Health Effects of the Fukushima Nuclear Disaster

Health Effects of the Fukushima Nuclear Disaster

Edited by

Kenji Kamiya
Vice President, Hiroshima University (HU) and Fukushima Medical University (FMU), Japan

Hitoshi Ohto
General Vice President, Fukushima Medical University (FMU), Fukushima, Japan

Masaharu Maeda
Professor and Chair, Department of Disaster Psychiatry, Fukushima Medical University (FMU), School of Medicine, Fukushima, Japan

ELSEVIER

ACADEMIC PRESS
An imprint of Elsevier

Academic Press is an imprint of Elsevier
125 London Wall, London EC2Y 5AS, United Kingdom
525 B Street, Suite 1650, San Diego, CA 92101, United States
50 Hampshire Street, 5th Floor, Cambridge, MA 02139, United States
The Boulevard, Langford Lane, Kidlington, Oxford OX5 1GB, United Kingdom

ISBN 978-0-12-824098-4

For information on all Academic Press publications
visit our website at https://www.elsevier.com/books-and-journals

Publisher: Andre G. Wolff
Acquisitions Editor: Elizabeth Brown
Editorial Project Manager: Timothy Bennett
Production Project Manager: Sreejith Viswanathan
Cover Designer: Christian Bilbow

Typeset by STRAIVE, India

Contents

Contributors

Numbers in parentheses indicate the pages on which the authors' contributions begin.

Keiya Fujimori (81), Radiation Medical Science Center for the Fukushima Health Management Survey; Department of Obstetrics and Gynecology, School of Medicine, Fukushima Medical University, Fukushima, Japan

Aya Goto (81, 319), Radiation Medical Science Center for the Fukushima Health Management Survey; Center for Integrated Science and Humanities, Fukushima Medical University, Fukushima, Japan

Mayumi Harigane (217), Radiation Medical Science Center for the Fukushima Health Management Survey, Fukushima Medical University; Department of Public Health, Fukushima Medical University School of Medicine, Fukushima, Japan

Arifumi Hasegawa (23), Department of Radiation Disaster Medicine, Fukushima Medical University School of Medicine, Fukushima, Japan

Kenichi Hata (81), Radiation Medical Science Center for the Fukushima Health Management Survey, Fukushima Medical University; Fukushima Society of Obstetrics and Gynecology, Fukushima, Japan

Fumikazu Hayashi (99), Radiation Medical Science Center for the Fukushima Health Management Survey, Fukushima Medical University; Department of Epidemiology, Fukushima Medical University School of Medicine, Fukushima, Japan

Arinobu Hori (357), Department of Psychiatry, Hori Mental Clinic, Minamisoma, Japan

Naoko Horikoshi (331), Radiation Medical Science Center for the Fukushima Health Management Survey, Fukushima Medical University, Fukushima, Japan

Mitsuaki Hosoya (99), Radiation Medical Science Center for the Fukushima Health Management Survey, Fukushima Medical University; Department of Pediatrics, Fukushima Medical University School of Medicine, Fukushima, Japan

Kayoko Ishii (81), Radiation Medical Science Center for the Fukushima Health Management Survey, Fukushima Medical University, Fukushima, Japan

Tetsuo Ishikawa (41), Department of Radiation Physics and Chemistry, Fukushima Medical University, Fukushima, Japan

Manabu Iwadate (69), Radiation Medical Science Center for the Fukushima Health Management Survey, Fukushima Medical University, Fukushima, Japan

Kenji Kamiya (81, 99), Radiation Medical Science Center for the Fukushima Health Management Survey, Fukushima Medical University, Fukushima; Research Institute for Radiation Biology and Medicine, Hiroshima University, Hiroshima, Japan

Yuka Karatani (195), Faculty of Urban Science, Meijo University, Nagoya, Aichi, Japan

Keiko Kawashima (195), Fukushima Collage, Fukushima, Japan

Tomoyuki Kobayashi (3, 281), Department of Disaster Psychiatry, Fukushima Medical University School of Medicine, Fukushima, Japan

Chihaya Koriyama (319), Department of Epidemiology and Preventive Medicine, Graduate School of Medical and Dental Sciences, Kagoshima University, Kagoshima, Japan

Hyo Kyozuka (81), Department of Obstetrics and Gynecology, School of Medicine, Fukushima Medical University, Fukushima, Japan

Masaharu Maeda (141, 331), Department of Disaster Psychiatry, Fukushima Medical University School of Medicine, Fukushima, Japan

Takashi Matsuzuka (69), Radiation Medical Science Center for the Fukushima Health Management Survey, Fukushima Medical University, Fukushima, Japan

Maho Momoi (331), Radiation Medical Science Center for the Fukushima Health Management Survey, Fukushima Medical University; Department of Disaster Psychiatry, Fukushima Medical University School of Medicine, Fukushima, Japan

Michio Murakami (247), Fukushima Medical University, Fukushima, Japan

Tsuyoshi Murata (81), Department of Obstetrics and Gynecology, School of Medicine, Fukushima Medical University, Fukushima, Japan

Masanori Nagao (99), Radiation Medical Science Center for the Fukushima Health Management Survey, Fukushima Medical University; Department of Epidemiology, Fukushima Medical University School of Medicine, Fukushima, Japan

Akihito Nakai (81), Radiation Medical Science Center for the Fukushima Health Management Survey, Fukushima Medical University, Fukushima; Nippon Medical School Tama Nagayama Hospital, Tokyo, Japan

Hironori Nakano (99), Radiation Medical Science Center for the Fukushima Health Management Survey, Fukushima Medical University; Department of Epidemiology, Fukushima Medical University School of Medicine, Fukushima, Japan

Tetsuya Ohira (81, 99), Radiation Medical Science Center for the Fukushima Health Management Survey, Fukushima Medical University; Department of Epidemiology, Fukushima Medical University School of Medicine, Fukushima, Japan

Hitoshi Ohto (81, 99), Radiation Medical Science Center for the Fukushima Health Management Survey, Fukushima Medical University, Fukushima, Japan

Kanako Okazaki (99), Radiation Medical Science Center for the Fukushima Health Management Survey, Fukushima Medical University; Department of Epidemiology, Fukushima Medical University School of Medicine; Department of Physical Therapy, Fukushima Medical University School of Health Sciences, Fukushima, Japan

Makiko Orita (247, 347), Atomic Bomb Disease Institute, Nagasaki University, Nagasaki, Japan

Misao Ota (81), Radiation Medical Science Center for the Fukushima Health Management Survey; Department of Midwifery and Maternal Nursing, School of Nursing, Fukushima Medical University, Fukushima, Japan

Kotaro Ozasa (123), Department of Epidemiology, Radiation Effects Research Foundation, Hiroshima, Japan

Akira Sakai (99), Radiation Medical Science Center for the Fukushima Health Management Survey, Fukushima Medical University; Department of Radiation Life Sciences, Fukushima Medical University School of Medicine, Fukushima, Japan

Hideki Sato (141), Department of Disaster Psychiatry, Fukushima Medical University School of Medicine, Fukushima, Japan

Naoya Sekiya (247), The University of Tokyo, Tokyo, Japan

Noriko Setou (69, 231), Radiation Medical Science Center for the Fukushima Health Management Survey; Department of Disaster Psychiatry, School of Medicine, Fukushima Medical University, Fukushima, Japan

Michio Shimabukuro (99), Radiation Medical Science Center for the Fukushima Health Management Survey, Fukushima Medical University; Department of Diabetes, Endocrinology, and Metabolism, Fukushima Medical University School of Medicine, Fukushima, Japan

Hiroki Shimura (69, 231), Radiation Medical Science Center for the Fukushima Health Management Survey; Department of Laboratory Medicine, School of Medicine, Fukushima Medical University, Fukushima, Japan

Kohta Suzuki (319), Department of Health and Psychosocial Medicine, Aichi Medical University School of Medicine, Nagakute, Japan

Kouta Suzuki (81), Radiation Medical Science Center for the Fukushima Health Management Survey, Fukushima Medical University, Fukushima; Department of Health and Psychosocial Medicine, School of Medicine, Aichi Medical University, Aichi, Japan

Satomi Suzuki (195), Institute of Counselling, Taisho University, Tokyo, Japan

Satoru Suzuki (69, 231), Radiation Medical Science Center for the Fukushima Health Management Survey, Fukushima Medical University, Fukushima, Japan

Satoshi Suzuki (69), Radiation Medical Science Center for the Fukushima Health Management Survey, Fukushima Medical University, Fukushima, Japan

Yasuyuki Taira (347), Atomic Bomb Disease Institute, Nagasaki University, Nagasaki, Japan

Noboru Takamura (347), Atomic Bomb Disease Institute, Nagasaki University, Nagasaki, Japan

Yoshitake Takebayashi (163), Department of Health Risk Communication, School of Medicine, Fukushima Medical University, Fukushima, Japan

Yui Takebayashi (141), Department of Disaster Psychiatry, Fukushima Medical University School of Medicine, Fukushima, Japan

Masaharu Tsubokura (265), Department of Radiation Health Management, Fukushima Medical University School of Medicine, Fukushima, Japan

Yuji Tsutsui (297), Fukushima University, Fukushima, Japan

Tokio Uchiyama (195), Clinical Psychology, Taisho University, Tokyo; Fukushima Collage, Fukushima; Institute of Counselling, Taisho University, Tokyo, Japan

Tatsuo Ujiie (297), The Open University of Japan, Nagoya, Japan

Akiko Yagi (177), Radiation Medical Science Center, Fukushima Medical University, Fukushima, Japan

Shun Yasuda (81), Radiation Medical Science Center for the Fukushima Health Management Survey; Department of Obstetrics and Gynecology, School of Medicine, Fukushima Medical University, Fukushima, Japan

Seiji Yasumura (3, 81, 99, 217), Radiation Medical Science Center for the Fukushima Health Management Survey, Fukushima Medical University; Department of Public Health, Fukushima Medical University School of Medicine, Fukushima, Japan

Susumu Yokoya (69), Radiation Medical Science Center for the Fukushima Health Management Survey, Fukushima Medical University, Fukushima, Japan

Kazuki Yoshida (281), Iryo Sosei University, Faculty of Nursing, Fukushima, Japan

Preface

People who once lived in the coastal area in Fukushima Prefecture will never be able to forget the painful days that followed the Fukushima Daiichi Nuclear Power Plant accident that occurred on March 11, 2011. In the initial phase, more than 160,000 people were evacuated to different areas across Japan, from Hokkaido to Okinawa. Many of these thought that their evacuation would last only a few days or several weeks at the longest. Now, however, more than 10 years since the disaster, more than 30,000 people continue living as evacuees. It has been very difficult for a large number of original residents who believed in the safety of the plant to expect such long-term effects at the time of the accident. Meanwhile, some experts expected serious, long-term health outcomes related to radiation exposure such as thyroid cancer while others voiced serious concerns about mental health and other nonradiological health effects. Heated arguments broke out about these effects among experts, policy makers, and the public, leading to splits and disagreements. Especially, emphasizing postdisaster nonradiological health effects was often regarded as displacing disputed points and confusing the issues as few international organizations, including the United Nations Scientific Committee on the Effects of Atomic Radiation, supported clear causality between the accident and its direct health effects. However, we now know that a great number of evacuees lost their lives as a result of different nonradiological health effects caused by long-term evacuation. For instance, among the evacuees, more than 2,000 deaths were officially recognized as disaster-related (not directly radiological) by national and/or local governments. Regardless of whether the effects were radiological or nonradiological, the Fukushima Daiichi Nuclear Power Plant accident certainly led to significant health effects that affected a large number of people.

This book aims to examine these various radiological and nonradiological health effects based on scientific evidence. The health effects described as topics in the book are broad-spectrum and multidimensional, from radiological effects such as thyroid cancer and problems related to pregnancy/childbirth to nonradiological effects such as problems related to mental health and lifestyle. Among such problems, an increase in the number of disaster-related suicides in Fukushima, which indicated the strong impact of the psychological trauma caused by the accident as well as culture-sensitive issues broadly seen in Asian countries, was among the most tragic consequences. In addition, in this book, we shed light on psychosocial agents that may involve individual health conditions.

Risk perception of radiation health effects, for example, might exert a great influence on not only individual behaviors such as relocation or returning home but also mental health conditions themselves. Unlike natural disasters, after the Fukushima disaster, intense social responses such as stigmatization were also elicited in communities, causing distress among many evacuees for a long time. Needless to say, traditional media and social networking services have been greatly affecting such social responses, often resulting in the formation of negative images toward the products, land, and people in Fukushima among the public. The chapters describing the psychosocial aspects of the period after the disaster allow readers to understand how people were distressed with intense social responses and unclear and sometimes groundless information and rumors. Furthermore, to identify the unique features of the health effects described above, this book refers to important historical nuclear events, including the atomic bombings of Hiroshima and Nagasaki in 1945 and the Chernobyl accident in 1986. Last, many medical and community health workers have been making great efforts to provide interventions and establish care networks for maintaining health conditions among a massive number of affected people. Their efforts have helped restore and foster resilience among affected people and communities, and as such, we believed that these significant challenges also warranted discussion in the book.

The authors of this book are experts who have been engaging in difficult, long-term rehabilitation work since the 2011 disaster. We believe that the book will provide an accurate picture of the various health effects experienced by those who were affected by the disaster in Fukushima as well as some useful suggestions for responding to future nuclear disasters.

Kenji Kamiya
Hitoshi Ohto
Masaharu Maeda

Section I

Event

Chapter 1

Development of Fukushima disaster

Tomoyuki Kobayashi[a] and Seiji Yasumura[b]
[a]*Department of Disaster Psychiatry, Fukushima Medical University School of Medicine, Fukushima, Japan,* [b]*Department of Public Health, Fukushima Medical University School of Medicine, Fukushima, Japan*

Accident

At 2:46 p.m. on March 11, 2011, an earthquake of magnitude 9.0 struck off the coast of Sanriku, Japan. The quake shook so hard and so long that it was impossible to remain standing; buildings collapsed, the earth thundered, and landslides occurred. The quake, which measured between 6 and 7 on the 7-point Japanese scale in Miyagi, Fukushima, Ibaraki, and Tochigi prefectures, spawned a vast tsunami (Japan Meteorological Agency, n.d.). The tsunami reached various parts of the Pacific Ocean side of Japan, mainly in the Tohoku and northern Kanto regions, with a height of more than five times that of the average Japanese man in Soma city, Fukushima Prefecture, and in Miyako city, Iwate Prefecture. The tsunami quickly engulfed cities and towns, swallowing people who had been unable to escape it. It swept away many buildings, even whole neighborhoods, and snarled trains into a tangled mess. Those who had reached higher ground could only watch as their cities and towns collapsed. Subsequent waves wreaked further damage, moving up the rivers and spreading the damage inland over a large area. Cars, household goods, and collapsed buildings were piled up in submerged disorder. Many fires broke out. More than 15,000 people were killed and a further 2500 went missing, mostly because of the tsunami, and more than 400,000 structures were damaged or destroyed (National Police Agency Emergency Disaster Security Headquarter, 2020).

The Great East Japan Earthquake spawned three disasters. Following the earthquake and tsunami came the nuclear power plant accident, the central topic of this book. The Fukushima Daiichi [Number 1] Nuclear Power Plant (FDNPP), the site of the accident, is located in the towns of Futaba and Okuma, Fukushima Prefecture, facing the Pacific Ocean to the east. In response to the earthquake, all the reactors in operation that day at the FDNPP were

Health Effects of the Fukushima Nuclear Disaster. https://doi.org/10.1016/B978-0-12-824098-4.00012-X

automatically shut down. As reported by the Tokyo Electric Power Company (TEPCO), the quake caused a loss of power to the site, and emergency diesel generators cut in to ensure a safe shutdown (Tokyo Electric Power Company, 2012). About 40 minutes after the earthquake, however, two huge tsunamis reached the plant. The entire main building area where the reactor and turbine buildings are located was inundated. With the exception of the emergency diesel generator in Unit 6, all power was lost. In response, at 3:42 p.m., the plant superintendent reported to government and emergency organizations that a "specific event" as defined in Section 1, Article 10 of the *Act on Special Measures Concerning Nuclear Emergency Preparedness* had occurred. In Unit 2, the reactor core isolation cooling system, which can supply emergency cooling water to reactors for a time, was manually activated, but as the power was lost, the operational status could not be confirmed, so the water supply condition was unknown. At 3:45 p.m., this was reported as a "special event" as stipulated in Section 1, Article 15 of the Act, and at 7:03 p.m., the Prime Minister issued a nuclear emergency situation declaration. The power plant was in darkness, status displays were dark, debris was scattered throughout, and roads were cut, making it extremely difficult to assess the situation and start recovery work. However, some of the status monitors, emergency fire pumps, and isolation condensers were recovered, and electricity and fire trucks were able to restore some control. If the containment vessel pressure is allowed to rise and the vessel bursts, a large amount of radioactive material can be released, so venting was attempted to reduce pressure by allowing some gas to be released. However, at 3:36 p.m. 0n March 12, a meltdown and a hydrogen explosion occurred in the Unit 1 reactor building. Although the response continued, explosions and fires were heard in the Unit 3 reactor building on March 14 and in the Unit 4 reactor building on March 15 (Fig. 1). As a result, a large amount of radioactive material was released.

Radiation effects

Radioactive materials emit radiation. The radioactivity of a particular material is measured in becquerels (Bq; particles disintegrated per second): the material emits radiation as a result of disintegration. The greater the radioactivity, the more energy emitted by the material. The extent to which this affects a person depends on the distance from the material. The amount of radiation a person is exposed to (radiation exposure) is measured in sieverts (Sv; joules of energy per kilogram of body mass weighted by type of radiation and body part exposed).

As radiation is invisible and has no smell, it cannot be perceived by human senses. Radiation occurs as both charged particles and electromagnetic waves emitted from nuclei of various elements when an imbalance in the numbers of protons and neutrons in the nuclei makes them unstable. An unstable nucleus can become stable by releasing the extra energy. The material it belongs to will continue to emit radiation until it all reaches a stable state. As the radioactive

FIG. 1 Unit 3 at the FDNPP, with white smoke, on March 15, 2011. The building behind it is Unit 4. *(Photo by Tokyo Electric Power Company.)*

content of the material decreases, the radioactivity becomes weaker, but the time period varies from seconds to billions of years, depending on the material.

Radiation exposure affects the human body, but its effects depend on the type of radiation, the parts exposed, and the degree of exposure. Radiation can be divided into particle radiation, in the form of alpha and beta rays, and electromagnetic radiation, in the form of gamma rays and x-rays. Alpha rays (protons) have a strong effect on the human body but do not penetrate far into the tissues. By external exposure, beta rays (electrons), which penetrate the skin and subcutaneous tissues, and gamma rays (high-energy photons), which penetrate deeply, are more problematic than alpha rays. On the other hand, internal exposure caused by the ingestion of radioactive materials is more likely to be a problem with alpha rays. Internal exposure occurs when radioactive materials are ingested or inhaled, absorbed through the skin or wounds, or administered as radiopharmaceuticals for medical treatment. The body remains exposed to radiation until the material is excreted from the body or decays with time.

The effects on the body also depend on the sensitivity to radiation and the ease of accumulation of the materials by the organs and the tissues of the body. In the case of external exposure to beta rays, for example, if a large amount of radioactive material adheres to the body surface and is not removed quickly, it may affect basal cells and capillary cells, which are highly sensitive to radiation, causing skin erythema and even alopecia. Gamma rays, which can reach the interior of the body, are more likely to affect the more sensitive internal organs, leading to dysfunction of organs and tissues. In the case of internal exposure, organs and tissues that are prone to accumulating radioactive materials will be exposed to higher doses, and if they are highly radiosensitive, the possibility of radiation-related effects is high. Since the Chernobyl nuclear accident, an increase in the number of thyroid cancers among children in Belarus and Ukraine has been attributed to the tendency for radioactive iodine to accumulate in the thyroid gland and the gland's higher radiation sensitivity in children than in adults (Eckerman, Harrison, Menzel, & Clement, 2012).

The effects of radiation on the body are classified as deterministic or stochastic according to the amount of radiation. Deterministic effects occur after a large exposure to radiation, which can kill or cripple a large number of cells of organs and tissues, resulting in effects such as skin damage, reduced blood cell count due to reduced hematopoiesis, and, in the womb, fetal damage and mental retardation. Deterministic effects occur only after exposure above a threshold. Stochastic effects, in contrast, result from mutations in genes, which can cause cancer and genetic abnormalities. They are called stochastic because the probability of their occurrence is not zero even at low dose levels. Epidemiologically, at least, the relationship between exposure to radiation and development of cancer is confirmed to increase almost linearly with dose above about 150 mSv, but even below 150 mSv, radiation has a theoretical effect on the development of cancer.

Radiation effects on the body mostly involve damage to DNA. Cells are capable of repairing DNA damage; when DNA is corrupted, repair enzymes rush in to repair it. In the case of limited damage, repair is successful and the cells return to normal, but in the case of extensive damage, the repair can fail and the cells die or degenerate. If many cells die, there may be clinical signs, but if only one or two cells die, no organ or tissue dysfunction is found, as other cells can take their place (deterministic effects). However, DNA repairs are often incomplete, and accidental mutations can occur. Most mutated cells are rectified or deleted, but some can survive, and the accumulation of additional gene mutations or changes in the gene expression levels in their progeny cells increases the probability of cancer cells (stochastic effects). Whereas deterministic effects can produce clinical symptoms within a few weeks, stochastic effects require years to decades before symptoms are seen because multiple processes take place between exposure and the development of cancer.

Evacuation in early phase

The earthquake and tsunami cut off much of the communication in Fukushima Prefecture. In the absence of sufficient information, residents began to evacuate for protection from the invisible threat of the radiation (Fig. 2). A number of documents and interviews available to us reveal what happened during the disaster but are often inconsistent with each other, illustrating the confusion at the time.

Before the Unit 1 reactor meltdown, the municipal offices were busy dealing with the aftermath of both the earthquake and the tsunami (Figs. 3 and 4). Administrators immediately began to assess the damage and guide the

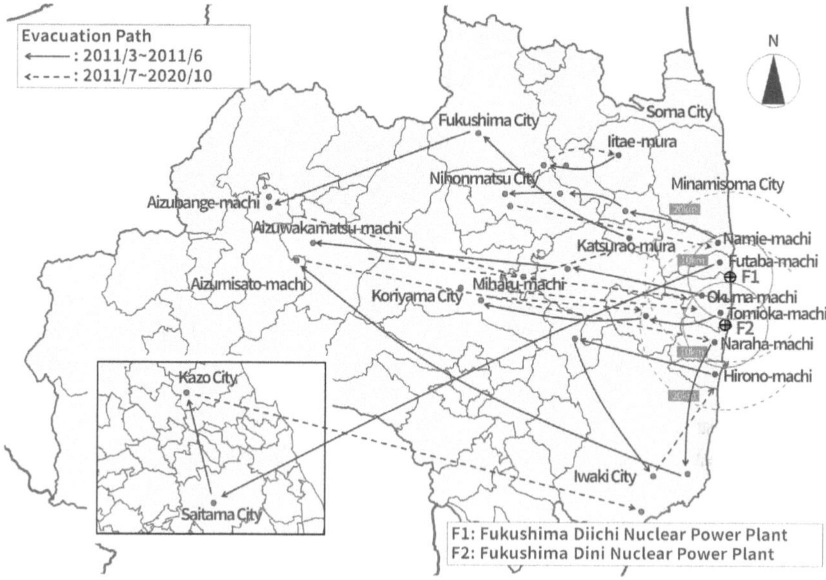

FIG. 2 Evacuation paths of nine cities in the evacuation zone.

FIG. 3 Namie town suffered extensive damage; March 12, 2011. *(Photo by The Fukushima Minpo Newspaper Publishing Company.)*

evacuation. To disseminate emergency response advice and help with evacuation, firefighters and disaster management staff drove to the coast in public information vehicles. As major aftershocks and tsunamis continued to hit, buildings were collapsing, roads were cracking, and the power and water supplies were cut off. Some emergency workers did not come back. Meanwhile,

FIG. 4 Volunteer firefighters carry victims found in the rubble in Minamisoma city, March 12, 2011. *(Photo by the Fukushima Minyu Shimbun.)*

tsunami evacuees gathered at schools and other shelters, where municipal office staff registered them.

In natural disasters, the primary source of information is local, but nuclear accidents are monitored centrally. If an emergency is detected at a TEPCO power plant, the information is sent to TEPCO's head office and the government in Tokyo (Articles 10 and 15 of the Act), and then to the prefectural and municipal governments where the power plant is located. However, the earthquake and tsunami crippled communications in Fukushima Prefecture. Most phone lines were cut, and only a few remained connected. The city's Internet and disaster radio systems were off. Moreover, the mobile phone service was not available in many areas, and even where it was available, the backup batteries in the base stations slowly went flat, and the service was gradually lost. The only way to get any information was to use an emergency power source to run televisions and radios, but government officials could not broadcast all the time as they had to run the emergency response.

Disruptions and obstacles caused by the earthquake and tsunami also occurred in the national government and other bodies. Government structure in a nuclear disaster is defined in detail in the *Act on Special Measures Concerning Nuclear Emergency Preparedness*, the *Basic Plan for Emergency Preparedness*, and the *Nuclear Emergency Response Manual*. The manual defines the main organizations that respond to a nuclear disaster as the Nuclear Emergency Response Headquarters (NERHQ) and the Local Nuclear Emergency Response Headquarters (Local NERHQ) of the national government. If the Prime

Minister issues a nuclear emergency situation declaration, the NERHQ, of which the Prime Minister is the director-general, is established in the Cabinet Office, and the Local NERHQ is established at an offsite center. The Local NERHQ, which is delegated some of the authority of the director-general of the NERHQ, responds to the accident such as by issuing evacuation orders with the support of the NERHQ and the support and cooperation of local governments and other organizations. However, the earthquake disrupted most of the circuits of terrestrial communication in Fukushima Prefecture. Consequently, communication among the government and Fukushima Prefecture's NERHQ (Prefectural NERHQ), the offsite center, and other institutions in the prefecture became extremely hard. General communication lines, including the mobile phone network, also experienced congestion in Tokyo owing to the surge in traffic immediately following the earthquake. Disaster priority circuits were strained, and communication among the agencies that relied on them was far from smooth. The earthquake also impaired and congested road and rail networks, significantly impeding traffic movement from Tokyo to Fukushima Prefecture and within the prefecture. Consequently, there was a long delay in setting up the offsite center.

In accordance with the *Basic Plan for Emergency Preparedness* and the *Fukushima Prefecture Disaster Prevention Plan*, the prefecture had set up 24 monitoring posts in the prefecture to collect data for the prefecture's Environmental Radioactivity Monitoring Center and published the data on its website. However, four of the 24 monitoring posts were swept away in the tsunami and another 19 lost their connection. Only one monitoring post remained working, so it was almost impossible for the government and Fukushima Prefecture to monitor atmospheric radiation doses. Although portable monitoring posts were set up beginning on March 12, the interruption of the mobile phone network and fuel shortages meant that it was not initially able to function properly. Therefore, owing to the disruption of the Local NERHQ and the inability to collect and share information, the secretariat of the NERHQ was delayed in considering and implementing a response to the accident.

Even though the Fukushima Prefectural Office was not affected by the tsunami, its main building was rendered unusable by the earthquake, and the office was moved to an alternative site to establish the Prefectural NERHQ. But as there was no adequate disaster response radio there, communications with the national government, municipalities, and disaster response agencies were fragile, making progress updates difficult. According to the official report of the Fukushima Nuclear Accident Independent Investigation Commission (National Diet of Japan Fukushima Nuclear Accident Independent Investigation Commission, 2012b), the prefectural government did not get the nuclear emergency situation declaration, which was issued at 7:03 p.m., until about an hour and a half later. But the Prefectural NERHQ was gathering information from TEPCO and was aware of the worsening situation at the FDNPP.

The prefectural government was aware that it did not have a clear legal basis for issuing evacuation orders, but alarmed by the slow pace of evacuation instructions from the national government, at 8:50 p.m., it issued an evacuation order for residents within a 2-km radius of the FDNPP. After the announcement at a press conference, prefectural police and firefighter radios were used to inform municipalities and residents about the evacuation. However, the extensive damage to equipment in some town and village offices made it extremely difficult to communicate. At 9:23 p.m., the Japanese government issued an evacuation order to residents within a 3-km radius of the FDNPP, and ordered residents within a 3–10-km radius of the plant to shelter indoors. These orders were conveyed mostly by the media, including TV.

TEPCO employees were dispatched to four towns–Naraha, Tomioka, Okuma, and Futaba (and nowhere else)–to explain the situation. But without sufficient expertise, the employees were unable to explain the full import of the danger. Okuma, for example, one of the towns where the plant is located, had relatively stable communication circuits and received evacuation orders from the national and prefectural governments. Nonetheless, with their attention focused on the tsunami and the search for missing persons scheduled to start the next morning (Sasaki, 2013), the residents did not comprehend the enormity of the danger.

In the absence of detailed information, the national government quickly expanded the evacuation zone. At 05:44 a.m. on March12, the order to evacuate within a 10-km radius of the plant was issued. Every municipality needed to evacuate its own residents and negotiated and secured its own shelters. Even through the evacuation orders were based on the distance from the power plant, many municipalities issued community-wide evacuation orders to encourage collaboration among residents. Some people left in their own vehicles, but many elderly or immobile residents–between several hundred and several thousand people, depending on the size of the town or village–left with municipal staff in buses and other vehicles. Long lines of cars were formed along the roads to the shelters. As the residents began to leave, the Unit 1 and Unit 3 reactors suffered meltdowns. The government's evacuation zone was extended at 5:39 p.m. on the same day to a 10-km radius from the Daini [Number 2] Plant, in the towns of Naraha and Tomioka, 11.5 km from Daiichi, and at 6:25 p.m. to a 20-km radius of the Daiichi Plant (Fig. 5).

On the night of March 11, following the earthquake and tsunami, more than 1000 residents of Naraha town, near the Fukushima Daini NPP, were evacuated to shelters within the town, some without blankets or food (Naraha-machi, 2014). The evacuation order for a 3-km radius issued at 9:23 p.m. did not cover all of Naraha, but the town authorities expected that it would be extended. At 7 a.m. the next morning, the disaster response headquarters in Naraha and Tomioka extended the zone to 30 km and asked Iwaki city, 30 km away, to accept the evacuees because it was impossible to predict the spread of radioactive materials on the wind and because the roads would be congested. At 8 a.m.,

FIG. 5 Lines of vehicles evacuating from coastal areas in Tomioka town, March 12, 2011. *(Photo by The Fukushima Minpo Newspaper Publishing Company.)*

the town authorities issued the order to evacuate the whole town. The staff and volunteer firefighters used public information vehicles to tell the residents to evacuate. Buses were used to relay children and the elderly first. Those who had their own transport were asked to go to designated shelters. By 4 p.m. that day, 7700 residents had left. In the shelters in Iwaki, water was still cut off and there was a lack of information because television coverage was limited. An increasing number of elderly people were especially suffering from respiratory symptoms. There were shortages of medicines, masks, blankets, kerosene, and other supplies, and the need for dialysis services and toilets for the disabled was identified. About 1000 residents of Naraha town then moved to Aizumisato town, with which it had originally had a sister city relationship, about 100 km away. Some evacuees were later taken by relatives and friends to other parts of the prefecture.

Kawauchi village, more distant from the power plant, gathered its own information and elected to evacuate. The earthquake moderately damaged about 20 buildings (Imai, 2014). The tsunami was not a threat. However, at around 7 a.m. on March 12, the village's mayor received a phone call from the mayor of neighboring Tomioka town with a request to take in residents from there because the situation at the plant was worsening (Tomioka-machi, 2015) (Fig. 6). Kawauchi immediately accepted the evacuees and opened 17 shelters, including elementary and junior high schools. Evacuees were already arriving. From Tomioka, 6000 people arrived in Kawauchi, with a population of fewer than 3000. Relief supplies such as blankets and stoves were in short supply. At this point, Kawauchi did not consider the situation dire. That perception

FIG. 6 Residents heading out of town with their luggage from Tomioka, March 12, 2011. *(Photo by The Fukushima Minpo Newspaper Publishing Company.)*

changed, however, on March 14, when the Unit 3 reactor exploded, and on March 15, when the Unit 4 reactor also exploded. Several villagers who worked at the plant had recommended that it was time to evacuate. At 7 a.m. on March 16, the village mayor ordered the whole village and the evacuees from Tomioka to evacuate. He called the director of the largest convention center in Fukushima Prefecture, *Big Palette Fukushima*, and got approval to send the evacuees there, as it was one of the largest shelters at the time. So 1300 residents of Tomioka and 700 residents of Kawauchi took shelter there.

Confusion reigned in both evacuated municipalities and refuge municipalities. Despite being devastated by the tsunami and the considerable confusion, Soma city did not evacuate, and instead it tried to accept as many evacuees as possible (Fig. 7). At 2:30 a.m. on March 12, it already housed 3849 survivors in 24 shelters. That night, evacuees began to arrive in the city, especially from Minamisoma city. Yet it had only about 200 blankets, there was a very high number of deaths, and numbers of coffins and cremations were restricted. The city compiled detailed reports of the situation and of the actions taken (Soma City Earthquake Response Headquarter, 2011). Staff asked to stop accepting evacuees from Minamisoma after 800, which they considered the maximum capacity. Mental health medications started to run out and the hospital was inundated with patients. Logistic companies were reluctant to come into the city, and supplies were scarce. Although Soma city is more than 30 km from the FDNPP, tensions persisted owing to the elevated atmospheric radiation dose in Iitate village, which is at a similar distance.

FIG. 7 Soma city, where fire broke out after buildings were swept away by the tsunami, March 11, 2011. *(Photo by the Fukushima Minyu Shimbun.)*

The lack of information caused distress. About 150,000 residents of Fukushima Prefecture were evacuated on order of the government (Figs. 8 and 9). However, only 20% of the residents in towns near the plant knew about the conditions at the plant when the order was given to evacuate within a 10-km radius (National Diet of Japan Fukushima Nuclear Accident Independent Investigation Commission, 2012a). In the shelters, digital broadcasts (1seg), car radios, and mobile phones were valuable sources of information for the evacuees. However, the mobile phone network gradually failed. Many residents had heard the evacuation orders on the municipality's emergency radios but had not heard about the severity of the accident or the expected evacuation period. A number of evacuees anticipated that the evacuation would last a few days and left with little more than the clothes on their backs. Some were evacuated to what were later determined to be high-dose areas. Some evacuees had to move several times. These problems created anxiety. Hospitals and nursing homes within a 20-km radius had a hard time securing shelters and safe transport, and at least 60 patients died by the end of March.

On March 15, residents within a radius of 20–30 km were ordered to shelter indoors. However, prolonged evacuation put a strain on lifelines and destroyed life infrastructure. Consequently, on March 25, residents were asked to evacuate. Once again, little information was provided by the government to help the residents decide whether they would.

FIG. 8 Survivors helping each other in shelters in Kawamata town, March 13, 2011. *(Photo by The Fukushima Minpo Newspaper Publishing Company.)*

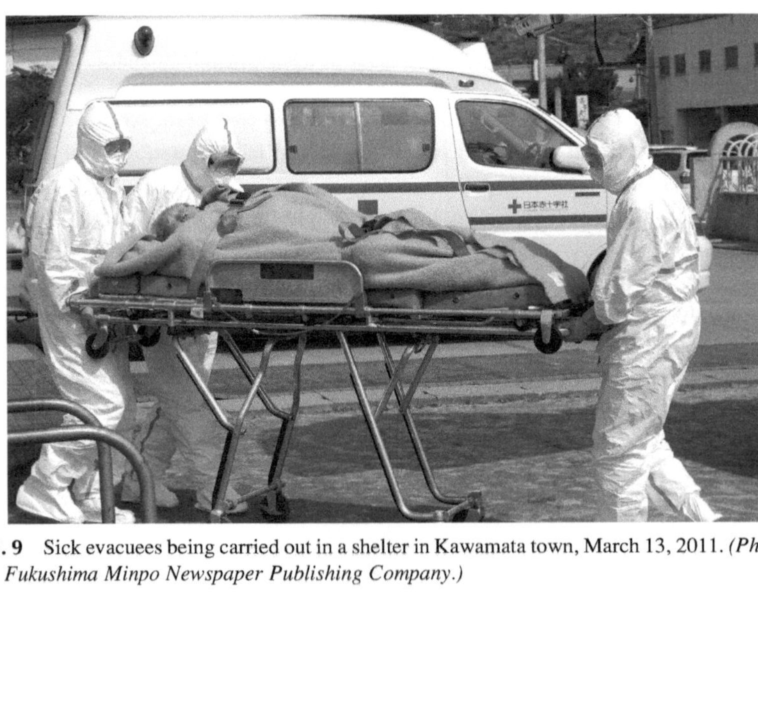

FIG. 9 Sick evacuees being carried out in a shelter in Kawamata town, March 13, 2011. *(Photo by The Fukushima Minpo Newspaper Publishing Company.)*

The issue of radiation is extremely technical. The majority of the Japanese people at the time did not know the technical terms. At press conferences, the government used a lot of phrases intended to reassure residents about the effects of radioactive materials on them, such as "We are taking all possible measures," "Just in case," and "There will be no immediate effects" (National Diet of Japan Fukushima Nuclear Accident Independent Investigation Commission, 2012b). However, there was no explanation of the basis for these statements that the residents could understand. Hence, they saw radiation as an invisible and unknown threat, dangerous to their lives and health, that would affect them for generations to come. Experts provided information and explanations about radiation exposure, but falsehoods were spread as well: people were told, for example, that the best way to reduce the damage caused by radiation was to gargle, and some actually did so.

In support of visiting medical teams, Fukushima Prefecture organized teams at each Public Health and Welfare Office and Public Health Center, and conducted emergency radiation exposure screenings over the entire prefecture beginning on March 13 (Fukushima Prefecture, n.d.-b) (Fig. 10). Because of the lack of personnel, the Prefectural Office requested local governments, universities, and research institutes to send personnel through the Ministry of Education, Culture, Sports, Science, and Technology and the Ministry of Health, Labor, and Welfare and asked them to bring survey meters and other devices. By the end of March, 114,488 people had been screened. The screening level for

FIG. 10 Evacuees being screened in Nihonmatsu city, March 13, 2011. *(Photo by the Fukushima Minyu Shimbun.)*

full-body decontamination was set at 100,000 cpm (counts per minute), and that for partial wipe-out decontamination was set at between 13,000 and 100,000 cpm. A total of 102 people were found to have been exposed to levels above 100,000 cpm. Even though many people were relieved after being screened, some broke down in tears when their screening values were very high. Screening continued after April, as the health concerns of the residents could not be alleviated and temporary entry to the evacuated areas began. Following many requests for a certificate of screening, screening certificates were issued to participants.

Fukushima Prefecture, furthermore, started emergency monitoring of foodstuffs and drinking water on March 19, owing to concerns about the health effects of ingested radioactive materials. In addition, internal radiation exposure tests by whole body counters were carried out, individual dosimeters were distributed, and experts held explanatory meetings to promote health management and allay residents' anxiety.

Initial evacuation zones were based on distance from the plant, but the government later based them on radiation dosages in April 2011, reorganizing them into three zones (Fukushima Prefecture, n.d.-a): the 20-km radius from the FDNPP was designated as the "evacuation order zone," the area outside this zone where radiation levels could reach 20 mSv per year was designated as a "planned evacuation zone," and additional areas within the 20–30-km radius were designated as an "emergency evacuation preparation zone." In June, households in areas with locally high radiation levels outside these zones were marked as specific sites recommended for evacuation. The evacuation order zone and the planned evacuation zone were reorganized in December, and three new zones were established to encourage people to return to their homes according to the radiation dose level: Areas where the annual cumulative radiation dose was estimated to be <20 mSv were marked as "evacuation order cancellation preparation zones," and residents of these zones prepared for a lifting of the evacuation order. Areas where the annual cumulative dose was between 20 and 50 mSv were marked as "restricted residence zones," allowing the residents to return home temporarily for maintenance. Areas where the annual cumulative dose was ≥ 50 mSv and did not fall below 20 mSv even after 5 years were marked as "difficult-to-return zones," and the evacuation continued. The zones covered 11 municipalities (Miyakoji-ku in Tamura city, Kawauchi village, Minamisoma city, Iitate village, Naraha town, Tomioka town, Okuma town, Futaba town, Katsurao village, Namie town, and Yamakiya in Kawamata town). The government originally aimed at rezoning the 11 municipalities to promote the return of residents on April 1, 2012, but as the amount of compensation available from TEPCO for the three zones changed because of a change in government criteria for defining the zones, it was difficult to reach consensus among residents, and the rezoning was delayed until August 2013.

Fukushima Health Management Survey

The accident at the FDNPP released large amounts of radioactive material, creating anxiety and confusion, and the evacuation forced residents to change their lifestyle, diet, exercise, and other personal habits. As of March 11, 2011, 210,189 individuals were living in government-marked evacuation zones or in high-air-dose areas. Concerned with potential health effects, Fukushima Prefecture established a long-term survey to screen and monitor residents and to support health management. The Fukushima Health Management Survey aims to track the health status of the residents over the long term, promote their future well-being, and determine the long-term effects of exposure to low-dose radiation on their health. The Radiation Medical Science Center for the Fukushima Health Management Survey was established at Fukushima Medical University to administer the survey.

The survey consists of a basic survey, detailed surveys, and follow-ups (Yasumura et al., 2012). With technical assistance from the National Institute of Radiological Sciences, the basic survey estimated individual external radiation dosages for approximately 2.05 million people, mainly Fukushima Prefecture residents and those who commuted to or lived there after the accident. The target people were sent questionnaires by mail, asking them where they lived, where they visited, how much time they spent indoors and outdoors, and how much they travelled between March 11 and July 11, when the atmospheric radiation dose was highest.

Past radiation-related tragedies, including the 1986 Chernobyl nuclear disaster, offer four important implications for health care after the FDNPP accident: the possibility of increased thyroid cancer in children, further disease due to radiation exposure or lifestyle changes, deterioration of mental health or lifestyle, and increased rates of abnormalities and deaths in children and babies. To address these issues, the Fukushima Health Management Survey included four detailed surveys and follow-ups. As thyroid cancer in children increased owing to internal exposure to radioactive iodine following the Chernobyl accident, it was decided to conduct thyroid ultrasound examinations of children in Fukushima Prefecture who were between 0 and 18 years old at the time of the accident to provide early detection and treatment and lifetime follow-up. About 360,000 people were enrolled in the survey in October 2011. When a nodule or cyst was found in the primary examination, it was confirmed at the Fukushima Medical University Hospital or another hospital accredited for advanced ultrasound examination. Detailed ultrasonography, blood testing, urine analysis, and aspiration biopsy cytology were performed as necessary. The comprehensive health checks are intended to review health information, assess the incidence of various diseases and risk factors of diseases, and improve the health status of the population.

Starting in January 2012, the same population was surveyed on mental health and lifestyle issues. Mental health is strongly challenged by the enormous impact of being separated from the familiarity of home by evacuation to an unfamiliar neighborhood as well as by the fear and distress associated with exposure to radiation. Mental health is part of the broad definition of health adopted by the World Health Organization, and effects can be long-lasting. Lifestyle-related diseases are also a point of interest in the comprehensive health check. The assessment of mental health and lifestyle and the provision of care were identified as essential to preventing avoidable deaths. Following the survey, a mental health support team, including clinical psychologists and public health nurses, provided telephone counselling based on the responses. When necessary, consultation with a registered psychiatrist, pediatrician, or radiologist was also provided.

The pregnancy and birth survey was designed to collect data that might improve obstetric and perinatal care and support women who had, or planned to deliver, a baby in Fukushima Prefecture. No significant increase in rates of induced abortion, miscarriage, low birth weight, or congenital malformations was observed after March 11, 2011. The nuclear accident, however, required pregnant women to change clinicians and hospitals, and many had insufficient prenatal care, which made it difficult for them to manage their own health and that of their children. The survey included 15,954 women who received a *Maternal and Child Health Handbook* from municipal offices in Fukushima Prefecture between Aug. 1, 2010, and July 31, 2011, or in a different prefecture but who gave birth in Fukushima Prefecture after March 11, 2011. Along with the survey, telephone and email hotlines were established for midwives and health-care workers to provide advice on health care, childcare, and anxiety. Midwives and nurses at Fukushima Medical University were also available. A mother's regular obstetrician or one from Fukushima Medical University could also be consulted, depending on the problems identified. For evacuees outside Fukushima Prefecture, a survey team, including the university's obstetrician, provided support.

The thyroid ultrasound examination started in October 2011, and an additional three detailed surveys began in January 2012.

Fukushima Prefecture created a Fukushima Resident Health Management File to collect and record the results of health surveys to allow residents to manage their own health care. The file is sent to residents in conjunction with the results of the basic survey. The prefecture also established a Radiation and Health Advisory Group in support of the health management program to provide expert advice to municipalities. The group holds training workshops for municipalities and prefectural public health centers as well as educators in charge of children's health management.

Goals of the book

It has now been 10 years since the accident. This book examines the effects of the accident on the health and lives of the population during that time. The physical and psychosocial issues raised in the aftermath of the accident are discussed in more detail by experts in each area in the following chapters. Physical issues are discussed in terms of the initial disruption of emergency medical treatment, radiation dosages and thyroid cancer, pregnancy, evacuation-related lifestyle diseases, and epidemiological studies of the health effects of exposure. Psychosocial issues include PTSD, suicide, and alcohol dependence; the mental health of children, the elderly, and people with thyroid tumors; risk perception; media; stigma; and community. Efforts made inside and outside Fukushima Prefecture to alleviate mental health problems are discussed. Nuclear disaster tragedies to date suggest that 10 years seems a long time but is still a short time for recovery. We hope that this book will prove helpful for ensuring people's future health and well-being and for lessening the tragedy of future nuclear disasters.

References

Eckerman, K., Harrison, J., Menzel, H.-G., & Clement, C. H. (2012). ICRP publication 119: Compendium of dose coefficients based on ICRP publication 60. *Annals of the ICRP*, *41*, 1–130. https://doi.org/10.1016/j.icrp.2012.06.038.

Fukushima Prefecture. (n.d.-a). Hinan kuiki no hensen nitsuite—Kaisetsu [Changes in evacuation zones—Commentary]. Fukushima Revitalization Station. Retrieved from https://www.pref.fukushima.lg.jp/site/portal/cat01-more.html (December 14, 2020).

Fukushima Prefecture. (n.d.-b). Kinkyu hibaku sukuriningu no katudou joukyou ni tsuite [Activities of the emergency radiation exposure screenings]. Fukushima Revitalization Station. Retrieved from https://www.pref.fukushima.lg.jp/site/portal/ps-iryou-screeningkatsudou.html (October 25, 2020).

Imai, A. (2014). *Jichitaisaiken: Genpatsu to idousurumura [Rebuilding municipality: the nuclear evacuees and moving villages]*. Chikuma Shinsho.

Japan Meteorological Agency. (n.d.). Information on the 2011 Great East Japan earthquake. Japan Meteorological Agency. Retrieved from https://www.jma.go.jp/jma/en/2011_Earthquake/Information_on_2011_Earthquake.html (October 25, 2020).

Naraha-machi. (2014). *Naraha-machi saigai kirokushi [The disaster record of Naraha-machi]*. Naraha-machi. https://www.town.naraha.lg.jp/saigai_kiroku/main/main.pdf.

National Diet of Japan Fukushima Nuclear Accident Independent Investigation Commission. (2012a). Overview of the damage and how it spread. In *The official report of the Fukushima Nuclear Accident Independent Investigation Commission*National Diet of Japan. https://warp.da.ndl.go.jp/info:ndljp/pid/3856371/naiic.go.jp/wp-content/uploads/2012/08/NAIIC_Eng_Chapter4_web.pdf.

National Diet of Japan Fukushima Nuclear Accident Independent Investigation Commission. (2012b). Problems with the nuclear emergency response. In *The official report of the Fukushima Nuclear Accident Independent Investigation Commission*National Diet of Japan.

National Police Agency Emergency Disaster Security Headquarter. (2020). *Heisei 23 nen (2011 nen) Tohoku-chiho Taiheiyoo oki jishin no keisatu katsudou to higai joukyou [Report of the police activities and the damage overview in the 2011 Tohoku-Pacific Coast Earthquake]*. https://www.npa.go.jp/news/other/earthquake2011/pdf/higaijokyo.pdf.

Sasaki, Y. (2013). Study on the relationship between information of Fukushima nuclear accident and evacuation of towns of Okuma and Tomioka. *Journal of Administration and Social Sciences*, *25* (4), 1–39.

Soma City Earthquake Response Headquarter. (2011). *Heisei 23 nen 3 gatsu 11 niti hassei higashi-nihon daishinsai no kiroku dai 1 kai chukan hokoku [March 11, 2011 Great East Japan Earthquake: The first interim report]*. http://infra-archive311.jp/data/doc/kiroku/souma01.pdf.

Tokyo Electric Power Company. (2012). *The investigation report of Fukushima nuclear accident* (p. 352). TEPCO. https://www.tepco.co.jp/cc/press/betu12_j/images/120620j0303.pdf.

Tomioka-machi. (2015). *Tomoioka-machi "higashi-nihon daishinsai/genshiryokusaigai" no kioku to kiroku [Memory and recod of Great East Japan Earthquake and the nuclear disaster In Tomioka-machi]*. Tomioka-machi. https://www.tomioka-town.jp/chosei/koho_kocho/2115.html.

Yasumura, S., Hosoya, M., Yamashita, S., Kamiya, K., Abe, M., Akashi, M., et al. (2012). Study protocol for the Fukushima Health Management Survey. *Journal of Epidemiology*, *22*(5), 375–383. https://doi.org/10.2188/jea.JE20120105.

Section II

Physical and radiological effects

Chapter 2

Initial turmoil in an emergency setting

Arifumi Hasegawa
Department of Radiation Disaster Medicine, Fukushima Medical University School of Medicine, Fukushima, Japan

Lessons learned from the Fukushima Daiichi Nuclear Power Plant accident

The accident at the Fukushima Daiichi Nuclear Power Plant (FDNPP) on March 11, 2011, which occurred in conjunction with the Great East Japan Earthquake, was a compound disaster in which a natural disaster and a nuclear disaster were superimposed. The occurrence of a compound disaster, in which an earthquake, tsunami, or other natural phenomenon would cause such a critical event as a nuclear power plant (NPP) accident, was perhaps inevitable in a seismically active country like Japan. Japan previously operated 54 NPPs along its coasts. At the time of the accident, the FDNPP was the only NPP that completely lost its core cooling system after the earthquake and tsunami, which caused serious damage to the core and released a substantial amount of radioactive material into the environment, leading to an International Nuclear and Radiological Event Scale Level 7 accident (Hayano & Adachi, 2013; Katata, Terada, Nagai, & Chino, 2012). One lesson to be learned from the FDNPP accident is that the effects of such an accident are noncentralized, being affected by meteorological conditions, and extensive. Another is that the effects of nuclear accidents are not limited to the health effects of radiation; they extend to social and psychological effects. The effects of the nuclear accident were more pronounced on people with greater vulnerability, such as inpatients in medical facilities and residents of nursing homes (Hasegawa, Ohira, Maeda, Yasumura, & Tanigawa, 2016). At the same time, we have learned that, in an acute crisis response, even the medical profession is influenced by emotions and delays in decision making. Healthcare professionals therefore need to understand this wide range of health risks that arise after nuclear accidents in order to properly address these issues.

Health Effects of the Fukushima Nuclear Disaster. https://doi.org/10.1016/B978-0-12-824098-4.00013-1

Initial turmoil in the general emergency setting

Causes and human factors related to the Fukushima accident

The Fukushima accident was the first major nuclear power plant accident in history to occur as part of a complex disaster: it resulted after a magnitude 9 earthquake occurred off the east coast of Japan. The earthquake generated massive tsunamis that severely damaged coastal areas and claimed more than 18,000 lives, but the human factor in the design of the emergency generator—which was submerged by the tsunami—is considered to have played an indirect though major role in the occurrence of the accident (The National Diet of Japan the Fukushima Nuclear Accident Independent Investigation Commission, 2012; National Police Agency, n.d; Hasegawa et al., 2015). During the magnitude 6.8 Chuetsu offshore earthquake in 2007, contaminated water leaked from the spent fuel pool at the Kashiwazaki-Kariwa NPP. Although the leak did not lead to criticality, in retrospect it is considered to have been a harbinger of the accident at the FDNPP in 2011 (Akashi, Kumagaya, Kondo, & Hirose, 2010).

Environmental dispersion of radioactive materials is weather-dependent, not concentric

The distribution of radioactive materials in the environment due to a large-scale accident is generally not concentric, but diffuses irregularly, depending on weather and topography, and varies over time (The National Diet of Japan the Fukushima Nuclear Accident Independent Investigation Commission, 2012) (5). The main cause of radiation exposure to the residents is the radioactive plume (an air mass containing radioactive noble gases, iodine, cesium, and other microradioactive materials from an NPP accident, which flows upward and outward in the atmosphere like smoke). The most significant release of radioactive material after the FDNPP accident occurred on March 15, 2011. At first, high air dose rates of radioactivity were measured in an area more than 40 km south of the plant, but then the wind direction changed and air dose rates in that area rapidly decreased. On the other hand, due to the change in wind direction to the northwest, high air dose rates were measured at about 60 km northwest on the evening of March 15 and throughout the night, and snowfall that night caused radioactive material to be deposited on the ground. Thus, radioactive cesium—with a long half-life—remained in a strip of area to the northwest of the plant. Even within the Precautionary Action Zone (PAZ), the temporal and geographic variability of air dose rates was substantial.

Lack of information

At the time of the accident, prompt or accurate information about the specifics of the accident and estimated radiation exposure were not provided to local authorities and residents, and only 10%–40% of residents were informed about the accident via local authorities or the police (The National Diet of Japan the Fukushima Nuclear Accident Independent Investigation Commission, 2012). The reason for evacuation was vaguely communicated as "just a precaution," and the order to evacuate was delayed, resulting in residents being forced to evacuate unprepared with no personal belongings and only the clothes they were wearing at the time. Furthermore, this uncoordinated response had a serious effect on the subsequent mental state of residents, who still harbor concerns about the timing of evacuation at the time of the accident (Hasegawa et al., 2018; The National Diet of Japan the Fukushima Nuclear Accident Independent Investigation Commission, 2012). In a nuclear facility accident, it should be assumed that it may be difficult to ascertain the damage to nuclear facilities from the perspective of radiation protection management, and that the situation may change from time to time, so that decision makers have to deal with unclear and indeterminate situations without sufficient or accurate information.

Disruption of command-and-control system

The command-and-control system for the management of a nuclear disaster was caught off guard and disorganized. Actually, the systems of dealing with a natural disaster and dealing with a nuclear disaster overlapped and were not sufficiently controlled in the disaster-affected area. Also, we were not familiar with the command-and-control system specific to a nuclear disaster. Therefore, at first, we did not know the names of organizations specific to nuclear disaster prevention, such as the Emergency Response Center (ERC) and the Nuclear Emergency Response Headquarters on scene (offsite center), nor did we know their roles. In addition, disruptions to electricity, water and gas supplies, communications, and radiation monitoring systems caused by the earthquake made it difficult to assess and address radiation health risks and forced the evacuation of the offsite center that had the responsibility for nuclear disaster management on the scene(The National Diet of Japan the Fukushima Nuclear Accident Independent Investigation Commission, 2012). Thus, the medical command-and-control system was in disarray as well.

Collapse of community health care

Medical functions in the Fukushima region were severely damaged, and regional medical care in the area around the FDNPP practically collapsed, even if only temporarily (The National Diet of Japan the Fukushima Nuclear Accident Independent Investigation Commission, 2012). On the second day after the accident, emergency evacuation was ordered within a 20-km radius of the plant, and all medical and welfare facilities were forced to evacuate without adequate preparation or support. In addition to supplies needed for daily life, medical products and medical equipment were in short supply in the 20–30 km zone where the evacuation order was issued on March 15. Hence, some medical facilities had to evacuate their inpatients, limiting them to outpatient care only, while others had to be evacuated completely (2012). In addition, medical staff at medical institutions in some areas outside the evacuation order zone evacuated themselves voluntarily because of concerns about radiation; this resulted in a decline in medical services, such as limiting the number of patients that could be accepted (Tanigawa, Hosoi, Hirohashi, Iwasaki, & Kamiya, 2011; Tominaga, Hachiya, Tatsuzaki, & Akashi, 2014). As a result, not only was patient care terminated or disrupted in and around the area where the evacuation order was issued, but inpatients at medical institutions and residents and staff of nursing care facilities also had to be evacuated (The National Diet of Japan the Fukushima Nuclear Accident Independent Investigation Commission, 2012). Therefore, we need to assume that, in the event of a nuclear accident or disaster, even if buildings or infrastructure sustain no damage, there might be shortages of supplies, staff, and medical services.

Disruption of medical care for patients contaminated by radioactive materials

The existing radiation emergency medical system (such as it was) did not work effectively. Hydrogen explosions occurred at Unit 1 on March 12 and at Unit 3 on March 14, injuring 16 people, including members of the Self-Defense Forces. In the responding medical facilities, it was difficult to assess the risk of human effects of radiation due to unclear information and because the staff lacked knowledge of radiation health effects. As a result, excessive protection was put in place, and in some facilities it was difficult to accept the injured and sick because of ignorance of the radioactive materials involved (The National Diet of Japan the Fukushima Nuclear Accident Independent Investigation Commission, 2012; Tominaga et al., 2014). At that time, it was difficult for persons to receive normal medical services because of the closure or disruption of local emergency medical facilities, and medical care for radiation-contaminated patients was even more difficult for the reasons mentioned above (The National Diet of Japan the Fukushima Nuclear Accident Independent Investigation Commission, 2012).

Public protection measures and emergency evacuation

After the Fukushima accident, emergency evacuation orders were given by the Japanese government and prefectural government as a public protection measure to sequentially evacuate areas within a 3-km, 10-km, and then a 20-km radius on the day after the accident, and the instructions were gradually extended to other areas on subsequent days (The National Diet of Japan the Fukushima Nuclear Accident Independent Investigation Commission, 2012). The International Atomic Energy Agency (IAEA) has proposed the following protective measures for the public in the event of a nuclear disaster: emergency evacuation; shelter in place; temporary relocation; blocking uptake by the thyroid gland with stable iodine; restricting intake of food, milk, and drinking water; decontamination;

etc. It is extremely important for public health that these protective measures be implemented in an effective way (International Atomic Energy Agency, 2011). Of these measures, emergency evacuation is the most important because it is the only way to completely avoid exposure to radioactive materials dispersed in the environment and thereby avoid definite and probable health effects (Kim et al., 2020). Therefore, the IAEA's technical documents consider emergency evacuation to be a basic protective measure, especially for people living in areas adjacent to a PAZ around a nuclear facility (International Atomic Energy Agency, 1997).

Unplanned emergency evacuations and preventable death during evacuation

Emergency evacuations without adequate medical support may prevent an unnecessarily high risk of exposure but may create a significant risk of life-threatening consequences for vulnerable people, such as hospitalized patients and residents of nursing care facilities who require continuous medical care. According to a survey after the FDNPP accident, more than 50 such persons died during emergency evacuations or at evacuation shelters due to inadequate preparation for dealing with medical needs during and after evacuation (Tanigawa, Hosoi, Hirohashi, Iwasaki, & Kamiya, 2012). The mortality rate among residents of long-term care health and welfare facilities was higher after the accident than before (Nomura et al., 2013), and the mortality rate among elderly persons who resided in long-term care health and welfare facilities after the accident remained high for about 6 months (Yasumura, Goto, Yamazaki, & Reich, 2013). The mortality rate among inpatients evacuated from hospitals was significantly higher than that among nursing home residents (Igarashi et al., 2018), and the risk associated with nursing home evacuation without adequate preparation was higher than the risk from radiation exposure due to the fact that emergency evacuation was carried out in the absence of adequate preparation (Tanigawa et al., 2011; Tanigawa et al., 2012). For effective radiation protection, it might be better in some areas to monitor changes in the radiation dispersion while sheltering indoors, even if the air dose rate increases temporarily. Indoor sheltering is an effective radiological protection measure and certainly reduces exposure to the radioactive plume. For vulnerable populations, among whom prophylactic emergency evacuation may increase health risks, even if they are in a PAZ, it is suggested that a combination of sheltering indoors and taking appropriate stabilized iodine might be reasonable (Nuclear Regulation Authority, 2020a, 2020b). It should be noted that, in contrast to the collapse of buildings due to earthquakes and tsunamis, there is no need to contemplate evacuation minute by minute; temporary exposure to the radioactive plume is not sufficient to cause acute radiation injury after a nuclear accident. The experience of the Fukushima accident taught us that evacuation plans made and actions taken should be calm and orderly while giving the highest priority to the well-being of the evacuees.

Worker health effects from radiation and other causes

No case of acute radiation syndrome or other radiation-related health effects has been identified among workers, but other work- and lifestyle-related fatalities occur at regular intervals among the workers at the FDNPP.

According to a report released by Tokyo Electric Power Company (TEPCO), worker radiation doses continue to decline year by year (TEPCO, n.d.-a, n.d.-b) (Fig. 1). As for the early phase of worker exposure, United Nations Scientific Committee on the Effects of Atomic Radiation (UNSECAR) reported that for 12 patients with ^{131}I internal thyroid exposure (2–12 Gy), the possibility of hypothyroidism could not be excluded, but the likelihood was low. The risk of cardiovascular disease due to radiation exposure for the workers with the highest exposure was extremely low, and the risk of cataracts could not be determined due to a lack of data on beta exposure of the lens. In regard to cancer in general, the majority of worker doses was below the level that epidemiological studies have demonstrated to be associated with a significantly increased cancer risk. There were 170 workers with committed effective doses above 100 mSv and approximately 2000 workers with thyroid absorbed doses in excess of 100 mGy, and so no epidemiological study would be expected to identify an increase in cancer incidence due to radiation in these workers (UNSCEAR, 2014). Fortunately, no case of acute radiation syndrome has been observed among workers to date as a result of the Fukushima accident.

On the other hand, 14 deaths happened onsite by July 2020, about half of which were endogenous diseases such as acute coronary syndrome, but none of them was attributable to radiation (Fig. 2). During the initial phase (first month) after the accident (March 2011), 67 workers, including those injured

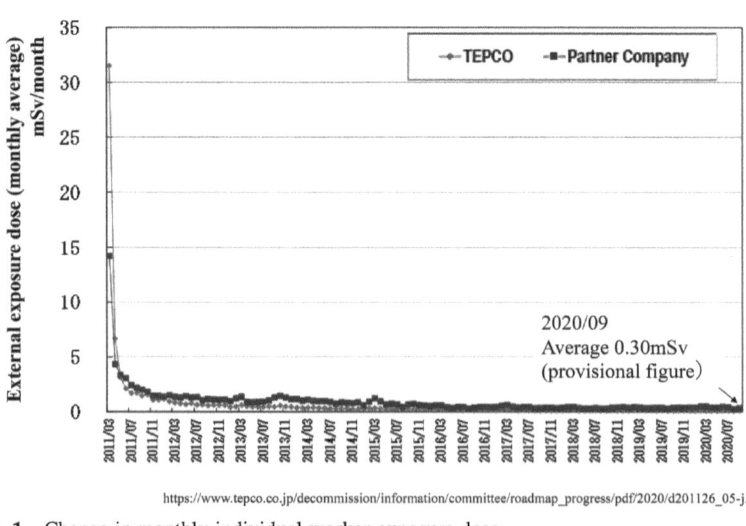

https://www.tepco.co.jp/decommission/information/committee/roadmap_progress/pdf/2020/d201126_05-j.pdf

FIG. 1 Change in monthly individual worker exposure dose.

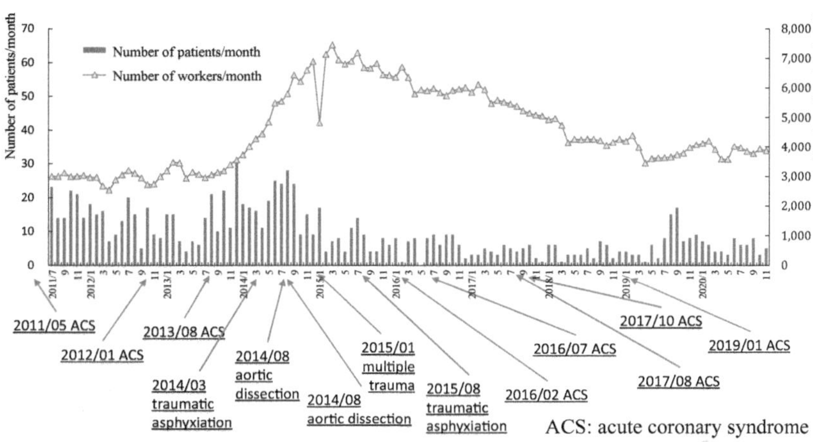

FIG. 2 Monthly distribution of the patients, workers, and lethal cases in the Fukushima Daiichi Nuclear Power Plant.

in two hydrogen explosions, were transported to an emergency room for treatment of injuries and sicknesses that occurred on the site of the FDNPP. The majority of these cases was caused by general fatigue and illness unrelated to radiation exposure. By June 30, 2012, a total of 264 workers had been treated at the site; three fatalities occurred—one each in April and August 2011 and another in January 2012. Two of these were cardiac arrests due to acute coronary syndrome and one a case of severe sepsis related to uncontrolled diabetes. Among the 264 cases, trauma accounted for 131 cases (49.6%). Regarding the severity of the trauma among them, only one case had an Injury Severity Score (ISS) greater than 15. Fortunately, no deaths due to trauma had been reported as of June 30, 2012, and only six out of 131 trauma cases showed evidence of radiation contamination, all of which occurred in March 2011 (Tanigawa & Hasegawa, 2013). For workers, heat stroke during the summer season was also a health concern because of the wearing of PPE to prevent the spread of contamination. In 2011, the incidences of heat stroke increased in May, June, and July, with a total of 44 workers suffering heat stroke, most of whom were treated in the onsite ER at the plant with only two of them requiring hospitalization. Beginning in June 2011, TEPCO prevented the occurrence of heat stroke by enforcing internal regulations for all workers at the FDNPP, including the wearing of cooling jackets under protective clothing, regular rest, adequate hydration and salt supplementation, and activity restrictions using the Wet Bulb Globe Temperature (WBGT) as a parameter (Radiation Emergency Medicine, 2019). Recent records suggest that the number of cases of heat stroke is under control, but it must certainly remain an important health risk factor for workers (Fig. 3). To deal with workers whose illnesses occurred in the plant, TEPCO set up an ER in the plant in July 2011. The most important role of the ER is to provide life-saving treatment and resuscitation for all persons related to the FDNPP

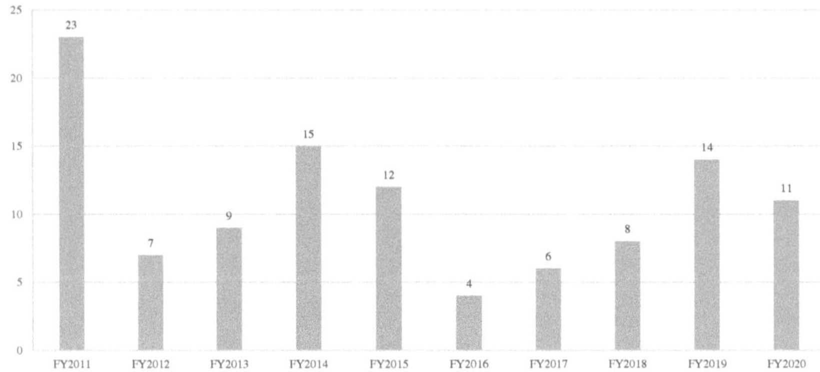

FIG. 3 Annual distribution of heatstroke patients in the FDNPP.

who suffered injury or sickness, and to identify appropriate medical institutions when more thorough treatment is required. This ER is in operation 24h a day with one emergency physician, one paramedic, one nurse, and one office staff person on duty, and it handles between 70 and 180 injured and sick people per year. In addition, automated external defibrillators (AEDs) have been installed throughout the plant to deal with cardiac arrests that occur onsite, and workers are provided with regular training in the use of these AEDs. Furthermore, flu vaccinations are offered free of charge in the winter season to substantially reduce the risk of outbreaks of infectious disease onsite (TEPCO) (Fig. 4).

Resident health effects from radiation and effects not attributable to radiation

One of the lessons learned from the Fukushima accident was that health effects not attributable to radiation can appear as a result of measures taken to avoid radiation exposure and contamination.

According to a report released by the Fukushima Health Management Survey Examination Committee in August 2020, the estimated external effective

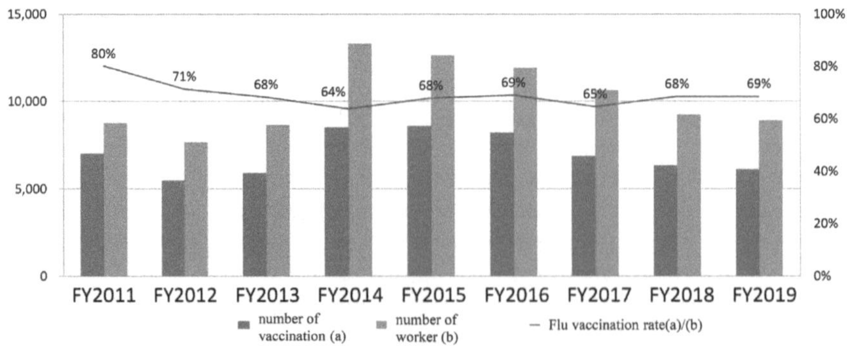

FIG. 4 Trends of the number of vaccinations, workers, and flu vaccination rates.

dose from March 11–July 11, 2011, was less than 2 mSv among 445,703 (93.8%) of the 568,632 people surveyed (average dose of 0.9 mSv) (Fukushima Health Management Survey Examination Committee, 2020). The UNSCEAR reported that no appreciable increases in incidence of cancers known to be associated with radiation exposure (leukemia, breast cancer, etc.) are expected in the population due to accidental radiation exposure and no congenital anomalies or genetic effects would be observed among residents. The most important health effects for residents were reported to be those related to mental and social well-being in relation to the severe effects of the earthquake, tsunami, and nuclear accident as well as fear and stigma associated with the recognized risks associated with radiation exposure (United Nations, 2014). According to the latest study (Ohba et al., 2020), which includes atmospheric dispersion simulations, dosimetry based on detailed evacuation behavior, dosimetry considering the effects of the diet of Japanese people and the effect of evacuation on that, and internal exposure from tap water, the average equivalent thyroid dose for a 1-year-old child in each of the seven evacuation areas was estimated to be 1.2–15 mSv. That is considerably lower than the UNSCEAR estimated average thyroid absorbed dose of 15–83 mGy.

As previously described, the accident at the FDNPP revealed health risks that are not attributable to direct radiation effects but are associated instead with the unplanned evacuation and relocation of vulnerable populations, such as hospitalized patients and elderly people who require nursing care, as well as the weakness of the emergency medical systems at nuclear power plants. Furthermore, the majority of disaster-related deaths in all three prefectures of the Tohoku region occurred in Fukushima Prefecture, 90% of which were among the elderly and were concentrated within a month of the accident. The reasons for this are assumed to be repeated transfer between shelters, significant changes in the medical and nursing care environments, and a decline in medical and nursing provision due to a shortage of medical and nursing staff caused by the withdrawal of employees (Ichiseki, 2013; Yasumura, 2014). It is therefore clear that health risks after a nuclear disaster include not only those caused by radiation, but also those related to protective measures (such as evacuation), and that vulnerable people (such as hospitalized patients in medical institutions and residents of nursing care facilities) are more strongly affected (Hasegawa et al., 2016). In addition, changes in living conditions due to relocation can be a factor in the subsequent deterioration of health among vulnerable people. The importance of a broader perspective and a sense of balance in reducing overall health risks, taking into account both the direct risks of radiation and the indirect risks posed by radiation protection measures, was an important lesson learned from the experience of the Fukushima nuclear accident (Carr, Maeda, Oughton, & Weiss, 2018; Hasegawa et al., 2015; Hasegawa et al., 2016). Health-care professionals are encouraged to understand that a wide range of health risks arise after nuclear accidents, so that it is necessary to address all these issues rather than focus on just the immediate and direct radiation-related risks.

Initial turmoil in the emergency setting at Fukushima Medical University Hospital

Japan's radiation emergency medical system was developed to deal with the situation subsequent to an industrial accident (Tominaga et al., 2014), not a situation compounded by a large-scale natural disaster such as the Fukushima accident (The National Diet of Japan the Fukushima Nuclear Accident Independent Investigation Commission, 2012). In the aftermath of the earthquake and tsunami, six hospitals that had been designated as primary radiological emergency medical facilities for initial medical response were closed or dysfunctional due to a combination of evacuation orders, damage to facilities, disruption of infrastructure, and exodus of medical staff fearful of the dangers of radiation (Koyama et al., 2011). Fukushima Medical University (FMU) Hospital, which was designated as a secondary radiation emergency hospital for the advanced treatment of radiation injuries, was the only hospital capable of responding to the emergency medical needs of people in and around the plant (Nakashima, Takamura, Suzuki, & Yamashita, 2012). Therefore, the load was concentrated on the FMU Hospital, which is the regional base hospital for dealing with disasters (The National Diet of Japan the Fukushima Nuclear Accident Independent Investigation Commission, 2012; Investigation Committee on the Accident at the Fukushima Nuclear Power Stations of Tokyo Electric Power Company, Investigation Report, n.d.). A national radiological emergency care network has been established to support the FMU's efforts (Morimura et al., 2013; Nakashima et al., 2012; Tanigawa, Hosoi, Hirohashi, et al., 2011; Tanigawa, Hosoi, Terasawa, et al., 2011). After the accident, FMU Hospital dealt with providing medical support to the FDNPP, as did many other facilities.

Role of FMU in the initial phase of the disaster

The roles of the FMU during the initial phase of the disaster included not only medical care for inpatients and persons out of the hospital, but also the health and welfare of residents in the Fukushima area. The FMU Hospital, located 57 km from the FDNPP, is the only university-affiliated hospital in Fukushima Prefecture. It had 778 beds and 30 departments, with an average of 630 inpatients and 1500 outpatients per day at the time of the accident. On the day of the earthquake, there were 1345 outpatients and 642 inpatients. Immediately after the earthquake, FMU Hospital set up a disaster response headquarters in the hospital, canceled general outpatient services, canceled or postponed elective surgical procedures, and discharged patients who were able to go home, reducing the number of hospitalized patients to about 70% of the previous number of inpatients. After the earthquake, the elevators at FMU Hospital were shut down immediately and the water supply was cut off for 8 days, forcing doctors to limit the number of surgical procedures and to stop dialysis, but electricity and gas continued to be supplied and there was no damage from the tsunami because it is located inland. During March 12–14, 2011, a total of 168 outpatients were admitted to the hospital, 30 of whom were assessed as critical. FMU Hospital received medical support from 35 of 180 acute disaster medical assistance teams (DMAT) during the same period. In the 10 days after the accident, 25 nonelective surgeries were performed at the hospital, including repairing broken bones and performing Cesarean sections (Murakawa, 2012). FMU Hospital was required to serve not only as a medical treatment center for patients injured in the natural and nuclear disasters, but also as a relay station for emergency evacuation, physical triage, and contamination screening. At the time the evacuation order was issued, there were approximately 2200 inpatients and residents in the affected area, and a government-organized emergency evacuation was carried out for them between March 12–18, 2011, with the FMU Hospital relaying 175 patients to other medical facilities outside the disaster area, 125 of whom initially required hospitalization in the FMU hospital so that they could be stabilized prior to movement (Murakawa, 2012).

Condition of FMU Hospital immediately after the nuclear disaster

After the Fukushima Daiichi Nuclear Power Plant accident became public knowledge, FMU Hospital was faced with the limits of disaster medical support, and was temporarily isolated and struggled to continue medical activities. Social systems in the area around the Fukushima Daiichi NPP were rendered completely useless during the acute phase of the accident. Emergency firefighting support from outside was restricted to the area outside the evacuation zone, so local firefighters must have been overloaded and forced to struggle alone (without help from outside). A no-fly zone was set up, making it difficult for aircraft, including helicopters, to fly within a 30-km radius of the FDNPP. Trucking associations restricted truck operations within a 50-km radius of the plant. The Japan Nurses Association restricted its support in Fukushima Prefecture due to uncertainties about the risks of radiation and how to deal with them. Acute medical support teams dispatched to Fukushima Prefecture had to be temporarily evacuated due to inadequate protective equipment against radioactive materials, and DMATs were evacuated from FMU Hospital on March 15. The totality of the above situation can be considered inevitable from a safety management perspective, but it has become a source of anxiety for those working on the front lines of disaster medicine (Fukushima Medical University Hospital Radiation Disaster Medical Center, 2013).

Difficulty in properly perceiving the radiation risk in the acute phase

FMU had difficulty making a proper radiation risk assessment because of a lack of knowledge, skill, and experience for dealing with a nuclear disaster and the radionuclides that proliferated from the FDNPP. Previous reports said that after the accident had become publicly known, a high expectation of what should be provided by hub hospitals during disasters was unavoidable and so a heavy demand on those hospitals was unavoidable. We were also expected to distribute accurate information to the public to help resolve concern about the radiation. However, knowledge at FMU was insufficient, which led to unnecessary concerns (The National Diet of Japan the Fukushima Nuclear Accident Independent Investigation Commission, 2012; Investigation Committee on the Accident at the Fukushima Nuclear Power Stations of Tokyo Electric Power Company, Investigation Report, n.d.). Not only did the staff at FMU lack knowledge and skills necessary for radiation detection and measurement, they also did not understand the actual impact on human health, what will happen next, and how much risk we might suffer; in other words, "the perceived value of risk." As a result, without knowing the specific biological effects, only the size of a risk number was seen by the public, which caused anxiety among staff, patients, and local residents. As a result, it was difficult to recognize the risk of radiation. The hazards of radiation and radioactive materials can be detected and measured, and the impact on human health can therefore be estimated. However, materials, accidents, and disasters associated with radiation tend to be considered more mysterious and dreadful than those associated with other threats to health (Slovic et al., 1991). The mystery of radiation is that it cannot be observed by the five senses, is difficult to notice by those exposed to the risk, the effects appear later, and the risks are subject to scientific uncertainties. Similarly, dread is associated with factors such as uncontrollability, tragedy involving a large number of people, difficulty mitigating risk, an increase in risk in recent years, and a sense of being captive involuntarily (Murakami, Kumagai, & Ohtsuru, 2018). These anxieties may not only concern ourselves but also our children and future generations and their present and future health status. The effects of nuclear accidents are not limited to the physical health effects of radiation but extend to social and psychological effects. As a medical institution for radiation emergency medicine and a disaster base hospital, FMU Hospital was expected to play an important role in relieving the anxiety of the local community as well as the role of the emergency and disaster center. However, it was later reported that they did not fully meet society's expectations because of a lack of knowledge about radiation and the anxiety felt by staff (2012) (Investigation Committee on the Accident at the Fukushima Nuclear Power Stations of Tokyo Electric Power Company, Investigation Report, n.d.). In a nuclear disaster, medical personnel are required to have basic, but to them not even recognized, knowledge and skills, such as knowledge of how to operate survey meters for radiation measurement and understanding of measurement units and their meaning. In addition, medical staff and hospital personnel who respond to disasters in high-risk situations need to calmly evaluate the crisis, perceive the risks appropriately, and recognize what they need to do as medical personnel.

Radiation emergency medicine at FMU Hospital

FMU Hospital, despite a lack of experience in radiation emergency medicine, had to provide treatment after the accident in an atmosphere of radiation-related anxiety and uncertain information; indeed, none of the medical facilities around the FDNPP, including FMU Hospital, had prior experience treating patients with radiation-related injuries and illnesses (Tominaga et al., 2014). Up until March 31, 2011, FMU Hospital treated a total of 14 patients for radiation-related problems: eight who were transported to the FDNPP due to injuries and illnesses that occurred on the site and six who were exposed or contaminated in other areas. FMU Hospital also had no previous experience treating patients contaminated by radioactive materials, so in the initial phase the emergency physicians treated them while referring to a textbook (Nakashima et al., 2012). From March 15 on, the SDF, the Japan Atomic Energy Agency (JAEA), and radiation emergency medical support teams consisting of Hiroshima and Nagasaki University staff, were stationed at FMU to provide human and material support, as if to replace the Hospital's DMAT. With their support, FMU established an in-hospital radiation medical team and gradually established an independent radiation emergency medical system by conducting multidisciplinary meetings, workshops, and drills. The basic goal of FMU was to maintain the quality of normal medical care at the same time as providing emergency (disaster-related) medical care. Therefore, the principle of the radiation emergency medical response at FMU Hospital was based on adding to the usual medical system such activities as radiation protection measures characteristic of treatment for injuries and illnesses involving radioactive materials, measures to prevent the spread of contamination, contamination measurement and decontamination, and dose assessment (Editorial Committee for the Fifth Revised Edition of the Guidelines for the Initial Treatment of Trauma, 2016). Support by SDF and JAEA personnel continued for 4 months, and support by initial members of the radiation emergency medical support teams continued for about a year after the accident; these personnel contributed greatly to the development of a nuclear disaster medical system at FMU Hospital.

Crisis communication

In a critical situation (including a nuclear disaster), crisis communication plays an important role in building consensus and fostering sound risk perception during the disaster response among front-line medical personnel. During the acute phase of the Fukushima accident, Japan's first severe nuclear accident, information required to implement disaster countermeasures was not provided to front-line medical staff (Nakashima et al., 2012). There was a lack of information in the FMU hospital; we had no idea what was happening at the nuclear power plant, what was going to happen near the plant, what the effects of radiation exposure would be, or what would happen in the future. Anxiety overwhelmed the public and the medical profession, and as a result the hospital faced a crisis, even if temporary, of business continuity (Nakashima et al., 2012). The hospital's crisis communication efforts to respond to the nuclear disaster were accomplished through three steps. As a first step, basic knowledge about radiation risks was provided by in-hospital experts to hospital staff. The second step was accomplished by the radiation emergency medical support team; dialogue and trust building were carefully performed as the team stayed at the hospital day and night. The third step was accomplished by an outside expert who provided us with discussion time to consider what we could do—how a hospital should deal with the crisis. More than 200 medical staff gathered in an auditorium and shared the time to consider countermeasures (Fig. 5). Crisis communication is a countermeasure to enable an organization to respond to a sudden and serious event, and it should have been prepared for and provided by the government. However, at that time in Fukushima, crisis communication was provided solely by private individuals who were aware of the hospital's crisis. After that, our perception of the nuclear crisis changed as we recognized it as something we had to solve by ourselves, not something that others could solve for us. It was like the process cancer patients undergo as they overcome despair, anguish, and anxiety and adjust to reality after being informed of their cancer (Holland & Rowland, 1990). The Fukushima experience suggests that an effective plan for crisis management and crisis communication with the population in a crisis is an absolute necessity.

Recent situation of radiation emergency medical preparedness at FMU Hospital

The radiation emergency medical system at FMU Hospital has now improved, and FMU Hospital is playing an important role in the Nuclear Disaster Medical System of Japan to help the nation prepare for future critical events. One reason for the difficulty in providing an appropriate medical response after the FDNPP accident was the lack of coordination among specialties within the existing medical system, such as emergency medicine and disaster medicine. In a huge

FIG. 5 Crisis communication.

compound disaster, the emergency medical system and disaster medical system did not function, nor did the radiation emergency medical system operate effectively, and the subsequent collapse led to overreliance on other specific facilities to take charge. Fortunately, no contaminated workers and residents suffered life-threatening injuries during this state of confusion; however, the collapse of medical systems might have—had things been different—resulted in preventable deaths among persons who suffered severe trauma or who were critically ill. At the governmental level, the regulation of atomic energy was previously included in the promotional organization before the FDNPP accident. After the accident, governmental amendments to nuclear management were created to newly establish the Nuclear Regulation Authority (NRA), in which nuclear regulation is integrated and independently maintained. As for the concept of medical care in the event of a nuclear disaster, it was examined in a project commissioned in 2013 by the NRA to improve the effectiveness of the radiation emergency medical system, and a new nuclear disaster medical system was launched in August 2016 (Nuclear Regulation Authority, 2020a, 2020b). In it, four facilities—including FMU Hospital—have been designated as Nuclear Emergency Medical Care and Integrated Support Centers to distribute knowledge and skill, and five facilities—also including FMU Hospital—have been designated as Advanced Medical Support Centers for Radiation Exposure to provide high-level radiation emergency medical care and to maintain a multilayered radiation emergency medical system. In FMU Hospital, a Radiation Disaster Medical Center was established as an organization directly supervised by the head of the hospital; it enacted a hospital-based system to proactively engage in nuclear disaster events and has been active since May 2012. Chemical, biological, radiological, nuclear, and explosive (CBRNE) disaster drills have been held every year since 2012 from the perspective of regarding radioactive materials as one of the potential hazards in our daily lives as a lesson from the FDNPP accident. Recently, disaster management drills have been officially designated as one of the job requirements of hospital staff. On the first Monday of every month, the person in charge from each department holds a meeting to plan education and training and to revise (as needed) the manual for dealing with contaminated patients in the event that they come to the hospital. Unexpectedly, these activities are none other than business continuity management in the Business Continuity Plan for a disaster event.

Conclusion

We at FMU Hospital were inadequately prepared for the accident; in other words, we lacked the knowledge, skills, experience, and awareness of the situation that arises from a nuclear disaster. On the other hand, what we learned from our experience in the crisis was the inescapable fact that front-line healthcare providers are called to provide the best possible healthcare services even in the face of an unexpected, unfamiliar, and uncertain situation. What we can do to prepare ahead of time for future crises is to take all possible measures in advance, and in doing so, reduce the number of unexpected, unfamiliar, and uncertain factors to the extent possible. In order to do so, it is essential that we continue each day to learn from the history and experiences of the past. On the other hand, the descriptions and records of past nuclear disasters have already reported many of the issues that arose during the Fukushima accident as if they were harbingers of the future. From the facts noted above, we should realize once again that we have not yet learned enough from our past experiences nor from the history of the FDNPP accident. The compass for Fukushima's future may actually be found in the history and experiences of the near past.

References

Akashi, M., Kumagaya, K., Kondo, H., & Hirose, Y. (2010). Concerns of disaster medical assistance team (DMAT) members about troubles at the nuclear power plant: Experience from the Niigata Chuetsu-Oki earthquake, 16 July 2007, in Japan. *Health Physics*, *98*, 804–809. https://doi.org/10.1097/HP.0b013e3181ac924a.

Carr, Z., Maeda, M., Oughton, D., & Weiss, W. (2018). Non-radiological impact of a nuclear emergency: Preparedness and response with the focus on health. *Radiation Protection Dosimetry*. https://doi.org/10.1093/rpd/ncy163.

Editorial Committee for the Fifth Revised Edition of the Guidelines for the Initial Treatment of Trauma, T. J. S.o. T. S. (2016). *Guidelines for the initial treatment of trauma: JATEC* (5th ed.). Herusu shuppan.

Fukushima Health Management Survey Examination Committee, Basic Survey (Radiation Dose Estimates) Reported on 31 March 2020. (2020). https://www.pref.fukushima.lg.jp/uploaded/attachment/386023.pdf. Retrieved from https://www.pref.fukushima.lg.jp/uploaded/attachment/386023.pdf.

Fukushima Medical University Hospital Radiation Disaster Medical Center. (2013). *Facing the radiation disaster: A message from the medical community in Fukushima*. Life Sciences Publishing Co.

Hasegawa, A., Ohira, T., Maeda, M., Yasumura, S., & Tanigawa, K. (2016). Emergency responses and health consequences after the Fukushima accident; evacuation and relocation. *Clinical Oncology (The Royal College of Radiologists)*, *28*(4), 237–244. https://doi.org/10.1016/j.clon.2016.01.002.

Hasegawa, A., Takahashi, M., Nemoto, M., Ohba, T., Yamada, C., Matsui, S., et al. (2018). Lexical analysis suggests differences between subgroups in anxieties over radiation exposure in Fukushima. *Journal of Radiation Research*, *59*(S2), 83–90. https://doi.org/10.1093/jrr/rry027.

Hasegawa, A., Tanigawa, K., Ohtsuru, A., Yabe, H., Maeda, M., Shigemura, J., et al. (2015). Health effects of radiation and other health problems in the aftermath of nuclear accidents, with an emphasis on Fukushima. *Lancet*, *386*(9992), 479–488. https://doi.org/10.1016/S0140-6736(15)61106-0.

Hayano, R. S., & Adachi, R. (2013). Estimation of the total population moving into and out of the 20 km evacuation zone during the Fukushima NPP accident as calculated using "Auto-GPS" mobile phone data. *Proceedings of the Japan Academy. Series B, Physical and Biological Sciences*, *89*, 196–199. Retrieved from http://www.ncbi.nlm.nih.gov/pubmed/23666090. http://www.ncbi.nlm.nih.gov/pmc/articles/PMC3722575/pdf/pjab-89-196.pdf.

Holland, J. C., & Rowland, J. H. (1990). *Handbook of psychooncology: Psycological care of the patient with cancer*. Oxford University Press.

Ichiseki, H. (2013). Features of disaster-related deaths after the Great East Japan Earthquake. *Lancet*, *381*(9862), 204. https://doi.org/10.1016/S0140-6736(13)60091-4.

Igarashi, Y., Tagami, T., Hagiwara, J., Kanaya, T., Kido, N., Omura, M., et al. (2018). Long-term outcomes of patients evacuated from hospitals near the Fukushima Daiichi nuclear power plant after the Great East Japan Earthquake. *PLoS ONE*, *13*(4). https://doi.org/10.1371/journal.pone.0195684, e0195684.

International Atomic Energy Agency. (1997). *Generic assessment procedures for determining protective actions during a reactor accident*. IAEA-TECDOC-955.

International Atomic Energy Agency. (2011). *Criteria for use in preparedness and response for a nuclear or radiological emergency: General safety guide (vol. no. GSG-2)*. International Atomic Energy Agency.

Investigation Committee on the Accident at the Fukushima Nuclear Power Stations of Tokyo Electric Power Company, Investigation Report. n.d., Retrieved from https://www.cas.go.jp/jp/seisaku/icanps/eng/final-report.html (12 December 2020).

Katata, G., Terada, H., Nagai, H., & Chino, M. (2012). Numerical reconstruction of high dose rate zones due to the Fukushima Dai-ichi Nuclear Power Plant accident. *Journal of Environmental Radioactivity*, *111*, 2–12. https://doi.org/10.1016/j.jenvrad.2011.09.011.

Kim, J., Rukundo, F., Musauddin, A., Tseren, B., Uddin, G. M. B., & Kiplabat, T. (2020). Emergency evacuation simulation for a radiation disaster: A case study of a residential school near nuclear power plant in Korea. *Radiation Protection Dosimetry*, *189*(3), 323–336. https://doi.org/10.1093/rpd/ncaa046.

Koyama, A., Fuse, A., Hagiwara, J., Matsumoto, G., Shiraishi, S., Masuno, T., et al. (2011). Medical relief activities, medical resourcing, and inpatient evacuation conducted by Nippon Medical School due to the Fukushima Daiichi Nuclear Power Plant accident following the Great East Japan Earthquake 2011. *Journal of Nippon Medical School*, *78*(6), 393–396. Retrieved from http://www.ncbi.nlm.nih.gov/pubmed/22197874. https://www.jstage.jst.go.jp/article/jnms/78/6/78_6_393/_pdf.

Morimura, N., Asari, Y., Yamaguchi, Y., Asanuma, K., Tase, C., Sakamoto, T., et al. (2013). Emergency/disaster medical support in the restoration project for the Fukushima nuclear power plant accident. *Emergency Medicine Journal*, *30*(12), 997–1002. https://doi.org/10.1136/emermed-2012-201629.

Murakami, M., Kumagai, A., & Ohtsuru, A. (2018). Building risk communication capabilities among professionals: Seven essential characteristics of risk communication. *Radiation Protection Dosimetry*. https://doi.org/10.1093/rpd/ncy140.

Murakawa, M. (2012). The Great East Japan Earthquake and Fukushima Daiichi Nuclear Power Plant Accident and Fukushima Medical University. *Japanese Journal of Reanimatology*, *31*(2), 82–84. https://doi.org/10.11414/jjreanimatology.31.82.

Nakashima, M., Takamura, N., Suzuki, K., & Yamashita, S. (2012). A new challenge of radiation health risk management. In *Proceedings of the 6th international symposium of Nagasaki University global COE program "Global Strategic Center for Radiation Health Risk Control"* Nagasaki Newspaper Publish.

National Police Agency, n.d. Press release. A damage situation and a police measure of the Great East Japan Earthquake 2011. https://www.npa.go.jp/news/other/earthquake2011/pdf/higaijokyo.pdf. Retrieved from https://www.npa.go.jp/news/other/earthquake2011/pdf/higaijokyo.pdf.

Nomura, S., Gilmour, S., Tsubokura, M., Yoneoka, D., Sugimoto, A., Oikawa, T., et al. (2013). Mortality risk amongst nursing home residents evacuated after the Fukushima nuclear accident: A retrospective cohort study. *PLoS ONE*, *8*, e60192. https://doi.org/10.1371/journal.pone.0060192. Epub 2013 March 26.

Nuclear Regulation Authority (2020a, 13 December). Estimation of emergency doses and effects of protective measures. Retrieved from www.nsr.go.jp/data/000055769.pdf.

Nuclear Regulation Authority. (2020, December 13). *Medical system in the event of a nuclear disaster*. Retrieved from http://www.nsr.go.jp/activity/bousai/measure/medicalsystem.html.

Ohba, T., Ishikawa, T., Nagai, H., Tokonami, S., Hasegawa, A., & Suzuki, G. (2020). Reconstruction of residents' thyroid equivalent doses from internal radionuclides after the Fukushima Daiichi nuclear power station accident. *Scientific Reports, 10*(1), 3639. https://doi.org/10.1038/s41598-020-60453-0.

Radiation Emergency Medicine. (2019). In A. Hasegawa (Ed.), *Vol. 43. The Japanese journal of acute medicine*. Tokyo: Herusu Shuppan. pp. 3p, 631–858.

Slovic, P., Layman, M., Kraus, N., Flynn, J., Chalmers, J., & Gesell, G. (1991). Perceived risk, stigma, and potential economic impacts of a high-level nuclear waste repository in Nevada. *Risk Analysis, 11*(4), 683–696. Retrieved from http://www.ncbi.nlm.nih.gov/pubmed/1780505.

Tanigawa, K., & Hasegawa, A. (2013). Medical perspective: 3.1 prehospital emergency medical response. In K. Tanigawa, & R. K. Chhem (Eds.), *Radiation disaster medicine: Perspective from the Fukushima nuclear accident*. New York: Springer.

Tanigawa, K., Hosoi, Y., Hirohashi, N., Iwasaki, Y., & Kamiya, K. (2011). Evacuation from the restricted zone of the damaged Fukushima nuclear power plant: Facing with the reality. *Resuscitation, 82*(9), 1248. https://doi.org/10.1016/j.resuscitation.2011.04.028.

Tanigawa, K., Hosoi, Y., Hirohashi, N., Iwasaki, Y., & Kamiya, K. (2012). Loss of life after evacuation: Lessons learned from the Fukushima accident. *Lancet, 379*(9819), 889–891. https://doi.org/10.1016/S0140-6736(12)60384-5.

Tanigawa, K., Hosoi, Y., Terasawa, S., Kondo, H., Asari, Y., Shishido, F., et al. (2011). Lessons learned from the Fukushima Daiichi nuclear power plant accident; the initial 5 days medical activities after the accident. *JJAAM, 22*(9), 782–791. https://doi.org/10.3893/jjaam.22.782.

TEPCO, n.d.-a. Implementation of heat stroke prevention measures at Fukushima Dai-ichi Nuclear Power Plant in FY2020. Retrieved from https://www.tepco.co.jp/decommission/information/committee/roadmap_progress/pdf/2020/d201126_12-j.pdf.

TEPCO, n.d.-b. Overview of decommissioning and countermeasures for contaminated water. Retrieved from https://www.tepco.co.jp/decommission/information/committee/roadmap_progress/pdf/2020/d201126_05-j.pdf.

The National Diet of Japan the Fukushima Nuclear Accident Independent Investigation Commission. (2012). The official report of the Fukushima Nuclear Accident Independent Investigation Commission: Main report. Retrieved from http://warp.da.ndl.go.jp/info:ndljp/pid/3856371/naiic.go.jp/en/report/ (20 December 2020).

Tominaga, T., Hachiya, M., Tatsuzaki, H., & Akashi, M. (2014). The accident at the Fukushima Daiichi Nuclear Power Plant in 2011. *Health Physics, 106*, 630–637. https://doi.org/10.1097/HP.0000000000000093.

UNSCEAR. (2014). *Sources, effects and risks of ionizing radiation: United Nations Scientific Committee on the Effects of Atomic Radiation: UNSCEAR 2013 report Volume I. Report to the General Assembly. Scientific annexe A: Levels and effects of radiation exposure due to the nuclear accident after the 2011 great east-Japan earthquake and tsunami (9789211422917, 9789211422931)*. Retrieved from http://ci.nii.ac.jp/ncid/BB15643898.

Yasumura, S. (2014). Evacuation effect on excess mortality among institutionalized elderly after the Fukushima Daiichi Nuclear Power Plant accident. Fukushima Journal of Medical Science Retrieved from http://www.ncbi.nlm.nih.gov/pubmed/25283975; https://www.jstage.jst.go.jp/article/fms/advpub/0/advpub_2014-13/_pdf.

Yasumura, S., Goto, A., Yamazaki, S., & Reich, M. R. (2013). Excess mortality among relocated institutionalized elderly after the Fukushima nuclear disaster. *Public Health, 127*, 186–188. https://doi.org/10.1016/j.puhe.2012.10.019.

Chapter 3

Radiation dose after the disaster

Tetsuo Ishikawa

Department of Radiation Physics and Chemistry, Fukushima Medical University, Fukushima, Japan

Release of radioactive materials into environment

Overview

Radioactive materials started to be released from the Fukushima Daiichi Nuclear Power Station (FDNPS) on March 12, 2011, and continued over an extended period. The most radiologically significant releases were isotopes of iodine and cesium, in particular, ^{131}I, ^{134}Cs, and ^{137}Cs. The releases to the atmosphere were estimated to be around 150 petabecquerels (PBq) for ^{131}I and 12 PBq for ^{137}Cs (Steinhauser, Brandl, & Johnson, 2014). The amounts of ^{134}Cs released were similar to those of ^{137}Cs. In addition, radioactive materials entered the marine environment directly from leaks from the damaged nuclear power plants (Table 1) (Aliyu, Evangeliou, & Mousseau, 2015).

The quantities of atmospheric releases were estimated to be about 10%–20% of the releases that occurred during the Chernobyl accident (United Nations Scientific Committee on the Effects of Atomic Radiation, 2011). Also, more than 80% of the atmospherically released radionuclides were estimated to have gone offshore from the FDNPS Fukushima, followed by deposition in the Pacific Ocean (Steinhauser et al., 2014). Because dilution in the ocean was found to occur quickly, the radiological consequences in countries other than Japan appeared negligible. Chernobyl, in contrast, is located in the center of the European continent and there was also significant contamination outside the evacuation zone, including in many European countries. The situations mentioned above resulted in differences in the sizes of contaminated areas between the Fukushima and Chernobyl accidents (Table 2) (Imanaka, Hayashi, & Endo, 2015).

Time series of atmospheric release

The pattern of atmospheric releases was complex, both temporally and spatially.

Health Effects of the Fukushima Nuclear Disaster. https://doi.org/10.1016/B978-0-12-824098-4.00016-7

TABLE 1 A comparison between Fukushima and Chernobyl in released amount of radionuclides.

	Released radioactivity ($\times 10^{15}$ Bq)		
	Fukushima		*Chernobyl*
Radionuclides	*Atmospheric*[a]	*Directly to ocean*[b]	*Total*[c]
^{133}Xe	14,000	–	6500
^{131}I	150	–	~1760
^{134}Cs	12	4–27	~47
^{137}Cs	12	4–27	~85
^{90}Sr	~0.02	0.1–2.2	~10
$^{239}Pu + ^{240}Pu$	$(1.0–2.4) \times 10^{-6}$	–	0.031

[a]*Steinhauser et al. (2014).*
[b]*Aliyu et al. (2015).*
[c]*United Nations Scientific Committee on the Effects of Atomic Radiation (2011).*

Fig. 1 shows temporal changes in released ^{131}I and ^{137}Cs radioactivities integrated for 3h each during the first 2 weeks (from March 12–26) (Terada et al., 2020). The rates of release varied considerably in magnitude, with marked increases associated with particular events in each unit (e.g., hydrogen explosions, venting, and leakage from the reactors and their containment systems). After the first week, the rates of release gradually declined. By the beginning of April, the release rates had fallen to a thousandth or less of the release rates during the first week of the accident.

The released radioactive materials formed a number of radioactive plumes (contaminated air masses). As noted in "Overview" section, most of the

TABLE 2 A comparison between Fukushima and Chernobyl in size of contaminated areas.

	Size of contaminated area (km^2)	
Soil contamination level	*Fukushima*	*Chernobyl*
>185 kBq/m^2	1700[a]	29,400[a]
555–1480 kBq/m^2	495[b]	7200[a]
>1480 kBq/m^2	272[b]	3100[a]

[a]*Steinhauser et al. (2014).*
[b]*Imanaka et al. (2015).*

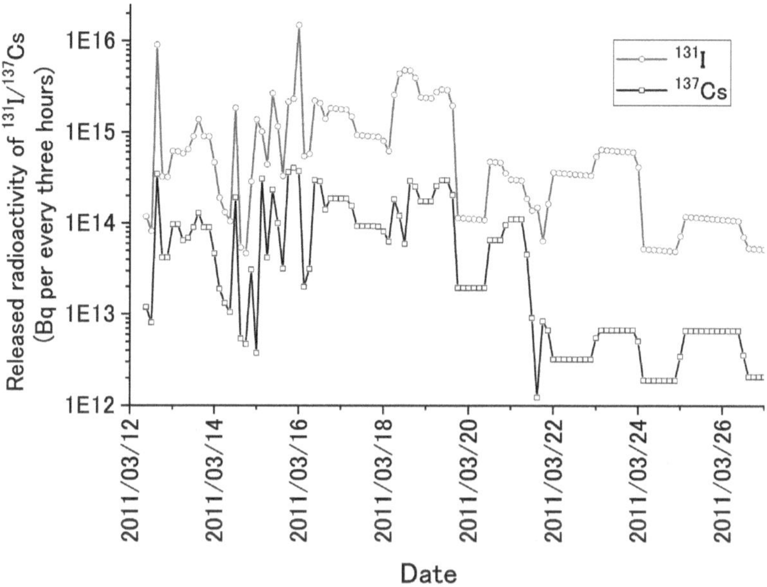

FIG. 1 Temporal changes in released amount of ^{131}I or ^{137}Cs during the first 2 weeks after the accident. The amount of radioactivity was integrated for 3-h intervals from hourly release rate data. *(No permission required.)*

radioactive plumes went to the ocean. Tsuruta, Oura, Ebihara, Ohara, and Nakajima (2014) categorized nine plumes that affected coastal land areas. Radionuclide concentrations during the plume events are important to estimate the inhalation dose, but very few measurement data were available. Terada et al. (2020) calculated radionuclide concentrations 1 m above the ground for each mesh area by an atmospheric dispersion model, the Worldwide version of the System for Prediction of Environmental Emergency Dose Information (WSPEEDI). Figs. 2–4 show ^{131}I concentrations calculated by WSPEEDI, when the first three plumes (P1–P3) among the nine hit coastal land areas. The first plume (P1) on the afternoon of March 12, which headed in a northerly direction, increased radionuclide concentrations in the northern coastal area (Fig. 2) while P2 on the morning of March 15, which headed in a southerly direction, increased radionuclide concentrations in the southern coastal area (Fig. 3).

P3 on the evening of March 15 headed in a northwesterly direction (Fig. 4).

It was observed that the air dose rate jumped for places where these plumes hit, as described in the next section.

Changes in air dose rate

There were a limited number of stations monitoring the ambient dose rate in Fukushima Prefecture before the accident.

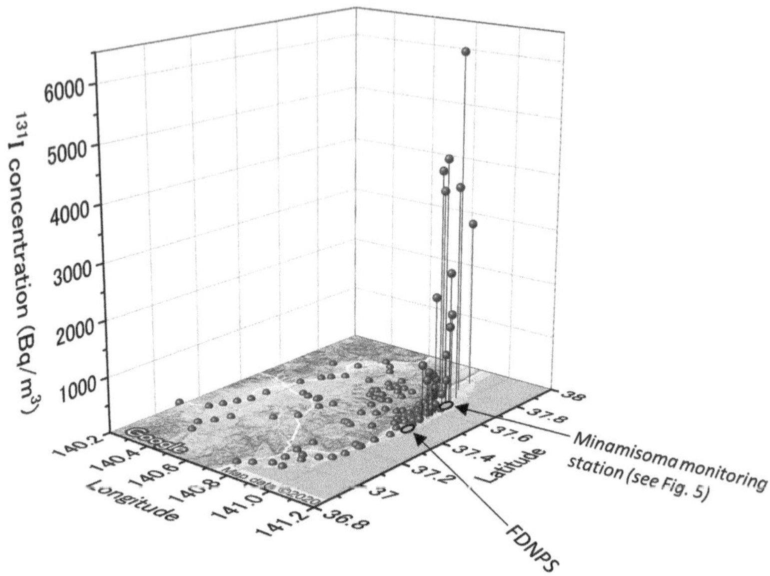

FIG. 2 Calculated ^{131}I concentrations (summed up for three chemical forms: particle, gas, and methyl) at 1 m height from land surface at 21:00 on March 12, 2011. These concentrations are depicted from WSPEEDI_2019DB results published by Terada et al. (2020), which demonstrate the temporal pattern of ^{137}Cs, ^{134}Cs, and ^{131}I concentrations (Bq/m^3) at 1 m height from the land surface. *(No permission required.)*

Fig. 5 plots reported data (Fukushima Prefecture, 2011) from such stations in seven municipalities and the map shows their locations. Typical dose rates before the accident at these stations were around 0.04–0.06 μSv/h. Corresponding to the times of the radioactive plumes, the air dose rate for each monitoring station jumped, as shown in Fig. 5.

For Minamisoma City, the air dose rate jumped the night of March 12, which was caused by plume P1 (Fig. 2). On the morning of March 15 when northeasterly winds prevailed, P2 headed in southern directions, and the air dose rate in Iwaki City, located south of the FDNPS (Fig. 3), increased. Radioactive materials were transported to metropolitan Tokyo areas such as Shinjyuku, located about 230 km south of the FDNPS, and an increase of ambient dose rate was observed. The dose rate increased to about 0.81 μGy/h on March 15, compared with an average value of 0.035 μGy/h before the accident (Konishi, Fujie, Ikushima, Hosaka, & Nakae, 2015).

Changes in air dose rate after being hit by the radioactive plumes were closely related with weather at the time the plumes arrived. For example, Minamisoma City and Iwaki City experienced no precipitation when the plumes hit, and the air dose rate rapidly decreased as they passed by. P3 on March 15 headed in a northwesterly direction because of southeasterly winds in the

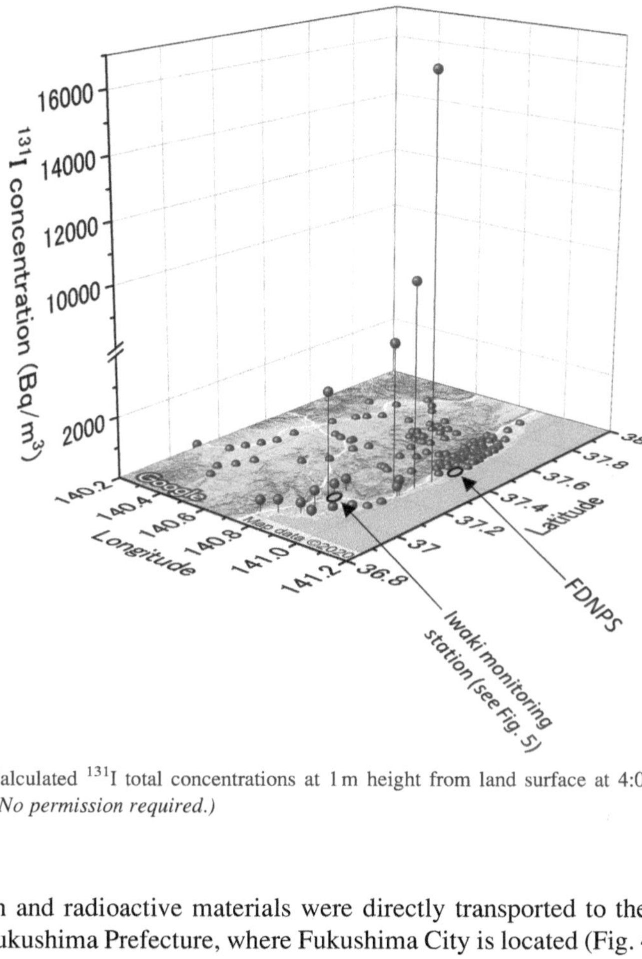

FIG. 3 Calculated ^{131}I total concentrations at 1 m height from land surface at 4:00 on March 15, 2011. *(No permission required.)*

afternoon and radioactive materials were directly transported to the northern part of Fukushima Prefecture, where Fukushima City is located (Fig. 4). At that time, it was raining in Fukushima City and deposition onto the ground was much larger compared with dry deposition (no precipitation). Thus, the decrease of air dose rate in Fukushima City was relatively slow, compared with that in the other two cities mentioned. As a result of P3 arriving at a time of precipitation, higher deposition of radionuclides was observed in the northwestern direction from the FDNPS. Fig. 6 is a map showing deposition of gamma-emitting radionuclides in eastern Japan.

Deposition was observed even in metropolitan Tokyo areas, although amounts were smaller than those in Fukushima Prefecture. By summing up the deposition-density data, the total ^{137}Cs deposition in eastern Japan was estimated to have been 2.48 PBq, which is around 20% of the atmospherically released ^{137}Cs (Saito, 2019).

After the deposition of released radionuclides onto the ground, during the rest of March the air dose rates decreased mainly due to the decay of short-lived

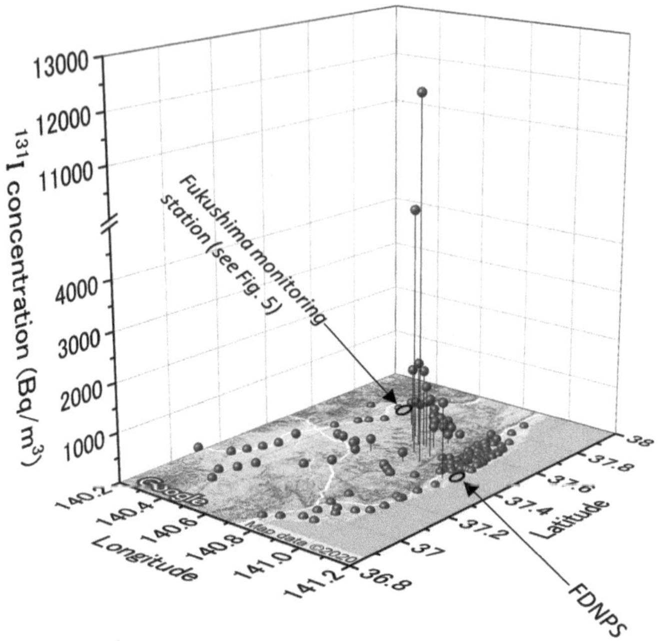

FIG. 4 Calculated ^{131}I total concentrations at 1 m height from land surface at 19:00 on March 15, 2011. *(No permission required.)*

radionuclides such as ^{131}I and ^{132}Te. After that, the dose rate gradually decreased according to the decay of ^{134}Cs (half-life: 2.1 years), weathering, and penetration of radiocesium into the ground. In later periods, decontamination also affected the decrease of air dose rates, which is discussed in "Decontamination" section.

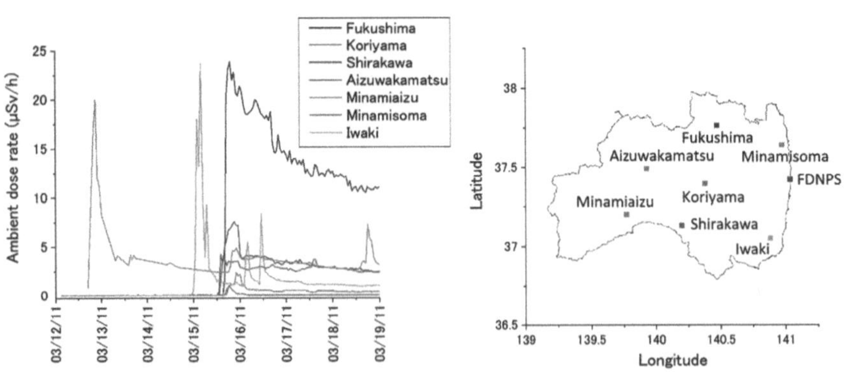

FIG. 5 Changes in air dose rate in seven municipalities (Fukushima, Minamisoma, Iwaki, Koriyama, Shirakawa, Aizuwakamatsu Cities, and Minamiaizu Village) (Fukushima Prefecture, 2011). *(No permission required.)*

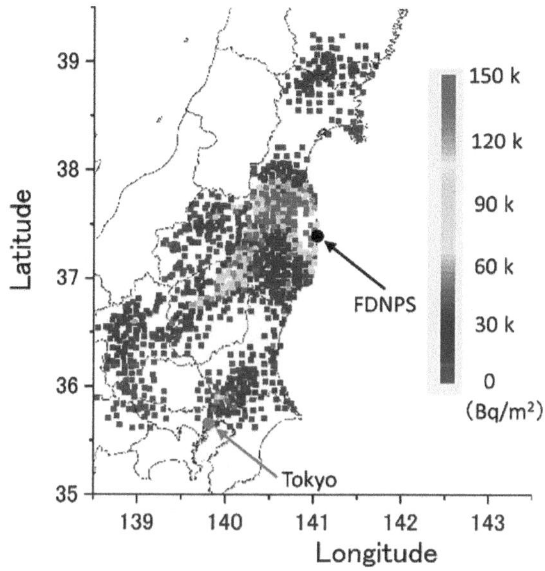

FIG. 6 Soil deposition densities of gamma-emitting nuclides measured from December 13, 2011, to May 29, 2012. The definition of "soil deposition" is the radioactivity per unit area, that is, the radioactivity integrated along the depth direction. Soil deposition measurements of gamma-emitting nuclides were conducted in the areas where the air dose rates were equal to or above 0.2 μSv/h, https://emdb.jaea.go.jp/emdb/. *(No permission required.)*

Radiation protection activities in acute phase

Soon after the release of radioactive materials, several protective actions were undertaken, including evacuations of residents and regulations preventing consumption of contaminated food and water. These protective actions ensured that radiation doses to the public were considerably reduced. "Acute phase" is here defined as within around 6 months following the beginning of the FDNPS accident (March 11, 2011).

Evacuation

Table 3 shows a timeline of evacuation orders.

At the time of the FDNPS accident, about 1900 residents lived in the 2-km zone, about 6000 residents in the 3-km zone, about 51,000 residents in the 10-km zone, and about 78,000 residents in the 20-km zone. Most people followed the evacuation orders and moved outside the 20-km zone (International Atomic Energy Agency, 2015).

The prompt evacuation was effective in reducing radiation doses, but the compulsory evacuation caused other problems. At least 12 of 27 patients with severe medical problems such as end-stage renal failure or stroke died, probably owing to their being transported more than 100 km under primitive conditions to

TABLE 3 A time series of evacuation orders.

Date	Time (JST)	Instructions on evacuation
March 11, 2011	20:50	Fukushima Prefecture issued an evacuation order for a 2-km radius around the FDNPS.
	21:23	The national government issued an evacuation order for a 3-km radius and a sheltering order for a 3–10-km radius around the FDNPS.
March 12, 2011	5:44	The national government issued an evacuation order for a 10-km radius around the FDNPS.
	18:25	The national government issued an evacuation order for a 20-km radius around the FDNPS.
March 15, 2011	11:00	The national government issued an order to shelter for residents within a 20–30-km radius of the FDNPS.

facilities by March 15 (Tanigawa, Hosoi, Hirohashi, Iwasaki, & Kamiya, 2012). Also, 50 of 438 elderly patients died in the compulsory evacuation or at evacuation centers.

On April 22, 2011, a Deliberate Evacuation Area was established beyond the 20-km evacuation zone to include areas where the projected dose criterion of 20 mSv in 1 year might be exceeded. This area was located to the northwest of the FDNPS. The national government issued an order to the effect that relocation of people from this area should be implemented in approximately 1 month.

As a result of the environmental monitoring conducted beyond the Restricted Area (i.e., the 20-km evacuation zone) and the Deliberate Evacuation Area, specific locations were identified with projected doses to residents above 20 mSv within 1 year after the accident occurrence. On June 16, the national government announced a guideline that specified that these locations should be designated as Specific Spots Recommended for Evacuation.

Food and water regulations

On the evening of March 15, initial results for radioactivity levels of agricultural products were available, indicating ^{131}I at levels ranging from 277 to 1230 kBq/kg and ^{137}Cs in the range of 31.1–169 kBq/kg in vegetation (weeds) at four locations, at a distance of 36 km and 46 km from the FDNPS (International Atomic Energy Agency, 2015).

Monitoring of food by the national and prefectural governments began on March 16, 2011. Selected foodstuffs (water, milk, vegetables, etc.) containing

TABLE 4 Provisional regulation values for radioiodine and radiocesium.

Radionuclide group	Category of food and water	Provisional regulation value (Bq/kg)
Radioiodine	Drinking water, milk, dairy products	300[a]
	Aquatic products and vegetables except coms, tubers, and roots	2000
Radiocesium	Drinking water, milk, dairy products	200
	Vegetables, cereals, meat, eggs, aquatic products, and other foodstuffs	500

[a] *100 Bq/kg was set as the provisional regulation value for powder milk for infants and milk for toddlers.*

radioactive materials that exceeded provisional regulation values (Table 4), as recommended on March 17, 2011, by the Ministry of Health, Labor, and Welfare (MHLW) of Japan, were prohibited from distribution on March 21, 2011, and from consumption on March 23, 2011.

Initially, the MHLW did not set a provisional regulation value for seafood. However, after detecting a high contamination of radioiodine in silver anchovies on April 4, 2011, the MHLW set 2000 Bq/kg as the provisional regulation value for seafood. The provisional regulation values for radioiodine were expressed as radioactive concentrations of an indicator radionuclide (^{131}I) by postulating the relative radioactive concentrations of coexisting radionuclides (^{132}I, ^{133}I, ^{134}I, ^{135}I, and ^{132}Te) (Hamada, Ogino, & Fujimichi, 2012).

After the Chernobyl accident, the continued consumption of contaminated milk caused internal exposure, especially to the thyroid. In the case of the Fukushima accident, however, regulations of food and water were started at a relatively early stage. Then, the level of ingestion dose was much smaller than expected from the deposited radionuclide (soil contamination) level, as described in "Internal dose" section.

Monitoring of water supplies was also started soon after the accident in various areas, and ^{131}I was detected in 12 out of 47 prefectures nationwide. Fig. 7 shows ^{131}I concentrations in water supplies sampled in Fukushima Prefecture and some other affected areas.

Radionuclide analyses of collected water samples were conducted using germanium semiconductor nuclide analysis devices. The highest concentrations of ^{131}I were detected at each of the respective locations from March 17–26, 2011, but ^{131}I concentrations began to decrease at many locations in the last week of March 2011. In and after April 2011, only small amounts of ^{131}I were detected at some of these locations. Radiocesium concentration was also monitored, but its concentration was relatively low compared with ^{131}I concentration. Well water

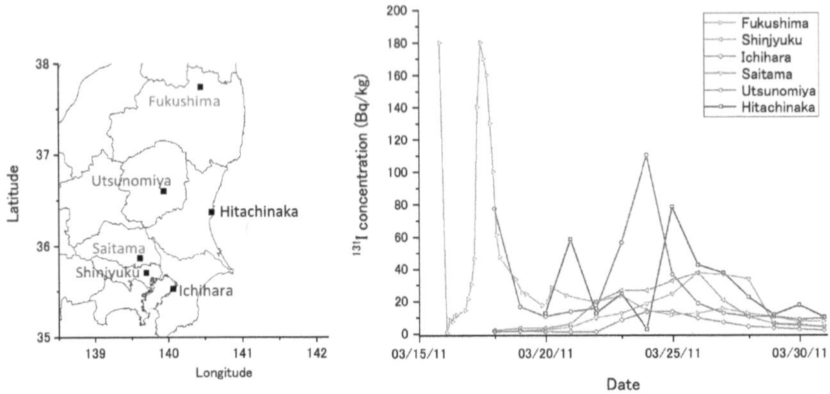

FIG. 7 ^{131}I concentration in drinking water (tap water) in Fukushima Prefecture, metropolitan Tokyo, and nearby prefectures. Water sampling locations are plotted on the map. https://emdb.jaea.go.jp/emdb/. *(No permission required.)*

for drinking was also monitored in many municipalities in Fukushima Prefecture, but cesium concentrations were below detection limits for almost all the samples.

Radiation protection activities in the chronic phase

A while after the accident, additional protective actions were started: air dose rate monitoring, decontamination, large-scale monitoring of agricultural and fishery products, and personal dose monitoring. These activities were in line with typical protection strategies to be implemented by authorities in a postaccident situation, as recommended by ICRP Publication 111 (International Commission on Radiation Protection, 2009). "Chronic phase" is here defined as after around 6 months following the FDNPS accident start.

Air dose rate monitoring

A series of airborne radiation monitoring surveys was started on April 6, 2011, using airplanes. These enabled health professionals and scientists to understand air dose rates in wide areas. Changes in ambient dose rate at 1 m height from the ground obtained by airborne monitoring are shown in Figs. 8–10.

The three maps corresponded to air dose rates on April 29, 2011, July 2, 2011, and September 28, 2013, respectively. Areas marked with red (light gray in print version) mostly had dose rates above 3.8 μSv/h, which corresponds to an annual dose of 20 mSv assuming that people spent 16 h indoors and 8 h outdoors. The size of the red areas (light gray areas in print version) gradually decreased with time. Most of the decrease was due to the physical decay of ^{134}Cs, but some could be attributed to weathering and decontamination effects.

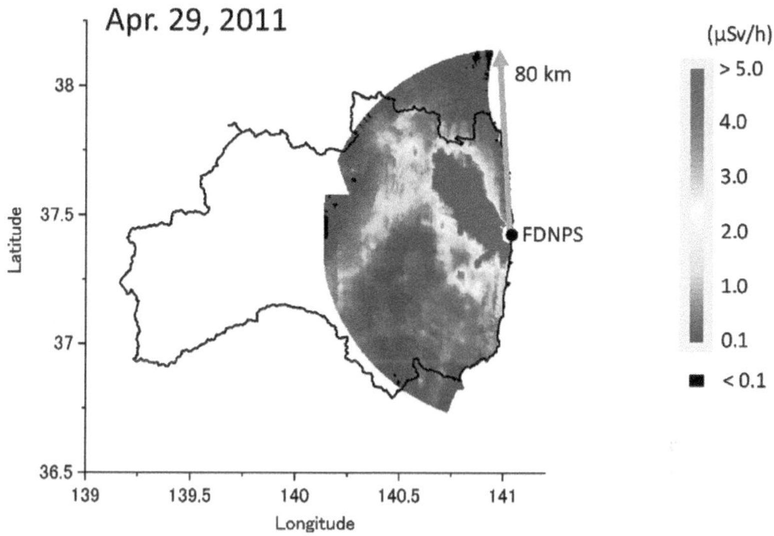

FIG. 8 Air dose rate map created by a series of airborne surveys for the 80-km zone at different times. Decay correction was made to April 29, 2011. https://emdb.jaea.go.jp/emdb/. *(No permission required.)*

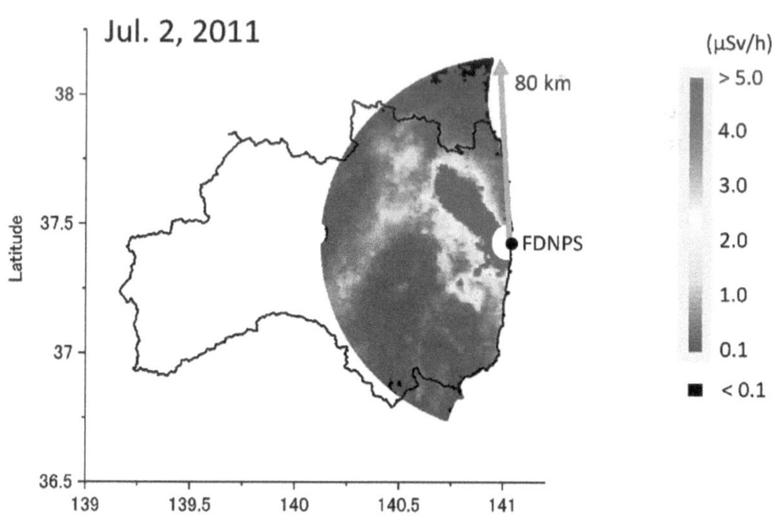

FIG. 9 Air dose rate map created by a series of airborne surveys for the 80-km zone at different times. Decay correction was made to July 2, 2011. https://emdb.jaea.go.jp/emdb/. *(No permission required.)*

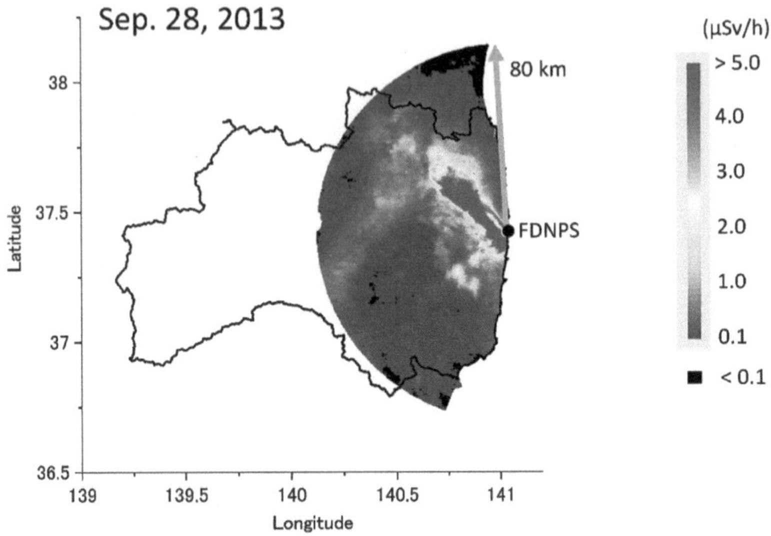

FIG. 10 Air dose rate map created by a series of airborne surveys for the 80-km zone at different times. Decay correction was made to September 28, 2013. https://emdb.jaea.go.jp/emdb/. *(No permission required.)*

During around 1 year after the accident, the national government set up about 2700 monitoring stations for real-time measurements of the ambient dose rate in Fukushima Prefecture (Fig. 11).

The number of stations later exceeded 3000. Most of them are still monitoring ambient dose rate levels, even in 2020.

Decontamination

Decontamination conducted after the accident followed three principles: removing, shielding, and keeping away from radioactive materials. These were used singly and in combination to reduce people's additional exposure doses. For removal, radioactive materials adhering to soil, vegetation, buildings, etc., were removed from people's living environment by such means as stripping topsoil; removing tree leaves, branches, and fallen leaves; and washing and cleaning the surfaces of buildings (Figs. 12 and 13).

For shielding, radioactive materials were covered with soil, etc., thereby reducing ambient doses and exposure doses accordingly. For keeping away from radioactive materials, advantage was taken of the radiation characteristic that the radioactivity intensity drops as the distance to the source increases.

Red areas (light gray areas in print version) in Fig. 8 mostly corresponded to the Returning is Difficult Zone (RDZ) and its surrounding areas named the Special Decontamination Areas (SDAs) (Ministry of the Environment, 2020). The national government was responsible for decontamination of SDAs. The

FIG. 11 Photo of a typical real-time monitoring postinstalled in Fukushima Prefecture. *(No permission required.)*

FIG. 12 Photo showing a worker engaged in decontamination of a house roof (Minamisoma City). *(From Ministry of the Environment Decontamination archive website (http://josen.env.go.jp/archive/).)*

FIG. 13 Photo showing a worker engaged in decontamination of a meadow by removing surface soil (Iitate Village). *(From Ministry of the Environment Decontamination archive website (http://josen.env.go.jp/archive/).)*

government, through the Ministry of the Environment, started decontamination of SDAs in 2012 and all decontamination work of SDAs was completed by the end of March 2017 (Ministry of Environment, 2020). Fig. 14 shows the air dose rate map decay corrected to November 16, 2017.

When compared with Fig. 8, it can be clearly seen that the size of the red areas (light gray areas in print version) was considerably decreased and had become a long and narrow area. This narrow area corresponded to the RDZ where no decontamination work was done.

In addition, areas with air dose rates above 0.23 μSv/h were designated as Intensive Contamination Survey Areas (ICSAs) in December 2011. These areas existed in almost 100 municipalities in Fukushima and neighboring prefectures. Each municipality with such areas was responsible for decontamination planning. The ICSA decontamination was completed in March 2018. Thus, Fig. 14 (the air dose rate map of November 16, 2017) shows the situation when the ICSA decontamination had almost been completed. A comparison between Fig. 8 and Fig. 14 shows a decrease in dose rates including areas outside the SDAs, which was partially due to the ICSA decontamination.

Monitoring of agricultural and fishery products

The provisional regulation values were based on the premise that the annual radiation dose from foods does not exceed 5 mSv. In April 2012, a new regulation, which limits the annual dose to 1 mSv, became effective. This value was stricter than regulation values set in other countries, as shown in Table 5 (Terada, Yamaguchi, Shimura, Svendsen, & Kunugita, 2018).

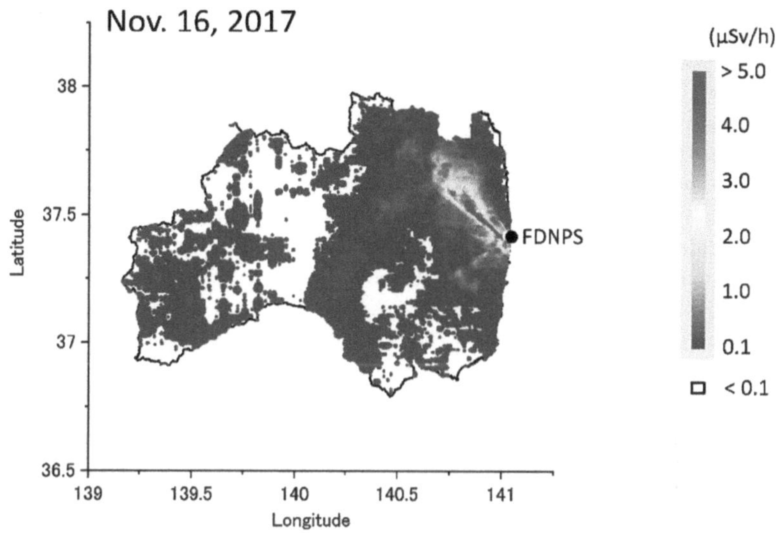

FIG. 14 Air dose rate map based on an airborne survey conducted after decontamination was almost finished. Decay correction was made to November 16, 2017. https://emdb.jaea.go.jp/emdb/. *(No permission required.)*

Foods were classified into five categories for the provisional regulation values, but were newly classified into four for these present standard limits. The standard limit for drinking water was set at 10 Bq/kg. The standard limit for milk, which children generally drink a lot of, was reduced to 50 Bq/kg. Additionally, a new category, "infant foods," was made for ensuring safety for infants

TABLE 5 Regulation values for radiocesium in Japan and comparisons with those in other countries.

Category of food and water	New regulation value in Japan (Bq/kg)	Regulation values in other countries (Bq/kg)		
		Codex committee (international food standards)	*EU*	*United States*
Drinking water	10	1000	1000	1200
Milk	50	1000	1000	1200
General foods	100	1000	1250	1200
Infant foods	50	1000	400	1200

and the standard limit therefore was set at 50 Bq/kg, the same as that for milk. The standard limit for other general foods was set at 100 Bq/kg. All foods other than infant foods were categorized as general foods based on the idea of minimizing gaps in additional doses caused by differences in individual eating habits. The food categories for the present standard limits are easier to understand for the general public and are also consistent with international views.

Local governments have been conducting drinking water and food monitoring to determine whether the radioactivity concentration in each product exceeds the regulation values. Table 6 shows monitoring results for agricultural and fishery products in a comparison between values of around the 1-year period just after the accident and fiscal year 2014, from April 2014 to March 2015 (Reconstruction Agency, 2020).

As shown by the numbers of inspected products, monitoring has been done with great effort. In particular, since 2012, Fukushima Prefecture has been inspecting all bags of rice harvested throughout the prefecture, not just the areas the national government set as requiring inspection, as an initiative by the prefecture. The proportion of products with cesium exceeding the regulation values has become very small. For example, in fiscal year 2014, the proportion was 0% for vegetables and fruits (the numbers of inspected items were 16,712 and 3302, respectively). Even for mushrooms and edible wild plants that tend to accumulate cesium, the proportion was 1.2% (the number of inspected items was 8557).

In addition to local government monitoring, independent monitoring by other authorities or individuals is also being conducted. For example, equipment for measuring cesium concentration in foods for self-consumption was installed in many municipalities in Fukushima Prefecture. People can check cesium concentration in products raised or obtained by themselves by using such equipment. In addition, independent monitoring is conducted by some agricultural cooperatives to check their own products.

Personal dose monitoring

While protective activities on environmental radiation and radioactivity were conducted as mentioned above, personal dose monitoring was also done as another protective activity. In the chronic phase, personal monitoring of internal and external exposures was done by whole-body counting (WBC) and personal dosimeters, respectively. Suitable feedback (including intervention and giving advice) based on people's measurement results could reduce their personal doses (Tsubokura et al., 2014). The monitoring methods and results are described in "Radiation dose to the public" section.

Radiation dose to the public

Due to the protective actions taken in the acute and chronic phases, the radiation dose to the public was reduced more than expected based on the environmental

TABLE 6 Monitoring results for agricultural and fishery products, comparing them between around 1 year just after the accident and FY2014 (fiscal year of 2014, from April 2014 to March 2015).

	From March 11, 2011 to the end of FY 2011 (March 2012)			FY2014 (From April 2014 to the end of March 2015)		
Category of food and water	*Number of products exceeding the regulation value*	*Proportion of exceeding (%)*	*Number of products inspected*	*Number of products exceeding the regulation value*	*Proportion of exceeding (%)*	*Number of products inspected*
Rice	592	2.2	26,464	2	0.00002	11,020,000
Wheat	27	4.8	557	0	0.0	383
Beans	16	2.3	689	4	0.1	3459
Vegetables	385	3.0	12,671	0	0.0	16,712
Fruits	210	7.7	2732	0	0.0	3302
Tea	192	8.6	2233	0	0.0	206
Eggs and meats	1058	1.3	79,181	0	0.0	1049
Mushrooms and edible wild plants	779	20.2	3856	103	1.2	8557
Aquatic products	1476	17.2	8576	100	0.5	20,922

contamination levels. Within a few years after the accident, international agencies such as the World Health Organization (WHO) and the United Nations Scientific Committee on the Effects of Atomic Radiation (UNSCEAR) reported public doses due to the accident (United Nations Scientific Committee on the Effects of Atomic Radiation, 2014; World Health Organization, 2012, 2013). However, these doses are generally considered to be overestimated. One of the reasons for this is that the protective actions actually taken were not fully considered in the WHO and UNSCEAR dose estimations. For example, WHO's reports assumed that people in Deliberate Evacuation Areas stayed there for the entire first 4 months although the inhabitants were subjected to relocations at different times during that period. UNSCEAR reported 10-year and lifetime doses, but it did not consider effects from decontamination in estimating these doses. External and internal doses described below were taken from results based on measurements that reflect the protective actions actually taken as much as possible.

External dose

Acute phase

As described before, people living within the 20-km zone were evacuated promptly. Most of the people in this zone moved through areas with different ambient dose rate levels. In the initial stage after the accident, personal dosimeters were not generally available to the public. That situation made external dose estimation difficult. Thus, individual external doses were estimated by obtaining people's behavior records (for the first 4 months) and superimposing them on the gamma ray dose rate maps. This was done as the Basic Survey that is a part of the Fukushima Health Management Survey (Yasumura et al., 2012). People's behavior data were obtained by sending a self-administered questionnaire to residents. According to the surveys on the deposition of radionuclides and air dose rate, potential exposure to residents was expected to be higher in areas to the northwest of the FDNPS (Fig. 8). Thus, two municipalities (Namie Town and Iitate Village) and one district (Yamakiya district of Kawamata Town) located in this direction were designated as "initially surveyed areas" of the Basic Survey, and the questionnaire was distributed earlier there than in other areas, beginning at the end of June 2011. Subsequently, the questionnaire was sent to all residents in Fukushima Prefecture, including those living in nonevacuated areas (Ishikawa, Yasumura, Sakai, et al., 2019).

The Basic Survey has estimated the first 4-month doses for about 466,000 residents in Fukushima Prefecture, as of the end of March 2019 (Ishikawa, Yasumura, Akahane, et al., 2019). The percentages of residents (excluding radiation workers) having various dose distributions were 62.2% less than 1 mSv; 93.9% less than 2 mSv; and 99.4% less than 3 mSv. The arithmetic mean and maximum doses were 0.8 and 25 mSv, respectively. The dose distributions were slightly different among the seven areas of Fukushima Prefecture; for example, in the Soso area, which was mostly designated as an Evacuation Area or Deliberate Evacuation Area, the percentages of residents (excluding radiation workers) having the following dose distributions were 77.3% <1 mSv; 94.9% <2 mSv; and 97.2% <3 mSv. Due to prompt evacuations, external doses were relatively low, even for residents in the evacuated areas. According to the Basic Survey, the age dependence of individual effective doses was not so large: the 4-month doses for infants (<6 years old) and children (6–15 years old) were 1.1 times and 1.04 times larger than that for adults (>15 years old) on average, respectively, in nonevacuated areas (Ishikawa et al., 2020).

The Basic Survey results showed that average 4-month doses for residents in the two municipalities (Iitate Village and Namie Town) among the highest deposition areas were around 4 mSv and 1 mSv for all age groups, respectively (Ishikawa, 2020). The difference could be mainly caused by evacuation timing. That is, while all of Iitate Village was beyond the 20-km zone and designated as a Deliberate Evacuation Area (evacuation within 1 month was recommended on April 22), most of Namie Town was included in the 20-km zone, where prompt evacuation was conducted.

Chronic phase

Around 6 months after the accident, measurements of external doses for residents by using integrating-type personal dosimeters (glass badges) were started on a large scale by local governments of Fukushima Prefecture. Generally, children and pregnant women were considered to have high priority for receiving those measurement devices at first. Later on, the target population was expanded to residents of all ages who wished to measure their personal doses. The dosimeters were usually retrieved after being used for a few months. According to the measurements for 22 municipalities in 2011, the estimated median annual external dose was reported to be <1 mSv. Some municipalities are continuing the measurements, even in 2020. Published results for personal dosimeter measurements were summarized elsewhere (Ishikawa, 2020).

As an example, monthly additional doses estimated from data obtained by wearing the glass badges for Koriyama City residents are shown in Fig. 15, together with the number of participants.

Koriyama City office started personal dosimeter measurements in October 2011 (Koriyama City, 2020). Since then, multiple measurements have been conducted per year. Four periodical series of measurements were scheduled for each year from 2013 to 2017. Measurement periods ranged from 33 to 70 days, but measurement for 70 days was typical. A background dose of 0.12 mSv per 70 days was subtracted from the measured dose to estimate the additional dose. The number of participants has gradually decreased with time. Thus, it is not possible to rigorously compare monthly additional doses to each

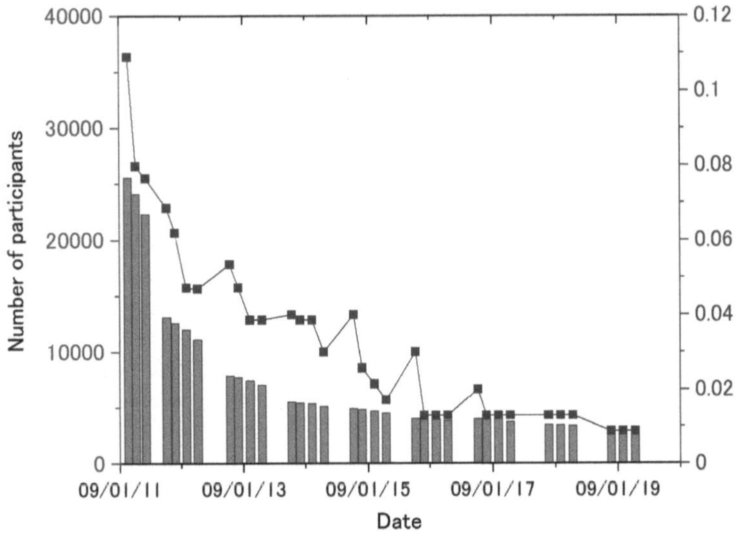

FIG. 15 Additional dose per month (right vertical axis) measured for children in Koriyama City who wore glass badges during periods from 2011 to 2019. The midpoint of each measurement period is plotted against the horizontal axis showing the date. The number of participants in each measurement is shown on the left vertical axis. *(No permission required.)*

other, but a decreasing tendency over the course of time can be seen from Fig. 15.

Although the glass badges are useful in large-scale personal monitoring, their disadvantages are that: (1) they generally need a measurement period of a few months and (2) it is not possible to determine when and where the person is exposed (at home, at work, etc.). Another type of personal dosimeter, the D-shuttle, overcoming these disadvantages was developed in 2013 (Naito, Uesaka, Kurosawa, & Kuroda, 2017). It can record hour-by-hour personal doses and it is useful for relating personal behavior with personal dose. Using the D-shuttle in combination with GPS systems makes it possible to identify exposure levels, locations, and personal doses. Identifying source contributions to the total dose is important in determining effective dose reduction measures. In this respect, the D-shuttle may be a good tool for risk communication.

In relation to the progress of decontamination work, the government started lifting evacuation orders in some areas. People began to return to or temporally stay in their hometowns in such areas. The D-shuttle was used for monitoring external doses for those people. A study on returnees to Iitate Village reported the additional external doses per year were below 3 mSv in 2017 when the evacuation order was lifted (Naito et al., 2017). Another study (Nomura et al., 2020) analyzed doses for 239 voluntary participants among returnees across 10 municipalities. Monte Carlo simulations used for quantifying both the uncertainty and population variability of observed data demonstrated that the mean of the annual dose in 2019 (including natural background doses) was 0.93 (95% uncertainty interval 0.53–1.76) mSv, with limited variation among municipalities. Even in the most affected areas, Iitate Village and Namie Town, the respective median annual doses were 1.16 and 0.86 mSv, including background doses.

Summary of external dose

The Basic Survey results showed that average 4-month doses for residents in the two most affected areas (Namie Town and Iitate Village) were around 4 mSv and 1 mSv, respectively, for all age groups. Because almost all the residents of these two municipalities had moved to nonevacuated areas after the first 4 months, their subsequent doses could be inferred from personal dosimeter measurements conducted in nonevacuated areas, as described elsewhere (Ishikawa, 2020). Municipality averages were around 0.5 mSv per 3 months in 2011, even in relatively higher ambient dose areas among nonevacuated areas. The 8-month dose could be roughly inferred as 1.5 mSv from this result. For Iitate Village and Namie Town, the average values for the first-year effective doses for residents in these areas could be 6 mSv and 3 mSv, respectively. Actually, some of the residents in these two areas evacuated to distant places with little contamination, including outside Fukushima Prefecture, and consequently, the average doses could be lower. For nonevacuated areas, the maximum municipality-average first-year effective dose could be around 3 mSv.

It is difficult to estimate the long-term integrated external dose, especially for residents of evacuated areas, because moves of residents were complex. The Basic Survey obtained behavior records for the first 4 months, but there were no available behavior data for subsequent periods. UNSCEAR estimated the first-year and 10-year effective doses for each municipality of nonevacuated areas. When these values were used, the difference in the two doses (i.e., the 9-year dose) was at most 4 mSv for nonevacuated areas. Because UNSCEAR did not consider the effects of decontamination, the dose could be a conservative value. Thus, an average 10-year dose for nonevacuated areas could be at most 7 mSv by adding 4 mSv to the first-year doses mentioned above (at most 3 mSv). In the case of Iitate Village, an average 10-year dose could be at most 10 mSv, even if residents continue to live in relatively high-dose-rate areas among the nonevacuated areas.

Some of the evacuated people have begun to return to Iitate Village. The reported additional external dose per year for the village was below 3 mSv in 2017 (Naito et al., 2017). The median external dose including background dose in 2019 was 1.16 mSv (Nomura et al., 2020). From these values, an average dose received by returnees after 2017 could be less than 5 mSv. Before returning, an average integrated dose up to 2016 could be less than 10 mSv. Thus, an average 10-year external dose for Iitate Village returnees might exceed 10 mSv, but it is not likely to reach 15 mSv. Also, in the case of Iitate Village (the evacuation order was lifted on March 31, 2017), the percent of returning people was around

30% in 2020. Thus, an average 10-year dose for all Iitate villagers is likely to be less than 15 mSv.

Internal dose

Acute phase

Thyroid exposure due to released radioiodine was a big concern, especially in children, after the Fukushima accident. In this respect, thyroid screenings (direct measurements of thyroid activity) in children were started around 2 weeks after the accident by the Nuclear Emergency Response Local Headquarters. In these measurements, the thyroid activities of 1080 children were estimated using survey meters without energy discrimination of gamma rays. As a result, the thyroid equivalent doses were estimated to be less than 30 mSv for most of the subjects. Because 100 mSv was regarded as a cutoff level in the screening, no measures were taken for persons below this level (Kim & Kurihara, 2020). Independently from the above measurements, Tokonami et al. (2012) estimated thyroid doses from ^{131}I activity measured using a NaI (Tl) spectrometer and found that the highest equivalent dose was 33 mSv among 62 persons from under 1 to 83 years of age. Although the doses in the two studies were much smaller than that received by evacuees of the Chernobyl accident (the average was estimated to be 490 mSv) (United Nations Scientific Committee on the Effects of Atomic Radiation, 2011), the number of subjects examined by the direct measurements of thyroid activity was limited compared with the population in the evacuated areas before the Fukushima accident. This encouraged studies on reconstruction of the thyroid dose, as described in "Reconstruction of thyroid dose" section.

In an early stage after the accident, there were no available whole-body counters (WBCs) to check internal contamination in Fukushima Prefecture. Then, WBCs installed in distant places were used in the acute phase and persons were brought there. The Fukushima Prefecture government organized WBC measurements by a committed institute, the National Institute of Radiological Sciences in Chiba, which started at the end of June 2011.

A total of 174 subjects mostly from contaminated areas (within 30-km areas) were measured from the end of June to the end of July 2011 (Kim, Kurihara, Kunishima, Nakano, et al., 2016). Radioiodine had already decayed at that time and only radiocesium could be measured. The 90th-percentile committed effective dose (CED) value for adults was around 0.1 mSv and the maximum CED (0.63 mSv) was found in an elderly male. Prompt evacuations and the regulations of food and water were considered to be effective in reducing the inhalation of radioactive particles and the ingestion of radiocesium.

On July 10, 2011, another committed institute, the Japan Atomic Energy Agency (JAEA), started WBC measurements (Momose et al., 2012). A total of 9927 subjects were measured until the end of January 2012. Although most of these subjects were residents of evacuated areas, the median CED values were 0.02 mSv and 0.025 mSv for subjects aged 13–17 years and for subjects aged >17 years, respectively. The results also showed low internal contamination levels by radiocesium. According to the UNSCEAR 1988 report on the exposures from the Chernobyl accident, the mean transfer factor from ^{137}Cs deposition density (kBq/m^2) to the first-year CED (μSv/year) for adults was about 20 (United Nations Scientific Committee on the Effects of Atomic Radiation, 1988). The soil deposition densities within the 30-km area of the Fukushima accident mostly exceeded 100 kBq/m^2. When the transfer factor was used, the CED should be more than 2 mSv (Hayano et al., 2013). The internal contamination level based on WBC measurements was much smaller than that expected from the soil contamination level.

Chronic phase

Internal dose due to cesium

Several months after the accident, WBC installation in Fukushima Prefecture was started. As of 2020, Fukushima Prefecture has more than 50 WBCs. The most common type is the standing-geometry type. The typical counting time is 2 min. Some of the standing-geometry type WBCs have been installed inside buses. These buses travel to distant prefectures for measuring evacuees who left Fukushima after the accident.

According to a report by Fukushima Prefecture (Fukushima Prefecture, 2020), around 345,000 residents (as of September 2020) had been measured using WBCs. The estimated doses were less than 1 mSv (CED) for almost all the subjects (more than 99.999%).

Results for WBC measurements by municipalities and other organizations independent from the program operated by Fukushima Prefecture have also been reported as scientific papers. They were summarized in a recent review (Ishikawa, 2020). All the papers reported very low doses in the chronic phase (an order of dozens of μSv).

Low-level internal contamination was also confirmed by measuring cesium contents in daily food prepared by ordinary families. Studies on ingestion dose estimation by food analysis were summarized elsewhere (Ishikawa, 2020). Two studies are outlined here. In the first, a research group collected 53 sets of 24-h food-duplicate samples in Fukushima Prefecture and neighboring regions in December 2011. Radioactive cesium was detected in 25 of 26 samples from Fukushima. However, the median dietary intake of radioactive cesium was 4.0 Bq/day (range <0.26–17 Bq/day). Assuming that the daily intake of radiocesium was constant throughout the year, the median estimated dose level was 23 μSv/year (range <2.6–99 μSv/year).

In the second, another research group conducted extensive analyses by the duplicate diet method. These samples were collected from November 2011 to March 2012 and from June 2012 to September 2012; each period covered 100 families selected from throughout Fukushima Prefecture. Among the 200 meals thus analyzed, only 12 were found to have ^{134}Cs and/or ^{137}Cs concentrations exceeding 1 Bq/kg. Even with the largest radiocesium value in this survey, the internal dose due to the ingestion of radiocesium did not exceed 0.1 mSv.

Reconstruction of thyroid dose

As described before, direct thyroid measurements were conducted for only about 1300 people, which was too small a number to accurately estimate the distribution of thyroid equivalent doses among evacuees who lived around the FDNPS at the time of the accident. Also, the Fukushima Prefectural Government and Fukushima Medical University (FMU) began an ultrasound thyroid examination campaign in October 2011 for about 360,000 residents who were less than 19 years old at the time of the accident (Yasumura et al., 2012). As of March 2019, 218 suspected or definite thyroid cancers had been found. There is an ongoing debate in Japan as to whether these observed thyroid cancers are radiation-induced. These situations have promoted studies on reconstruction of the thyroid dose.

Kim, Kurihara, Kunishima, Momose, et al. (2016) estimated the 90th percentile of 1-year infant thyroid dose due to inhalation exposure as a representative value for each municipality in Fukushima Prefecture by using three methodologies: simulation of the atmospheric dispersion of radionuclides, WBC results of cesium content and the intake ratio of iodine to cesium, and data on direct measurements of the thyroid. The maximum thyroid dose was 30 mSv for three municipalities. Considering results obtained by subsequent studies, Kim and Kurihara (2020) considered that the thyroid doses were unlikely to exceed 50 mSv, excluding some extreme cases (e.g., persons who continued to consume highly contaminated items or did not evacuate from areas close to the FDNPS).

Efforts at reconstruction of the thyroid dose due to internal exposure are continuing even in 2020, almost 10 years after the accident. Ohba et al. (2020) published the thyroid dose with reduced uncertainties. The reduction of uncertainties was achieved by (a) improving the simulation of atmospheric dispersion of radionuclides, (b) reevaluating evacuation scenarios, (c) introducing a decontamination factor for sheltering in Japanese houses, and (d) introducing a correction factor for dose coefficient to reflect a lower iodine uptake ratio by the thyroid in the Japanese. As a result, estimated thyroid equivalent doses via inhalation were very close to doses estimated from direct thyroid measurements. The median and 95th percentile of thyroid equivalent doses in the most critical group (1-year-old children) for seven municipalities designated as evacuated areas ranged from 0.6 to 16 mSv, and 7.5 to 30 mSv, respectively.

Summary of internal dose

The internal effective dose due to radiocesium was mostly less than 0.1 mSv, even in the first year. It could be much smaller in subsequent years. The effective dose is the tissue-weighted sum of the equivalent doses in all specified tissues and organs of the human body. Thus, the thyroid equivalent dose mainly given by radioiodine also contributes to the effective dose. However, the median thyroid dose for 1-year-old children was up to 16 mSv, which contributes 0.64 mSv as the effective dose, considering the tissue weighting factor of the thyroid (0.04). That for adults was around half, 8 mSv, for the thyroid equivalent dose, which contributes as 0.32 mSv to the effective dose. Thus, the CED due to internally deposited radiocesium and radioiodine could be mostly below 1 mSv for all age groups, even in evacuated areas.

As described above, the thyroid equivalent dose for 1-year-old children (the most critical group) was at most around 30 mSv.

Summary

The Fukushima nuclear accident is often compared with the Chernobyl nuclear accident. However, the amount of radioactive materials released in the Fukushima accident was around 10%–20% of that released at Chernobyl. Also, for the former, most of the atmospherically released radioactive materials have gone to the ocean. In addition, prompt protective actions such as evacuations and regulations of food and water were effective in reducing external and internal doses to the Japanese public. The subsequent protective actions taken in the rehabilitation phase, such as the decontamination of inhabited areas and the monitoring of agricultural and fishery products, contributed to reducing external and internal doses due to radiocesium. Also, individuals making personal dosimetry measurements by the D-shuttle and monitoring by WBC measurements have also been effective in dose reduction, when suitable advice from trained staff has been given.

Due to such protective actions, actual individual doses to the public in the Fukushima accident were smaller than those estimated in an early phase after the accident. As examples of two affected areas, Iitate Village and Namie Town, the actual average values for the first-year effective doses for residents could be up to 6 mSv and 3 mSv, respectively. For nonevacuated areas, the maximum municipality-average first-year effective dose could be around 3 mSv. Although the thyroid dose estimation still had a large uncertainty, the most critical group (1-year-old infants) had doses up to around 30 mSv, according to the latest estimation.

External exposure due to deposited radiocesium continues at a very low level. It is difficult to estimate the long-term integrated external dose, especially for residents in the evacuated areas, because moves of residents were complex. According to a rough estimate, 10-year external doses for residents of Iitate Village (most affected areas) could slightly exceed 10 mSv, but were not likely to reach 15 mSv.

References

Aliyu, A., Evangeliou, N., & Mousseau, T. (2015). An overview of current knowledge concerning the health and environmental consequences of the Fukushima Daiichi Nuclear Power Plant (FDNPP) accident. *Environment International*, *85*, 213–228.

Fukushima Prefecture. (2011). *Environmental radiation monitoring data in seven regions in Fukushima Prefecture (March, 2011)* (in Japanese) https://www.pref.fukushima.lg.jp/sec/16025d/h22-24-7houbu.html.

Fukushima Prefecture. (2020). *Results of whole-body counting* (in Japanese) http://www.pref.fukushima.lg.jp/site/portal/ps-wbc-kensa-kekka.html.

Hamada, N., Ogino, H., & Fujimichi, Y. (2012). Safety regulations of food and water implemented in the first year following the Fukushima nuclear accident. *Journal of Radiation Research*, *53*, 641–671.

Hayano, R., Tsubokura, M., Miyazaki, M., Satou, H., Sato, K., Masaki, S., et al. (2013). Internal radiocesium contamination of adults and children in Fukushima 7 to 20 months after the Fukushima NPP accident as measured by extensive whole-body-counter surveys. *Proceedings of Japan Academy Series B, Physical and Biological Sciences*, *89*, 157–163.

Imanaka, T., Hayashi, G., & Endo, S. (2015). Comparison of the accident process, radioactivity release and ground contamination between Chernobyl and Fukushima-1. *Journal of Radiation Research*, *56*(S1), i56–i61.

International Atomic Energy Agency. (2015). *The Fukushima Daiichi accident: Technical volume 4/5 radiological consequences*. International Atomic Energy Agency.

International Commission on Radiation Protection. (2009). Application of the commission's recommendations to the protection of people living in long-term contaminated areas after a nuclear accident or a radiation emergency (ICRP 111). *Annals of the ICRP*, *39*.

Ishikawa, T. (2020). Individual doses to the public after the Fukushima nuclear accident. *Journal of Radiation Protection and Research*, *45*(2), 53–68.

Ishikawa, T., Yasumura, S., Akahane, K., Yonai, S., Ohtsuru, A., Sakai, A., et al. (2020). Age dependence of individual external doses in an early stage after the Fukushima accident. *Radiation Protection Dosimetry*, *188*(2), 238–245.

Ishikawa, T., Yasumura, S., Akahane, K., Yonai, S., Ohtsuru, A., Sakai, A., et al. (2019). The latest update of individual external doses in an early stage after the Fukushima accident. *Radiation Protection Dosimetry*, *187*(3), 402–406.

Ishikawa, T., Yasumura, S., Sakai, A., Ohtsuru, A., Miyazaki, M., Hosoya, M., et al. (2019). Experience in individual dose estimation after the Fukushima nuclear accident using self-administered questionnaires—activities to encourage responses to the questionnaires and resulting response rate. *Radiation Environment and Medicine*, *8*(2), 118–126.

Kim, E., & Kurihara, O. (2020). Thyroid doses in children from radioiodine following the accident at the Fukushima Daiichi nuclear power plant. *Journal of Radiation Protection and Research*, *45*, 2–10.

Kim, E., Kurihara, O., Kunishima, N., Momose, T., Ishikawa, T., & Akashi, M. (2016). Internal thyroid doses to Fukushima residents—estimation and issues remaining. *Journal of Radiation Research*, *57*(S1), i118–i126.

Kim, E., Kurihara, O., Kunishima, N., Nakano, T., Tani, K., Hachiya, M., et al. (2016). Early intake of radiocesium by residents living near the TEPCO Fukushima Daiichi Nuclear Power Plant after the accident. Part 1: Internal doses based on whole-body measurements by NIRS. *Health Physics*, *111*, 451–464.

Konishi, H., Fujie, S., Ikushima, K., Hosaka, M., & Nakae, D. (2015). Time-dependent change of ambient dose rate and gamma-ray spectrum in Shinjuku, Tokyo, after the Fukushima Daiichi Nuclear Power Plant accident. *Radioisotopes*, *64*, 185–195.

Koriyama City. (2020). *Individual cumulative dose measurement of children and students.* https://www.city.koriyama.lg.jp/bosai_bohan_safecommunity/shinsai_hoshasentaisaku/4/gaibukenko/10870.html.

Ministry of the Environment. (2020). *Environmental decontamination.* http://josen.env.go.jp/en/decontamination/.

Momose, T., Takada, C., Nakagawa, T., Kanai, K., Kurihara, O., Tsujimura, N., et al. (2012). Whole-body counting of Fukushima residents after the TEPCO Fukushima Daiichi Nuclear Power Station accident. In O. Kurihara, K. Akahane, S. Fukuda, N. Miyahara, & S. Yonai (Eds.), *The 1st NIRS symposium on reconstruction of early internal dose in the TEPCO Fukushima Daiichi nuclear power station accident* (pp. 67–82). National Institute of Radiological Sciences.

Naito, W., Uesaka, M., Kurosawa, T., & Kuroda, Y. (2017). Measuring and assessing individual external doses during the rehabilitation phase in Litate village after the Fukushima Daiichi nuclear power plant accident. *Journal of Radiological Protection, 37*, 606–622.

Nomura, S., Murakami, M., Naito, W., Yasutaka, T., Sawano, T., & Tsubokura, M. (2020). Low dose of external exposure among returnees to former evacuation areas: A cross-sectional all-municipality joint study following the 2011 Fukushima Daiichi nuclear power plant incident. *Journal of Radiological Protection, 40*, 1–18.

Ohba, T., Ishikawa, T., Nagai, H., Tokonami, S., Hasegawa, A., & Suzuki, G. (2020). Reconstruction of residents' thyroid equivalent doses from internal radionuclides after the Fukushima Daiichi nuclear power station accident. *Scientific Reports, 10*, 3639. https://doi.org/10.1038/s41598-020-60453-0.

Reconstruction Agency. (2020). *Basic information on radiation risk* (In Japanese) https://www.reconstruction.go.jp/english/topics/RR/index.html.

Saito, K. (2019). Temporal change in radiological environments on land after the Fukushima Daiichi Nuclear Power Plant accident. *Journal of Radiation Protection and Research, 44*, 128–148.

Steinhauser, G., Brandl, A., & Johnson, T. E. (2014). Comparison of the Chernobyl and Fukushima nuclear accidents: A review of the environmental impacts. *Science of the Total Environment, 470–471*, 800–817.

Tanigawa, K., Hosoi, Y., Hirohashi, N., Iwasaki, Y., & Kamiya, K. (2012). Loss of life after evacuation: Lessons learned from the Fukushima accident. *Lancet, 379*, 889–891.

Terada, H., Nagai, H., Tsuduki, K., Furuno, A., Kadowaki, M., & Kakefuda, T. (2020). Refinement of source term and atmospheric dispersion simulations of radionuclides during the Fukushima Daiichi Nuclear Power Station accident. *Journal of Environmental Radioactivity, 213*, 106104. https://doi.org/10.1016/j.jenvrad.2019.106104.

Terada, H., Yamaguchi, I., Shimura, T., Svendsen, E., & Kunugita, N. (2018). Regulation values and current situation of radioactive materials in food. *Journal of National Institute of Public Health, 67*(1), 21–33.

Tokonami, S., Hosoda, M., Akiba, S., Sorimachi, A., Kashiwakura, I., & Balonov, M. (2012). Thyroid doses for evacuees from the Fukushima nuclear accident. *Scientific Reports, 2*, 507.

Tsubokura, M., Kato, S., Nomura, S., Gilmour, S., Nihei, M., Sakuma, Y., et al. (2014). Reduction of high levels of internal radio-contamination by dietary intervention in residents of areas affected by the Fukushima Daiichi Nuclear Plant disaster: A case series. *PLoS One, 9*(6), e100302.

Tsuruta, H., Oura, Y., Ebihara, M., Ohara, T., & Nakajima, T. (2014). First retrieval of hourly atmospheric radionuclides just after the Fukushima accident by analyzing filter-tapes of operational air pollution monitoring stations. *Scientific Reports, 4*, 6717. https://doi.org/10.1038/srep06717.

United Nations Scientific Committee on the Effects of Atomic Radiation. (1988). *UNSCEAR 1988 report: Sources, effects and risks of ionizing radiation, annex D: Exposures from the Chernobyl accident.* United Nations.

United Nations Scientific Committee on the Effects of Atomic Radiation. (2011). *UNSCEAR 2008 report annex D: Health effects due to radiation from the Chernobyl accident*. United Nations.
United Nations Scientific Committee on the Effects of Atomic Radiation. (2014). *UNSCEAR 2013 report annex A: Levels and effects of radiation exposure due to the nuclear accident after the 2011 Great East-Japan Earthquake and Tsunami*. United Nations.
World Health Organization. (2012). *Preliminary dose estimation from the nuclear accident after the 2011 Great East Japan Earthquake and Tsunami*.
World Health Organization. (2013). *Health risk assessment from the nuclear accident after the 2011 Great East Japan Earthquake and Tsunami based on a preliminary dose estimation*.
Yasumura, S., Hosoya, M., Yamashita, S., Kamiya, K., Abe, M., & Akashi, M. (2012). Study protocol for the Fukushima Health Management Survey. *Journal of Epidemiology*, *22*, 375–383.

Chapter 4

Thyroid ultrasound examination program to address health concerns of Fukushima residents after the nuclear accidents

Takashi Matsuzuka, Satoru Suzuki, Satoshi Suzuki, Manabu Iwadate, Noriko Setou, Susumu Yokoya, and Hiroki Shimura
Radiation Medical Science Center for the Fukushima Health Management Survey, Fukushima Medical University, Fukushima, Japan

Thyroid examination in the Fukushima Health Management Survey

In the Fukushima Prefecture, the March 2011 earthquake and tsunami followed by the Tokyo Electric Power Company (TEPCO) Fukushima Daiichi Nuclear Power Plant accident forced about 165,000 people to evacuate (Ohto et al., 2017). Although there were no direct deaths due to radiation exposure, it was the most severe nuclear accident since the 1986 Chernobyl nuclear power plant accident, and the only other disaster to receive the Level 7 event classification of the International Nuclear Event Scale. Hence, as with Chernobyl, there have been concerns regarding health damage for the residents of the Fukushima Prefecture, one of which is childhood thyroid cancer.

During the Chernobyl nuclear accident, the amount of ^{131}I released was 1.8 million tera-Becquerel and adequate precautions were not taken against diets such as milk contaminated with radioactivity, resulting in a large amount of internal exposure of the thyroid among the inhabitants. Thyroid cancer in children reportedly began to be diagnosed 4–5 years after the nuclear accident, with most of the 6000 or more thyroid cancers among children and adolescents at the time of the accident being caused by exposure (United Nations Scientific Committee on the Effects of Atomic Radiation [UNSCEAR], 2011). According to a report in Belarus, the thyroid cancer morbidity rate has continued to rise for more than 20 years after the accident, mainly among people aged 0–5 years at

Health Effects of the Fukushima Nuclear Disaster. https://doi.org/10.1016/B978-0-12-824098-4.00017-9

the time of the accident. By contrast, there was no rise in morbidity rate for people born more than 5 years after the accident (Demidchik, Saenko, & Yamashita, 2007).

During the Fukushima Daiichi Nuclear Power Plant accident, the amount of ^{131}I released was reported to be 0.16 million tera-Becquerel (Nuclear Emergency Response Headquarters, 2011). It was assumed that the evacuation of residents and food safety regulations would be implemented and that the external and internal doses would be much smaller than those of the Chernobyl accident. However, there have been concerns about the same health hazards as the Chernobyl accident. Therefore, the Fukushima Prefecture and the Fukushima Medical University have been conducting a thyroid ultrasound examination (TUE) program since October 2011 as part of the Fukushima Health Management Survey that aims to observe children's health over the long term. The Fukushima Medical University (FMU) has been collaborating with medical institutions inside and outside the prefecture for these long-term observations.

TUE program outline

The preliminary baseline survey (PBLS), which was the first-round survey, started in October 2011. Examinations thereafter have been conducted as full-scale surveys (FSS). The first FSS (second-round survey) started in April 2014, the second FSS (third-round survey) in April 2016, and the third FSS (fourth-round survey) in April 2018. These surveys were conducted once every 2 years until participants were >20 years old, and then once every 5 years when participants were >25 years old. Fig. 1 shows the examination flow of the TUE program. In response to the possibility of detecting even harmless thyroid

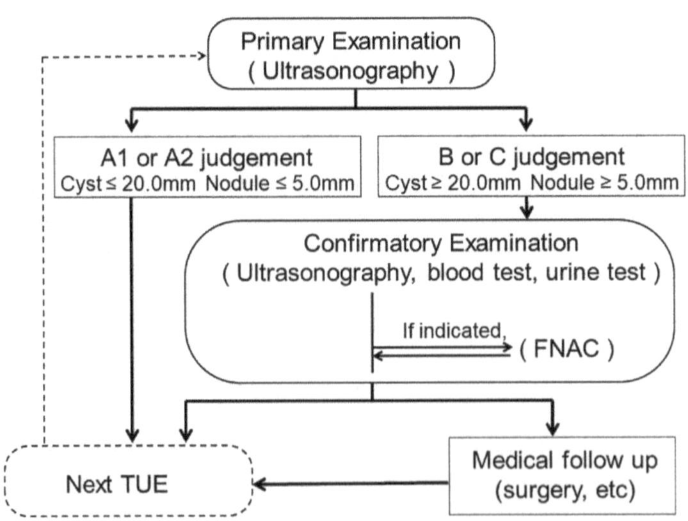

FIG. 1 The thyroid ultrasound examination (TUE) program procedure within the Fukushima Health Management Survey. *FNAC*; fine needle aspiration cytology.

cancer that otherwise would not show any symptoms or affect the quality and duration of life, measures were taken to minimize the diagnosis of lesions that do not require treatment.

Primary examination

An ultrasound examination of the thyroid gland was conducted among all participants to assess whether a more detailed examination, which is defined as the secondary confirmatory examination, was required. Immediately after the accident, evacuated local residents were strongly requested to undergo the examination as soon as possible because of anxieties about radiation exposure. In October 2011, the first part of the PBLS started at the medical office of Fukushima Medical University Hospital for those who lived in the 13 evacuation zone municipalities. Since FY2012, the original primary examination system for each municipality was established in descending order of radiation dose. Because many of the targets to be examined are students, they and their parents were requested to undergo the primary examination during school hours. Therefore, in-school examinations were introduced during the implementation of the primary examination to comply with demand. As of March 2020, 84 medical facilities in the Fukushima prefecture and 124 medical institutions outside the prefecture were registered. Those who moved after the earthquake can be examined at medical institutions in each prefecture. Examinations at the specified medical institutions or at general venues such as public facilities were recommended to those who have not undergone in-school examinations. Examinations at the general venues were held during long school holidays. During the primary examination, a probe with a frequency of $\geq$10 MHz and an ultrasonic device with color Doppler imaging were used. The presence or absence of nodules and cysts in the thyroid gland was confirmed by a committee during the judgment of sonographic findings at the Fukushima Medical University. A "B" judgment is defined as a nodule size of $\geq$5.1 mm or a cyst size of $\geq$20.1 mm. A "C" judgement is defined as immediate requirement of a secondary examination based on findings from the nodule. During B and C judgments, the examinee is recommended to undergo a secondary confirmatory examination. If no nodule or cyst is present, it is defined as an "A1" judgment, and if the nodule is $\leq$5.0 mm or the cyst is $\leq$20 mm, it is defined as an "A2" judgment. During A1 and A2 judgments, the next primary examination is recommended.

Confirmatory examination

The secondary confirmatory examinations were conducted at five facilities in the Fukushima Prefecture, including the Fukushima Medical University, and at 37 facilities outside the prefecture. Secondary examinations included blood and urine examinations and a detailed ultrasound examination. The decision to make a cytologic diagnosis is based on nodule size and ultrasonographic findings (Suzuki, 2016). The Japan Society of Ultrasonics in Medicine first published the ultrasound diagnostic criteria for thyroid nodules in 1991, and the third edition was published in 2011 based on a multicenter study (Shimura et al., 2005). Ultrasound findings were divided into primary and secondary findings. Moreover, ultrasound findings that are obviously useful are treated as primary findings, whereas those with low statistical significance are regarded as secondary findings. Based on these criteria, the Japan Association of Breast and Thyroid Sonology (JABTS) published the criteria for conducting fine needle aspiration cytology (FNAC) of thyroid nodules. FNAC is recommended for nodules of 5.1–10.0mm in diameter with strongly suspicious sonographic findings for malignancy; nodules of 10.1–20.0mm with any suspicious findings; and all nodules >20mm. All nodules <5.1mm are not subjected to FNAC because the program does not diagnose malignant lesions for which the treatment need is low.

Informed consent for the TUE program

Recently, thyroid cancer overdiagnosis has become a major issue worldwide. Cancer screening should improve the quality of life with low-invasive treatment, reduced medical costs, and reassurance in true negatives through early detection; moreover, it must not lead to overdiagnosis. Overdiagnosis is a major disadvantage of cancer screening; nevertheless, if unnecessary screening is not performed, unnecessary treatment is not performed and psychological stress does not occur. Some thyroid cancers, such as latent thyroid cancers or papillary thyroid cancers, are very low risk, tend not to progress or progress slowly, and do not cause cancer death. In the Korean National Cancer Screening Program, breast cancer screening by ultrasonography was started in 1999. Although thyroid cancer screening was not included in the program, providers frequently offered additional ultrasonographic screening, including for thyroid cancer. Hence, the overall prevalence of thyroid cancer increased by about 25%, most of which was <10mm. However, the mortality rate remained unchanged (Ahn, Kim, & Welch, 2014). Therefore, this Korean thyroid screening has been regarded as an overdiagnosis (Ahn & Welch, 2015; Lee & Shin, 2014).

Nodules diagnosed in children have a higher risk of malignancy than those in adults, including a higher frequency of local lymph node metastases, extrathyroidal extension, and lung metastases. However, there are insufficient data to determine whether early detection of nonpalpable tumors will significantly improve the quality and longevity of life among children. Therefore, thyroid ultrasonography in asymptomatic children is not recommended by the American Thyroid Association (Francis et al., 2015). The TUE program might be fundamentally different from general cancer screening, and careful consideration was required to reduce the risk of overdiagnosis. Therefore, in this program, we have adopted the Japanese guideline for the implementation of FNAC that corresponds to the risk of overdiagnosis and have disclosed this potential risk to the subjects. Through 2019, examinees and their families were informed of the disadvantages as well as examination objectives prior to each primary examination, and examinations were conducted after obtaining informed consent. During the prefectural expert meeting in 2019, it was recommended that more detailed explanations should be offered. Hence, the notification "Information on Thyroid Ultrasound Examination" was revised, with a special focus on the advantages and disadvantages of the examination. Easy-to-understand leaflets were also created for elementary school pupils and junior high school students. In this revised information, the advantages and disadvantages considered were as follows.

Advantages of the examination

(1) If no abnormality is found through the examination, a patient's concerns over radiation effects on health may be relieved and the quality of life may be enhanced accordingly.
(2) Early diagnosis and early treatment may reduce the risk of surgery complications, risk of side effects due to treatment, and risk of recurrence.
(3) Information concerning the existence or nonexistence of radiation effects can be provided through the analyses of examination results not only to the examinees and their families but also broadly to people in and outside the Fukushima Prefecture.

Disadvantages of the examination

(1) Diagnosis may lead to treatment of cancers that otherwise would not show any symptoms or affect the quality and duration of life.
(2) If cancer or suspected cancer is diagnosed early, the treatment and follow-up period can be prolonged and may increase psychological burdens or cause social and economic disadvantages.
(3) Nodules or cysts that do not require treatment may be found, and secondary examination or cytology may be recommended even in cases of benign nodules. This may impose physical burdens and anxieties to the examinee and his/her family.

We also informed the examinees about countermeasures for the above-mentioned disadvantages.

For disadvantage (1): We take measures to avoid diagnoses of nodules with little need for treatment by making it a rule to exclude nodules of ≤ 5.0 mm during the secondary confirmatory examination and to decide the need for FNAC for nodules ≥ 5.1 mm by considering imaging findings in line with the guidelines of the JABTS.

For disadvantage (2): As a support program for the Fukushima Health Management Survey thyroid examination, the Fukushima Prefecture offers financial support for medical costs necessary for treatment and follow-up after the TUE.

For disadvantages (2) and (3): The Fukushima Medical University has established the Mental Care Support Team, and its specialized staff responds to the anxieties of those who have been subjected to the confirmatory examination. Additionally, FMU offers telephone consultation services to respond to medical questions concerning the TUE results and thyroid disorders, to respond to psychological distress, and to hold explanation sessions at schools and other places.

Mental care support

Most people have little interest and knowledge about the thyroid gland and thyroid cancer. As a result, when nodules and cysts are found despite not requiring treatment, and if secondary examination or cytology is recommended even with benign nodules, the examinees and his/her family are physically burdened and anxious. For individuals, we send a booklet before an examination and periodic newsletters two times a year to explain the details about ultrasound examinations, including benefits as well as possible harms. As an approach to the psychosocial impact of thyroid examination, we started the dialog sessions mainly for parents and faculty members in 2013 and the onsite lectures targeting elementary through high school students in 2015. At these opportunities, doctors and medical technologists provide basic knowledge about the thyroid gland and explanations on thyroid examinations (Midorikawa, Tanigawa, Suzuki, & Ohtsuru, 2017).

It takes about 2–3 months to receive these results through the mail, giving participants and their families a lot of time to worry about the results of the primary examination. This latency made them anxious. After finishing the primary examination, we gave leaflets to participants. This leaflet explained the details of cysts or nodules, the reason why thyroid cysts should not be worrisome, and that cytologic diagnosis would not be performed in all cases with nodules. Since 2015, small booths have been set up at the primary examination site so doctors can explain tentative results by presenting just images from the day of examination to participants and their families. For questions and consultations from the examinees and their families who were examined where there were no small booths, we have also prepared a medical hotline and e-mail address. These remote devices are connected to medical professionals to answer thyroid examination results and questions related to medical issues, and provide counseling for mental problems.

Participants who are recommended to have the secondary confirmatory examination and their families might have anxieties regarding their upcoming examination. Therefore, we send leaflets about the confirmatory examination beforehand. This leaflet explains details of the secondary confirmatory examination, the medical hotline, and FNAC, which will not be performed for all cases with nodules. The Thyroid Support Team was organized at FMU to address anxiety during the confirmatory examination as well as provide medical care. The team consists of clinical psychologists, medical social workers, and nurses specialized in pediatric oncology, who sit closer to participants and their families, listen to their anxieties, and help in their communications with doctors. They provide mental support for those with worries and anxieties through counseling (see Chapter 13).

When a malignancy or a suspected malignancy is diagnosed earlier than the usual course, the treatment and follow-up period can be prolonged; this may increase psychological burdens or cause social and economic disadvantages. In response, the Fukushima Prefecture supports the medical expenses necessary for medical treatment after thyroid examination through the Prefectural Health Survey Thyroid Examination Support Project.

Regular results of the TUE program (Table 1)

1. The first-round surveys (PBLS) were performed between October 2011 and March 2014 to determine the thyroid status of the subjects within the period prior to the discovery of thyroid cancer in the Chernobyl nuclear accident. The subjects of this survey were 367,637 Fukushima citizens aged $\leq$18 years at the time of the earthquake. By March 2018, 300,472 residents underwent the primary examination. In total, 154,605 residents (51%) had the A1 judgment, 143,573 (48%) had the A2 judgment, and 2293 (0.8%) had the B and C judgments. The primary examination consisted of 2275 nodules (0.8%) with a size $\geq$5.1 mm, 1713 nodules (0.6%) with a size $\leq$5.0 mm, 12 cysts with a size $\geq$20.1 mm, and 143,899 cysts (48%) <20.1 mm. Because this is an unprecedented cohort study, it was difficult to interpret the significance of these prevalence rates. Therefore, members of the JABTS conducted ultrasonic examinations on children in three municipalities outside the Fukushima Prefecture (Hirosaki, Kofu, and Nagasaki City) to investigate the frequency of nodules and cysts among children. No difference was found between the prevalence rates in the Fukushima Prefecture and this survey (Hayashida et al., 2013; Taniguchi et al., 2013). Confirmatory examinations were performed on 2130 residents, of which 547 (26%) underwent cytology, and 116 (5.4%, 0.04% of those who underwent primary examination) were diagnosed as malignant or suspicious for malignancy (Fukushima Revitalization Station, 2018).
2. In the second-round survey, the first FSS was performed between April 2014 and March 2016 for 381,237 Fukushima citizens and subjects for PBLS plus children born in the next fiscal year following the earthquake. By March 2021, 270,552 residents underwent primary examination. In total, 108,726 residents (40%) had the A1 judgment, 159,596 (59%) had the A2 judgment, and 2230 (0.8%) had the B judgment. The primary examination consisted of 2219 nodules (0.8%) with a size $\geq$5.1 mm, 1570 nodules (0.6%) with a size $\leq$5.0 mm, six cysts with a size $\geq$20.1 mm, and 160,363 cysts

TABLE 1 Regular results of the thyroid ultrasound examination program.

	Survey type	Examination period (fiscal year)	Subjects	Primary examination					Confirmatory examination		
					Nodules	*>5mm*	*Cysts*	*>20mm*		*Cytology*	*Diagnosed as malignant or suspicious for malignancy*
1.	PBLS (1st round)	2011–2014	367,637	300,472	3988 (1.4%)	1713 (0.6%)	143,911 (48%)	12	2130	547 (26%)	116 (5.4%)
2.	1st FSS (2nd round)	2014–2016	381,237	270,552	3789 (1.4%)	1570 (0.6%)	160,369 (59%)	6	1877	207 (11%)	71 (3.8%)
3.	2nd FSS (3rd round)	2016–2018	336,667	217,922	2328 (1.1%)	1499 (0.7%)	140,678 (65%)	3	1104	79 (7.2%)	31 (2.8%)
4.	3rd FSS (4th round)	2018–2020	294,237	183,352	2053 (1.1%)	1386 (0.7%)	120,962 (66%)	5	1021	87 (8.5%)	36 (3.5%)

FSS, full-scale survey; *PBLS*, Preliminary Baseline Survey.

(59%) <20.1 mm. Confirmatory examinations were performed on 1877 residents, of which 207 (11%) underwent cytology, and 71 (3.8%, 0.03% of those who underwent primary examination) were diagnosed as malignant or suspicious for malignancy (Fukushima Revitalization Station, 2021a).

3. In the third-round survey, the second FSS was performed between May 2016 and March 2018. By March 2021, 217,922 residents underwent primary examination. In total, 76,431 residents (35%) had the A1 judgment, 139,989 (64%) had the A2 judgment, and 1502 (0.7%) had the B judgment. The primary examination consisted of 1499 nodules (0.7%) with a size ≥5.1 mm, 829 nodules (0.4%) with a size ≤5.0mm, three cysts with a size ≥20.1 mm, and 140,675 cysts (65%) <20.0mm. Confirmatory examinations were performed on 1104 residents, of which 79 (7.2%) underwent cytology, and 31 (2.8%, 0.01% of those who underwent primary examination) were diagnosed as malignant or suspicious for malignancy (Fukushima Revitalization Station, 2021a).
4. A fourth-round survey, the third FSS, was performed between May 2019 and 2020. By June 2021, 183,352 residents underwent primary examination. In total, 61,691 residents (34%) had the A1 judgment, 120,256 (66%) had the A2 judgment, and 1391 (0.8%) had the B judgment. The primary examination consisted of 1386 nodules (0.8%) with a size ≥5.1 mm, 667 nodules (0.4%) with a size ≤5.0mm, five cysts with a size ≥20.1 mm, and 120,957 cysts (66%) <20.1 mm. The confirmatory examination was performed on 1021 residents, of which 87 (8.5%) underwent cytology, and 36 (3.5%, 0.02% of those who underwent primary examination) were diagnosed as malignant or suspicious for malignancy (Fukushima Revitalization Station, 2021b).

An FSS is planned to be conducted every 2 years until the resident is >20 years old, and every five years after the age of 25.

Obvious findings of the TUE program

The association between radiation exposure and thyroid cancer in children found in the TUE program has attracted attention. The following research results have been reported.

1. Although the Chernobyl distribution shows a high risk related to young age at radiation exposure, with a steep fall with increasing age, the Fukushima distribution in the first 3 years after the accident shows no increased risk among the youngest, and the increasing incidence with age is similar to that in normal cohorts (Williams, 2015).
2. In childhood papillary thyroid cancer, regardless of the history of radiation exposure, rearrangements during transfection/papillary thyroid carcinoma (RET/PTC) are a major event. Conversely, in adult-onset papillary thyroid cancer among the general population, the most common molecular event is

the v-raf murine sarcoma viral oncogene homolog B1 (BRAF) point mutation, but not RET/PTC rearrangements. Compared to atomic bomb survivors and cases who have not been exposed to atomic bomb radiation, it was reported that RET/PTC rearrangements showed significantly increased frequency with increased radiation dose and the BRAF mutation was less frequent in cases exposed to higher radiation doses (Hamatani et al., 2008). Post-Chernobyl tumors are characterized by the frequent occurrence of chromosomal rearrangements, such as RET/PTC, whereas point mutations of BRAF and other genes are much less common in this population (Fukushima Revitalization Station, 2021a). We also analyzed genetic alterations in thyroid cancers and found that the BRAFV600E mutation was highly prevalent and was found in 69.6% of cases. Chromosomal rearrangements such as those involving RET were found in 16.7% of cases (Iwadate et al., 2020). The genetic pattern with a high prevalence of the BRAF point mutation and low frequency of chromosomal rearrangements was completely different from post-Chernobyl papillary thyroid carcinomas. Moreover, many post-Chernobyl thyroid cancers were diagnosed as solid variant papillary carcinoma. By contrast, among 115 of our cases with papillary thyroid carcinomas, only four were diagnosed as the solid variant of papillary thyroid cancer (Suzuki et al., 2019).

3. To examine the association between external radiation dose and thyroid cancer in the Fukushima Prefecture, a cross-sectional study design was divided into three groups based on individual external doses, with 300,476 participants who underwent the first-round survey (PBLS) included. There were no significant associations between individual external doses and thyroid cancer prevalence. Within the first 4 years after the nuclear accident, we concluded that the external radiation dose was not associated with thyroid cancer prevalence among Fukushima children (Ohira et al., 2016).
4. UNSCEAR estimated absorbed doses in the thyroid gland for individuals during the first year following the Fukushima nuclear power plant accident (UNSCEAR, 2013). In the TUE program, we grouped participants from each municipality in the Fukushima Prefecture according to estimated absorbed doses based on this report, and identified that no dose-dependent pattern emerged from the geographical distribution of absorbed doses and the detection of thyroid cancer in the first- and second-round surveys (Ohira et al., 2020).

The results of the PBLS and the first FSS have been analyzed at the Prefectural Oversight Committee Meetings, and it has been concluded that there may be no relationship between the detection of childhood thyroid cancer and radiation exposure. However, compared to the duration of the post-Chernobyl nuclear power plant accident, the ongoing examinations correspond to the time when the incidence of thyroid cancer began to increase. Although it may be difficult to derive the effects of radiation exposure, we plan to continue the investigation and monitor the health damage of the nuclear power plant accident for a long time in response to the anxiety of Fukushima residents about radiation exposure.

References

Ahn, H. S., Kim, H. J., & Welch, H. G. (2014). Korea's thyroid-cancer "epidemic"—screening and overdiagnosis. *The New England Journal of Medicine*, *371*, 1765–1767.

Ahn, H. S., & Welch, H. G. (2015). South Korea's Thyroid-Cancer "Epidemic"—Turning the Tide Korea's thyroid-cancer "epidemic"—Turning the tide. *The New England Journal of Medicine*, *373*, 2389–2390.

Demidchik, Y. E., Saenko, V. A., & Yamashita, S. (2007). Childhood thyroid cancer in Belarus, Russia, and Ukraine after Chernobyl and at present. *Arquivos Brasileiros de Endocrinologia e Metabologia*, *51*, 748–762.

Francis, G. L., Waguespack, S. G., Bauer, A. J., Angelos, P., Benvenga, S., et al. (2015). Management guidelines for children with thyroid nodules and differentiated thyroid cancer. *Thyroid*, *25*, 716–759.

Fukushima Revitalization Station. (2018). *Dai 31 kai Fukushima ken "kenminkenkouchosa" kentou iinkai (heisei 30 nen 6 gatsu 18 niti) nit suite [The 31st meeting for Fukushima Health Management Survey (June 18, 2018)]*. http://www.pref.fukushima.lg.jp/uploaded/attachment/461404.pdf (Accessed 25 October 2021, in Japanese).

Fukushima Revitalization Station. (2021a). *Dai 42 kai Fukushima ken "kenminkenkouchosa" kentou iinkai (reiwa 3 nen 7 gatsu 26 niti) nit suite [The 42rd meeting for Fukushima Health Management Survey (July 26, 2021)]*. http://www.pref.fukushima.lg.jp/uploaded/attachment/461564.pdf (Accessed 26 October 2021, in Japanese).

Fukushima Revitalization Station. (2021b). *Dai 43 kai Fukushima ken "kenminkenkouchosa" kentou iinkai (reiwa 3 nen 10 gatsu 15 niti) nit suite [The 43rd meeting for Fukushima Health Management Survey (October 15, 2021)]*. http://www.pref.fukushima.lg.jp/uploaded/attachment/475143.pdf (Accessed 28 October 2021, in Japanese).

Hamatani, K., Eguchi, H., Ito, R., Mukai, M., Takahashi, K., et al. (2008). RET/PTC rearrangements preferentially occurred in papillary thyroid cancer among atomic bomb survivors exposed to high radiation dose. *Cancer Research*, *68*, 7176–7182.

Hayashida, N., Imaizumi, M., Shimura, H., Okubo, N., Asari, Y., et al. (2013). Thyroid ultrasound findings in children from three Japanese prefectures: Aomori, yamanashi and Nagasaki. *PLoS ONE*, *8*, e83220.

Iwadate, M., Mitsutake, N., Matsuse, M., Fukushima, T., Suzuki, S., et al. (2020). The Clinicopathological results of thyroid cancer with BRAFV600E mutation in the young population of Fukushima. *Journal of Clinical Endocrinology and Metabolism*, *105*, e4328–e4336.

Lee, J. H., & Shin, S. W. (2014). Overdiagnosis and screening for thyroid cancer in Korea. *Lancet*, *384*, 1848.

Midorikawa, S., Tanigawa, K., Suzuki, S., & Ohtsuru, A. (2017). Psychosocial issues related to thyroid examination after a radiation disaster. *Asia-Pacific Journal of Public Health*, *29*, 63S–73S.

Nuclear Emergency Response Headquarters. (2011). *Genshiryoku anzen ni kansuru IAEA kakuryou kaigi ni taisuru nihonkoku seifu no houkokusyo: Tokyo denryoku Fukushima genshiryoku hatudensyo no jiko ni tsuite [Report of the Japanese government of Japan to the IAEA Ministerial Conference on Nuclear Safety: The accident at the Fukushima Nuclear Power Plant of Tokyo Electric Power Co.]*. https://www.kantei.go.jp/jp/topics/2011/pdf/06-kankyo.pdf (in Japanese).

Ohira, T., Shimura, H., Hayashi, F., Nagao, M., Yasumura, S., et al. (2020). Absorbed radiation doses in the thyroid as estimated by UNSCEAR and subsequent risk of childhood thyroid cancer following the Great East Japan Earthquake. *Journal of Radiation Research*, *61*, 243–248.

Ohira, T., Takahashi, H., Yasumura, S., Ohtsuru, A., Midorikawa, S., et al. (2016). Comparison of childhood thyroid cancer prevalence among 3 areas based on external radiation dose after the Fukushima Daiichi nuclear power plant accident: The Fukushima health management survey. *Medicine*, *95*, e4472.

Ohto, H., Yasumura, S., Maeda, M., Kanuma, H., Fujimori, K., et al. (2017). From devastation to recovery and revival in the aftermath of Fukushima's nuclear power plants accident. *Asia-Pacific Journal of Public Health*, *29*, 10S–17S.

Shimura, H., Haraguchi, K., Hiejima, Y., Fukunari, M., Fujimori, K., et al. (2005). Distinct diagnostic criteria for ultrasonographic examination of papillary thyroid carcinoma: A muticenter study. *Thyroid*, *15*, 251–258.

Suzuki, S. (2016). Childhood and adolescent thyroid cancer in Fukushima after the Fukushima Daiichi nuclear power plant accident: 5 years on. *Clinical Oncology*, *28*, 263–271.

Suzuki, S., Bogdanova, T. I., Saenko, V. A., Hashimoto, Y., Ito, M., et al. (2019). Histopathological analysis of papillary thyroid carcinoma detected during ultrasound screening examinations in Fukushima. *Cancer Science*, *110*, 817–827.

Taniguchi, N., Hayashida, N., Shimura, H., Okubo, N., Asari, Y., et al. (2013). Ultrasonographic thyroid nodular findings in Japanese children. *Journal of Medical Ultrasonics*, *40*, 219–224.

United Nations Scientific Committee on the Effects of Atomic Radiation. (2011). *Health effects due to radiation from the Chernobyl accident. Vol. II* (pp. 45–219). United Nations: United Nations Scientific Committee on the Effects of Atomic Radiation.

UNSCEAR. (2013). *REPORT Vol. I. Sources Eff Risks Ionizing Radiat.* https://www.unscear.org/unscear/en/publications/b2013_1.html (Accessed 15 November 2019).

Williams, D. (2015). Thyroid growth and cancer. *European Thyroid Journal*, *4*, 164–173.

Chapter 5

Trends in pregnancy and birth after the Great East Japan earthquake and Fukushima Daiichi nuclear power plant accident in the Fukushima prefecture: A 7-year survey*

Keiya Fujimori[a,b], Kayoko Ishii[a], Hyo Kyozuka[b], Shun Yasuda[a,b], Tsuyoshi Murata[b], Aya Goto[a,c], Seiji Yasumura[a,d], Misao Ota[a,e], Kenichi Hata[a,f], Kouta Suzuki[a,g], Akihito Nakai[a,h], Tetsuya Ohira[a,i], Hitoshi Ohto[a], and Kenji Kamiya[a]

[a]*Radiation Medical Science Center for the Fukushima Health Management Survey, Fukushima Medical University, Fukushima, Japan,* [b]*Department of Obstetrics and Gynecology, School of Medicine, Fukushima Medical University, Fukushima, Japan,* [c]*Center for Integrated Science and Humanities, Fukushima Medical University, Fukushima, Japan,* [d]*Department of Public Health, Fukushima Medical University School of Medicine, Fukushima, Japan,* [e]*Department of Midwifery and Maternal Nursing, School of Nursing, Fukushima Medical University, Fukushima, Japan,* [f]*Fukushima Society of Obstetrics and Gynecology, Fukushima, Japan,* [g]*Department of Health and Psychosocial Medicine, School of Medicine, Aichi Medical University, Aichi, Japan,* [h]*Nippon Medical School Tama Nagayama Hospital, Tokyo, Japan,* [i]*Department of Epidemiology, Fukushima Medical University School of Medicine, Fukushima, Japan*

Introduction

In the Fukushima prefecture, pregnant women were greatly affected physically and mentally by the Great East Japan Earthquake that occurred on March 11, 2011, and the subsequent radioactive contamination caused by the accident at the Tokyo Electric Power Company's (TEPCO's) Fukushima Daiichi Nuclear Power Plant (NPP). The specific effects of these disasters included the temporary cessation of medical treatment and deliveries due to damage

* Pregnancy and Birth Survey Group of the Fukushima Health Management Survey.

Health Effects of the Fukushima Nuclear Disaster. https://doi.org/10.1016/B978-0-12-824098-4.00001-5

to clinics, disruption of lifelines caused by a lack of resources and services, and evacuation due to concerns over radioactive contamination.

Soon after the disasters, the Fukushima prefectural government launched a prefecture-wide cohort survey called the Fukushima Health Management Survey (FHMS) to investigate the health effects of long-term low-dose radiation and disaster-related stress caused by the accidents (Yasumura et al., 2012). The FHMS consists of a basic survey (to estimate the external radiation exposure) and four detailed surveys. The detailed surveys include a thyroid ultrasound examination for all Fukushima children aged 18 or younger, a comprehensive health checkup with an assessment of mental health and lifestyles for all residents from the evacuation zones, and the pregnancy and birth survey for all mothers who were given a *Maternal and Child Health handbook* in the Fukushima prefecture (Fujimori et al., 2014; Yasumura et al., 2012). The purpose of the pregnancy and birth survey is to accurately determine the mental health, physical condition, opinions, and requests of pregnant women seeking to have children and mothers raising children in the Fukushima prefecture as well as to reduce their anxiety and provide the necessary care (Fujimori et al., 2014; Yasumura et al., 2012). This survey has been conducted annually since 2011 and is still ongoing. In the first year (2011), we reported no significant adverse outcomes following the disaster over the entire Fukushima prefecture (Fujimori et al., 2014). However, evacuation and concerns about radioactive contamination were significantly associated with depressive symptoms (Goto et al., 2017).

This chapter assesses and presents the trends in pregnancy and birth outcomes over 7 years in the Fukushima prefecture based on self-administered questionnaires (the pregnancy and birth survey) after the two disasters. Meanwhile, the mental health issues among the mothers identified by this survey are deeply discussed in Chapter 18.

Materials and methods

Study design

In this analysis, we used the results of the maternal survey questionnaire that was part of the FHMS. The methods used for the FHMS and maternal survey have been reported previously (Yasumura et al., 2012). In brief, the maternal survey is a population-based study conducted as a part of the FMHS launched by the Fukushima prefecture government in 2011. We divided the Fukushima prefecture into six districts and included women who received the Maternal and Child Health handbooks since August 1, 2010. This handbook describes the unique perinatal healthcare system in Japan and helps maintain a record of the women's antenatal and postnatal check-ups by physicians. The self-administered questionnaire was sent to the women by mail beginning on January 18, 2012. The mothers were asked to refer to their handbooks when completing the questionnaire. Pregnant women who delivered between March 11, 2011, and December 31, 2017, and their newborns were included in this study. Women who delivered before

March 11, 2011, and received the handbook outside the Fukushima prefecture were excluded from the analysis. Moreover, for 2016 and 2017, we did not send a questionnaire to women who had an abortion, stillbirth, or whose neonates could not be confirmed to be alive.

This survey was approved by the local ethics review committee of the Fukushima Medical University, guided by local policy, national law, and the World Medical Association Declaration of Helsinki (approval #13047).

Maternal information and obstetrics outcomes

The self-administered questionnaire aimed to collect maternal information such as the geographic district where the pregnant women received the *Maternal and Child Health handbook*, year of delivery, maternal age at delivery, single or multiple gestational pregnancies, gestational weeks at delivery, method of pregnancy, mode of delivery, and neonatal information such as neonatal birth weight, gender of the newborn, and anomalies in the newborn. The Fukushima prefecture was divided into six geographic districts: Kenpoku, Kenchu, Kennan, Soso, Iwaki, and Aizu, as shown in Fig. 1 (Fujimori et al., 2014). Stillbirth was defined as fetal death after 22 completed weeks of gestation. Delivery before 37 weeks of gestation was defined as preterm birth (PTB). The PTB rate was defined as the number of PTBs (singleton or multiple) divided by the total number of live births based on the World Health Organization (WHO) definition (WHO, 1977). A birth weight less than 2500 g was defined as low birth weight (LBW). The incidence of LBW in a population is defined as the percentage of live births that weigh less than 2500 g out of the total live births during the same period (UNICEF, 2004). The method of pregnancy was categorized as natural pregnancy or with the aid of some kind of fertility treatment, such as ovulation, artificial insemination, or in vitro fertilization. The mode of delivery was categorized into vaginal delivery or cesarian section. The major anomalies

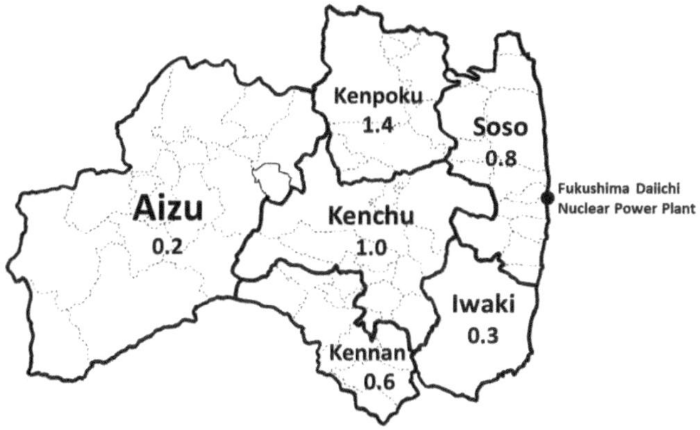

FIG. 1 Map of the Fukushima prefecture. Shown is the average estimated external dose by area (mSv).

in a newborn were the following: cataracts, cardiac malformations, kidney or urinary tract malformations, spina bifida, microcephaly, hydrocephalous, cleft lip or palate, intestinal atresia (esophagus, duodenum, ileum), imperforate anus, and poly or syndactylism.

Statistical analysis

The maternal or neonatal characteristics were categorized into seven groups according to the year of birth. The subjects' demographic and perinatal outcome data are presented as mean ± standard deviation or as incidences where appropriate. There were discrepancies in the total number of valid respondents because of missing data in each subject area. The Mantel-Haenzel test for trend or chi-square test was used to analyze the annual or geographical trends, respectively. The Jonckheere-Terpstra trend test was used to compare the continuous variables by year. SPSS ver 24 (IBM Japan, Tokyo) was used for data analysis. The P-value $<.05$ was considered significant.

Results

Over the past 7 years, a questionnaire was sent to 102,425 pregnant women in the Fukushima prefecture. The response rate was 49.8%, with 50,996 women responding (Fig. 2). A total of 258 invalid responses (47 no answer, 1 death of the participant, 41 multiple responses, 168 inconsistent answers, 1 canceled registration after reply) were excluded, and 50,738 subjects were included in the study. Of these, 49,716 subjects participated in this analysis, after excluding 459 deliveries before the disaster, 165 pregnant while answering, 46 induced and 350 spontaneous abortions, and 2 triplets (Fig. 2).

In the first year (2011), the number of participants was 15,972, which decreased to 14,420 in 2012, and then temporarily increased to 15,108 in 2013. After that, the numbers gradually decreased every year (Table 1). The response rate was relatively higher at 58.2% in the first year but fluctuated around 50% in the subsequent years. The rates differed significantly based on the region and year. Kenpoku constantly had a higher response rate than the other areas. Soso had a higher response rate only in 2011 and has been lower since then (Table 1).

Table 2 shows the characteristics of participants based on the year of delivery. Significant differences were seen in maternal age, gestational week at delivery, mode of pregnancy, and gender of the neonates over the 7 years. While maternal age has increased significantly, the natural pregnancy rates have declined over time (Table 2). However, no significant differences were seen in the rate of stillbirths over the years.

As shown in Table 3, no significant differences were seen in the incidence of preterm deliveries based on region and year. Regional differences were

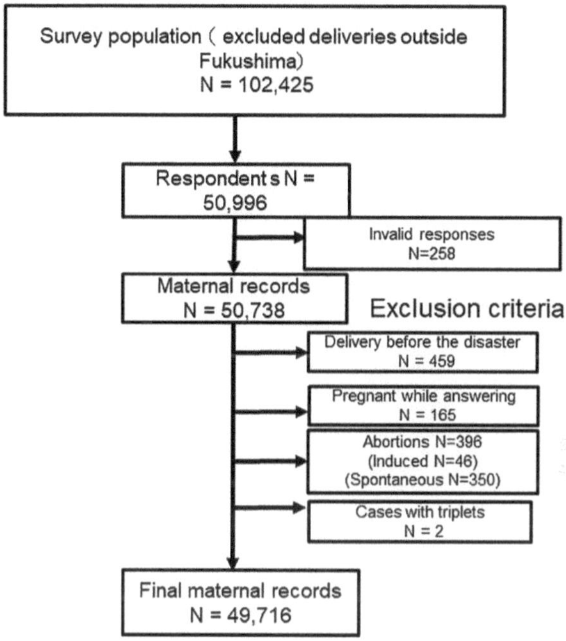

FIG. 2 Flowchart showing study enrollment.

recognized only in 2012 when the incidence was lower in Kenopoku and higher in Aizu.

Table 4 shows the incidence of LBW infants by region and year. No significant differences were observed in the Fukushima prefecture as a whole. Regional differences were seen only in 2011 and 2012. In 2011, the incidence was lower in Kenopoku and higher in Iwaki while in 2012, it was lower in Kenopoku and higher in Aizu.

Table 5 shows the incidence of congenital anomalies in newborns in each district over time. It ranged from 2.21% to 2.85% in the entire Fukushima prefecture, with no significant change over time. The different regions also showed no significant changes over time.

Table 6 shows changes in the incidence of each anomaly over time. Cardiac malformation was the most common anomaly in the entire Fukushima prefecture, with an incidence rate of 0.61%–1.01%. No significant changes were seen in the incidence of all anomalies over time.

The onset of microcephaly and spina bifida is considered to be caused by exposure to radiation. Tables 7 and 8 show the incidence of microcephaly and spina bifida, respectively, for each district over the 7 years. In 2011, 2012, and 2014, sporadic cases of microcephaly in newborns were recognized in Kenchu. While 3–5 cases of spina bifida in newborns occur every year in the Fukushima prefecture, no chronological or regional fluctuations have been observed.

TABLE 1 Changes in response rate over time.

Response rate	2011	2012	2013	2014	2015	2016	2017	*P*-value[a]
All of Fukushima (%)	58.2	49.1	47.3	46.8	47.8	51.3	47.1	<.001
(Res/send)	(9299/15972)	(7085/14420)	(7152/15108)	(7024/15017)	(6913/14454)	(7191/14019)	(6332/13435)	
Kenpoku (%)	62.8[b]	55.5[b]	53.2[b]	52.4[b]	52.3[b]	55.9[b]	50.9[b]	<.001
(Res/send)	(2289/3647)	(1857/3347)	(1936/3637)	(1841/3515)	(1806/3453)	(1875/3352)	(1634/3212)	
Kenchu (%)	59.3	48.7	44.5[b]	44.8[b]	45.2[b]	49.8[c]	46.8	<.001
(Res/send)	(2858/4819)	(2067/4243)	(1982/4453)	(1961/4376)	(1924/4261)	(2065/4150)	(1862/3980)	
Kennan (%)	50.2[b]	48.1	48.5	46.5	47.9	51.1	45.1	.209
(Res/send)	(631/1256)	(560/1164)	(588/1213)	(553/1188)	(560/1168)	(571/1118)	(473/1048)	
Soso (%)	65.6[b]	43.7[b]	45.4	42.2[b]	44.2[b]	43.6[b]	40.5[b]	<.001
(Res/send)	(963/1468)	(500/1145)	(535/1178)	(512/1213)	(523/1183)	(511/1171)	(442/1091)	
Iwaki, %	55.9[b]	47.8	45.1[c]	45.8	46.6	50.1	45.5	<.001
(Res/send)	(1515/2711)	(1203/2516)	(1195/2649)	(1213/2648)	(1148/2461)	(1192/2377)	(1054/2317)	
Aizu (%)	50.4[b]	44.8[b]	46.3	45.5	49.4	52.8	48.5	.032
(Res/send)	(1043/2071)	(898/2005)	(916/1978)	(944/2077)	(952/1928)	(977/1851)	(867/1787)	
P-value[d]	<.001	<.001	<.001	<.001	<.001	<.001	<.001	

[a]*P-value was calculated by the Mantel-Haenzel test for trend.*
[b]*Indicates* $P < .01$.
[c]*Indicates* $P < .05$.
[d]*P-value was calculated by the chi-square test.*

TABLE 2 Characteristics of respondent based on the year of delivery.

	2011	2012	2013	2014	2015	2016	2017	*P*-value
Maternal age, mean ± SD	30.7 ± 5.0	31.0 ± 5.0	31.3 ± 5.0	31.4 ± 5.1	31.6 ± 5.0	31.4 ± 5.0	31.6 ± 5.0	<.001[a]
n	8598	6940	7044	6940	6815	7091	6288	
Single pregnancy, *n* (%)	8517 (99.1)	6877 (99.1)	6967 (98.9)	6868 (99.0)	6757 (99.1)	7031 (99.2)	6240 (99.2)	.264[b]
Gestational week, mean ± SD	38.9 ± 1.7	38.9 ± 1.7	38.9 ± 1.7	38.9 ± 1.8	38.8 ± 1.8	38.9 ± 1.7	38.8 ± 1.7	.001[a]
n	8566	6926	7015	6930	6801	7086	6286	
Neonatal weight, mean ± SD	3013 ± 418	2993 ± 435	3003 ± 435	2996 ± 455	2994 ± 445	2999 ± 432	3007 ± 432	.770[b]
n	8603	6935	7088	6999	6859	7122	6319	
Mode of pregnancy								
Natural pregnancy, *n* (%)	8203 (95.7)	6541 (94.6)	6590 (94.0)	6462 (93.4)	6324 (93.1)	6499 (92.1)	5753 (92.0)	<.001[b]
Sex of newborn								
Male, *n* (%)	4423 (51.4)	3642 (52.5)	3574 (50.4)	3595 (51.3)	3435 (50.1)	3535 (50.3)	3163 (50.7)	.022[b]
Stillbirth, *n* (%)	22 (0.25)	20 (0.29)	24 (0.34)	15 (0.21)	17 (0.25)	15[c] (0.21)	14[c] (0.22)	.304[b]
Mode of delivery								
Cesarean delivery, *n* (%)	1745 (20.7)	1446 (21.7)	1409 (20.3)	1440 (21.0)	1511 (22.3)	1489 (21.2)	1364 (21.8)	.084[b]

[a]*P-value was calculated by the Jonckheere-Terpstra trend test.*
[b]*P-value was calculated by the Mantel-Haenzel test for trend.*
[c]*Not included confirmed abortions and stillbirths, or neonates not confirmed to have survived before sending the questionnaires.*

TABLE 3 Incidence of preterm birth in the different districts over time.

	2011	2012	2013	2014	2015	2016	2017	*P*-value[a]
All areas, *n* (%)	395 (4.6)	393 (5.6)	373 (5.3)	375 (5.4)	386 (5.6)	380 (5.3)	335 (5.3)	.099
Kenpoku, *n* (%)	91 (4.3)	82 (4.5)[b]	94 (4.9)	96 (5.2)	105 (5.9)	91 (4.9)	72 (4.4)	.356
Kenchu, *n* (%)	114 (4.3)	127 (6.2)	95 (4.8)	107 (5.5)	119 (6.2)	103 (5.0)	103 (5.6)	.198
Kennan, *n* (%)	28 (4.7)	23 (4.1)	33 (5.7)	25 (4.6)	34 (6.1)	30 (5.3)	27 (5.7)	.247
Soso, *n* (%)	38 (4.3)	31 (6.5)	35 (6.6)	31 (6.1)	30 (5.8)	29 (5.7)	21 (4.8)	.673
Iwaki, *n* (%)	71 (5.1)	65 (5.5)	69 (5.9)	72 (5.9)	46 (4.1)	63 (5.3)	62 (5.9)	.926
Aizu, *n* (%)	53 (5.4)	65 (7.3)[b]	47 (5.2)	44 (4.7)	52 (5.5)	64 (6.6)	50 (5.8)	.945
P-value[c]	.633	.020	.501	.718	.215	.498	.481	

[a] *P-value was calculated by the Mantel-Haenzel test for trend.*
[b] *Indicates* $P < .05$.
[c] *P-value was calculated by the chi-square test.*

TABLE 4 Incidence of infants with low birth weight in the different districts over time.

	2011	2012	2013	2014	2015	2016	2017	*P*-value[a]
All of Fukushima, *n* (%)	736 (8.6)	640 (9.3)	681 (9.6)	683 (9.8)	650 (9.5)	659 (9.3)	584 (9.3)	.212
Kenpoku, *n* (%)	160 (7.6)[b]	138 (7.6)[c]	171 (8.9)	168 (9.2)	155 (8.7)	151 (8.1)	155 (9.5)	.061
Kenchu, *n* (%)	217 (8.3)	208 (10.3)	191 (9.7)	192 (9.8)	208 (10.9)	187 (9.2)	166 (9.0)	.523
Kennan, *n* (%)	47 (8.0)	51 (9.2)	58 (9.9)	49 (9.1)	54 (9.8)	56 (9.9)	38 (8.1)	.737
Soso, *n* (%)	70 (7.9)	43 (9.3)	47 (8.9)	57 (11.3)	56 (10.9)	46 (9.1)	50 (11.3)	.047
Iwaki, *n* (%)	143 (10.3)[c]	101 (8.6)	120 (10.3)	119 (9.8)	99 (8.7)	118 (10.0)	98 (9.3)	.680
Aizu, *n* (%)	99 (10.2)	99 (11.3)[b]	94 (10.4)	98 (10.4)	78 (8.4)	101 (10.4)	77 (8.9)	.174
P-value[d]	.031	.025	.761	.736	.099	.349	.638	

[a]*P-value was calculated by the Mantel-Haenzel test for trend.*
[b]*Indicates* $P < .05$.
[c]*Indicates* $P < .01$.
[d]*P-value was calculated by the chi-square test.*

TABLE 5 Incidence of anomalies in newborns in the different districts over time.

	2011	2012	2013	2014	2015	2016	2017	*P*-value[a]
All of Fukushima, *n* (%)	238 (2.85)	161 (2.35)	170 (2.44)	163 (2.36)	150 (2.21)	176 (2.49)	153 (2.44)	.156
Kenpoku, *n* (%)	55 (2.66)	39 (2.17)	44 (2.32)	54 (2.99)	31 (1.75)	41 (2.22)	40 (2.47)	.524
Kenchu, *n* (%)	78 (3.07)	49 (2.43)	50 (2.60)	49 (2.54)	48 (2.54)	57 (2.81)	34 (1.85)	.101
Kennan, *n* (%)	24 (4.20)	13 (2.39)	9 (1.56)	11 (2.04)	18 (3.26)	12 (2.14)	9 (1.92)	.113
Soso, *n* (%)	18 (2.09)	13 (2.86)	9 (1.75)	9 (1.79)	7 (1.36)	13 (2.60)	8 (1.82)	.651
Iwaki, *n* (%)	38 (2.81)	26 (2.23)	35 (3.04)	22 (1.83)	27 (2.41)	26 (2.24)	33 (3.15)	.977
Aizu, *n* (%)	25 (2.63)	21 (2.40)	23 (2.58)	18 (1.93)	19 (2.04)	27 (2.80)	29 (3.36)	.399
P-value[b]	.268	.970	.411	.261	.181	.787	.099	

[a]*P-value was calculated by the Mantel-Haenzel test for trend.*
[b]*P-value was calculated by the chi-square test.*

TABLE 6 Incidence of major anomalies in newborns over time.

	2011	2012	2013	2014	2015	2016	2017	*P*-value[a]
All anomalies	238/ 8345	161/ 6858	170/ 6954	163/ 6915	150/ 6790	176/ 7064	153/ 6278	.156
%	2.85%	2.35%	2.44%	2.36%	2.21%	2.49%	2.44%	
Cataract	1	2	1	0	0	1	1	.645
%	0.01%	0.03%	0.01%	0.00%	0.00%	0.01%	0.02%	
Cardiac malformation	77	54	70	53	53	66	38	.139
%	0.92%	0.79%	1.01%	0.77%	0.78%	0.93%	0.61%	
Kidney/urinary tract malformation	22	14	12	24	14	16	25	.181
%	0.26%	0.20%	0.17%	0.35%	0.21%	0.23%	0.40%	
Spina bifida	5	5	3	5	4	3	4	.819
%	0.06%	0.07%	0.04%	0.07%	0.06%	0.04%	0.06%	
Microcephaly	1	1	0	1	0	0	0	.185
%	0.01%	0.01%	0.00%	0.01%	0.00%	0.00%	0.00%	
Hydrocephalus	1	3	4	2	1	2	1	.661
%	0.01%	0.04%	0.06%	0.03%	0.01%	0.03%	0.02%	
Cleft lip/palate	17	14	14	14	14	11	8	.258
%	0.20%	0.20%	0.20%	0.20%	0.21%	0.16%	0.13%	

Continued

TABLE 6 Incidence of major anomalies in newborns over time—cont'd

	2011	2012	2013	2014	2015	2016	2017	*P*-value
Intestinal atresia	6	7	7	9	4	10	7	.398
%	0.07%	0.10%	0.10%	0.13%	0.06%	0.14%	0.11%	
Imperforate anus	4	2	2	5	2	2	3	.949
%	0.05%	0.03%	0.03%	0.07%	0.03%	0.03%	0.05%	
Poly/syndactylism	22	15	22	14	14	18	16	.809
%	0.26%	0.22%	0.32%	0.20%	0.21%	0.25%	0.25%	
Others	104	61	55	63	58	64	59	.106
%	1.25%	0.89%	0.79%	0.91%	0.85%	0.91%	0.94%	

[a]*P-value was calculated by the Mantel-Haenzel test for trend.*

TABLE 7 Incidence of microcephaly in newborns in the different districts over time.

	2011	2012	2013	2014	2015	2016	2017
All of Fukushima, *n* (%)	1 (0.01)	1 (0.01)	0 (0.00)	1 (0.01)	0 (0.00)	0 (0.00)	0 (0.00)
Kenpoku, *n* (%)	0 (0.00)	0 (0.00)	0 (0.00)	0 (0.00)	0 (0.00)	0 (0.00)	0 (0.00)
Kenchu, *n* (%)	1 (0.04)	1 (0.05)	0 (0.00)	1 (0.05)	0 (0.00)	0 (0.00)	0 (0.00)
Kennan, *n* (%)	0 (0.00)	0 (0.00)	0 (0.00)	0 (0.00)	0 (0.00)	0 (0.00)	0 (0.00)
Soso, *n* (%)	0 (0.00)	0 (0.00)	0 (0.00)	0 (0.00)	0 (0.00)	0 (0.00)	0 (0.00)
Iwaki, *n* (%)	0 (0.00)	0 (0.00)	0 (0.00)	0 (0.00)	0 (0.00)	0 (0.00)	0 (0.00)
Aizu, *n* (%)	0 (0.00)	0 (0.00)	0 (0.00)	0 (0.00)	0 (0.00)	0 (0.00)	0 (0.00)

TABLE 8 Incidence of spina bifida in newborns in the different districts over time.

	2011	2012	2013	2014	2015	2016	2017	*P*-value[a]
All of Fukushima, *n* (%)	5 (0.06)	5 (0.07)	3 (0.04)	5 (0.07)	4 (0.06)	3 (0.04)	4 (0.06)	.819
Kenpoku, *n* (%)	2 (0.10)	1 (0.06)	0 (0.00)	3 (0.17)	1 (0.06)	0 (0.00)	1 (0.06)	.577
Kenchu, *n* (%)	2 (0.08)	0 (0.00)	1 (0.05)	1 (0.05)	1 (0.05)	1 (0.05)	0 (0.00)	.536
Kennan, *n* (%)	0 (0.00)	1 (0.18)	0 (0.00)	0 (0.00)	0 (0.00)	1 (0.18)	0 (0.00)	.954
Soso, *n* (%)	0 (0.00)	1 (0.22)	1 (0.19)	0 (0.00)	1 (0.19)	0 (0.00)	0 (0.00)	.765
Iwaki, *n* (%)	1 (0.07)	1 (0.09)	0 (0.00)	1 (0.08)	1 (0.09)	0 (0.00)	1 (0.10)	.925
Aizu, *n* (%)	0 (0.00)	1 (0.11)	1 (0.11)	0 (0.00)	0 (0.00)	1 (0.10)	2 (0.23)	.259
P-value[b]	.836	.543	.378	.586	.752	.439	.332	

[a]*P-value was calculated by the Mantel-Haenzel test for trend.*
[b]*P-value was calculated by the chi-square test.*

Discussion

The current air radiation dose rates in Fukushima prefecture have decreased significantly since April 2011, and are almost at the same level as major cities overseas (Fukushima Revitalization Station, 2021). However, the basic survey, which is a part of the FHMS, reported the individual external gamma ray doses for residents, and in the first 4 months after the disaster, the distribution was as follows: 62.0% < 1 mSv, 94.0% < 2 mSv, and 99.4% < 3 mSv (Ishikawa et al., 2015). In terms of the regions affected, the radioactive plume from the damaged NPP was carried north by wind and rain toward the prefecture's most populated Kenpoku area (Nagata, Kimura, & Ishii, 2012). For that reason, it is also reported that the external gamma ray doses were the highest in the Soso area (the evacuation area or deliberate evacuation area), followed by the Kenpoku and Kenchu areas. The average doses for the Soso, Kenpoku, Kenchu, and Iwaki areas were 0.8 mSv, 1.4 mSv, 1.0 mSv, and 0.3 mSv, respectively (Ishikawa et al., 2015). The average dose for all the respondents was 0.8 mSv. The average dose for the Aizu area was smaller (0.2 mSv) than that for other areas due to its greater distance from the Fukushima Daiichi NPP (Fig. 1) (Ishikawa et al., 2015). Though regional differences in radioactive doses were observed, we found no significant adverse outcomes and no regional differences in the pregnancy and birth survey. Moreover, no differences over the 7 years have been observed in the entire Fukushima prefecture.

The response rate in the current study was approximately 50%. The response rates varied significantly over the years and between the regions. While Kenpoku constantly had a higher response rate than other areas, Soso had a higher rate only in 2011. This difference may be related to the concern that pregnant women in Kenpoku and Soso areas were exposed to relatively higher radiation doses.

In the entire Fukushima prefecture, the postdisaster incidences of stillbirth (0.21%–0.34%), PTB (4.6%–5.6%), LBW (8.6%–9.8%), and congenital anomalies (2.21%–2.85%) over the 7-year study period are similar to the recent averages in Japan.

The incidences of stillbirth (over 22 completed gestational weeks) are gradually decreasing in Japan as the quality of perinatal care is improving. They have decreased from 0.33% in 2011 (Mother's and Children's Health Organization, 2013) to 0.28% in 2017 (Mother's and Children's Health Organization, 2019c). Though the incidence of high-risk pregnancies has increased in Japan due to advanced maternal age and complicated pregnancies, the incidences of PTB (5.7%, Mother's and Children's Health Organization, 2019b) and LBW (9.4%, Mother's and Children's Health Organization, 2019a) have been almost stable during the postdisaster period.

The incidences of congenital anomalies in the entire Fukushima prefecture range from 2.21% to 2.85%, indicating no regional differences over time. It is widely reported that 2%–3% of all newborns have a major congenital abnormality, detectable at birth (Cragan & Gilboa, 2009; Dolk, Loane, & Garne, 2010). In Japan, from 2011 to 2016, the incidences of birth defects were 2.43%–2.59%, according to a report of the International Clearinghouse for Birth Defects Surveillance and Research (ICBDSR) Japan Center (ICBDSR, 2020). It was also reported that the prevalence per 10,000 pregnancies was 298.6, including miscarriages, stillbirths, and live births, of major congenital anomalies at delivery or in 1-month-old infants, using data from the Japan Environment and Children's Study (Mezawa et al., 2019).

Disasters potentially influence perinatal health, including maternal mental health (Harville, Xiong, & Buekens, 2010; Zotti, Williams, Robertson, Horney, & Hsia, 2013). The Chinese Wenchuan earthquake, which caused nearly 70,000 deaths in 2008, was associated with many birth complications, including higher rates of LBW, PTB, and birth defects and lower Apgar scores in a large study of survivors (Tan et al., 2009). A hospital-based cohort study showed that the Taiwan earthquake of 1999 was associated with high rates of LBW and psychiatric distress but no significant changes in birth length (Chang, Chang, Lin, & Kuo, 2002). Pregnant women who had experienced the 1994 Northridge earthquake in California gave birth earlier than expected, particularly those exposed in the first trimester (Glynn, Wadhwa, Dunkel-Schetter, Chicz-Demet, & Sandman, 2001). An increase in preterm deliveries was seen in one hospital after two of the five Israeli earthquakes (Weissman, Siegler, Neiger, Jakobi, & Zimmer, 1989). Several major environmental or technical disasters have been related to congenital anomalies. These include the nuclear reactor accidents at Chernobyl, Ukraine, in 1986 and at Three Mile Island, Pennsylvania, in 1979. Chernobyl involved a much larger radiation leak and affected many more people than Three Mile Island or Fukushima. Reviews on the Chernobyl disaster's effects indicate increased microcephaly and neural tube defects (Wertelecki, 2010; Wertelecki et al., 2017, 2018). However, most birth defects did not increase in most European countries (Dolk & Nichols, 1999; Hoffmann, 2001; Little, 1993). Although data for the most exposed areas are limited, a suggestion of an increase in congenital anomalies in Chernobyl's immediate vicinity has been reported (Feshchenko, Schröder, Müller, & Lazjuk, 2002).

Based on these previous observations and our findings, we concluded that the Great East Japan Earthquake and the accident at TEPCO's Fukushima Daiichi NPP did not have any significant adverse outcomes. Moreover, our results suggest that routine pregnancies in Fukushima received appropriate prenatal care after the disaster.

Our study has several major limitations. First, the response rate was only around 50%, limiting the generalizability of our findings. Second, as this study used self-administered questionnaires, it is assumed that the mothers answered correctly, especially regarding fetal anomalies. Third, this survey analyzes each district and does not investigate the relationship with individual radiation exposure doses. Finally, although it is possible to underestimate incidences when using a self-administered questionnaire with variable response rates, we found no significant adverse outcomes based on the pregnancy and birth survey over the whole Fukushima prefecture after the disaster.

In conclusion, the Great East Japan Earthquake and the accident at TEPCO's Fukushima Daiichi NPP do not appear to have caused significant adverse outcomes in the Fukushima prefecture in 7 years since the disasters.

References

Chang, H. L., Chang, T. C., Lin, T. Y., & Kuo, S. S. (2002). Psychiatric morbidity and pregnancy outcome in a disaster area of Taiwan 921 earthquake. *Psychiatry and Clinical Neurosciences*, *56*(2), 139–144.

Cragan, J. D., & Gilboa, S. M. (2009). Including prenatal diagnoses in birth defects monitoring: Experience of the metropolitan Atlanta congenital defects program. *Birth Defects Research. Part A, Clinical and Molecular Teratology*, *85*(1), 20–29.

Dolk, H., Loane, M., Garne, E., & de la Posada Paz, M. (2010). The prevalence of congenital anomalies in Europe. In S. Groft (Ed.), *Vol 686. Rare diseases epidemiology. Advances in experimental medicine and biology*. Dordrecht: Springer. https://doi.org/10.1007/978-90-481-9485-8_20.

Dolk, H., & Nichols, R. (1999). Evaluation of the impact of Chernobyl on the prevalence of congenital anomalies in 16 regions of Europe. EUROCAT working group. *International Journal of Epidemiology*, *28*(5), 941–948.

Feshchenko, S. P., Schröder, H. C., Müller, W. E., & Lazjuk, G. I. (2002). Congenital malformations among newborns and developmental abnormalities among human embryos in Belarus after Chernobyl accident. *Cellular and Molecular Biology*, *48*(4), 423–426.

Fujimori, K., Kyozuka, H., Yasuda, S., Goto, A., Yasumura, S., Ota, M., et al. (2014). Pregnancy and birth survey after the great East Japan earthquake and Fukushima Daiichi nuclear power plant accident in Fukushima prefecture. *Fukushima Journal of Medical Science*, *60*(1), 75–81.

Fukushima Revitalization Station. (2021). *Radioactivity measurement map*. Fukushima Prefectural Government. http://www.pref.fukushima.lg.jp/site/portal-english/.

Glynn, L. M., Wadhwa, P. D., Dunkel-Schetter, C., Chicz-Demet, A., & Sandman, C. A. (2001). When stress happens matters: Effects of earthquake timing on stress responsivity in pregnancy. *American Journal of Obstetrics and Gynecology*, *184*(4), 637–642.

Goto, A., Bromet, E. J., Ota, M., Ohtsuru, A., Yasumura, S., Fujimori, K., et al. (2017). The Fukushima nuclear accident affected mothers' depression but not maternal confidence. *Asia-Pacific Journal of Public Health*, *29*(2_suppl), 139S–150S.

Harville, E., Xiong, X., & Buekens, P. (2010). Disasters and perinatal health: A systematic review. *Obstetrical and Gynecological Survey*, *65*(11), 713–728.

Hoffmann, W. (2001). Fallout from the Chernobyl nuclear disaster and congenital malformations in Europe. *Archives of Environmental Health*, *56*(6), 478–484.

ICBDSR. (2020). *International clearinghouse for birth defects surveillance and research*. Japan Center. https://icbdsr-j.jp/data.html. in Japanese, Accessed June 2020.

Ishikawa, T., Yasumura, S., Ozasa, K., Kobashi, G., Yasuda, H., Miyazaki, M., et al. (2015). The Fukushima health management survey: Estimation of external doses to residents in Fukushima prefecture. *Scientific Reports*, *5*, 12712. https://doi.org/10.1038/srep12712.

Little, J. (1993). The Chernobyl accident, congenital anomalies and other reproductive outcomes. *Paediatric and Perinatal Epidemiology*, *7*(2), 121–151.

Mezawa, H., Tomotaki, A., Yamamoto-Hanada, K., Ishitsuka, K., Ayabe, T., Konishi, M., et al. (2019). Prevalence of congenital anomalies in the Japan environment and children's study. *Journal of Epidemiology*, *29*(7), 247–256.

Mother's & Children's Health Organization. (2013). Foetal deaths and Foetal death rates at 22 completed weeks and over of gestation by prefecture. In *Maternal and child health statistics of Japan* (p. 90).

Mother's & Children's Health Organization. (2019a). Live births and percentages by birthweight and sex, 1951–2011. In *Maternal and child health statistics of Japan* (p. 44).

Mother's & Children's Health Organization. (2019b). Live births and percentages by period of gestation, 1980-2011. In *Maternal and child health statistics of Japan* (p. 49).

Mother's & Children's Health Organization. (2019c). Foetal deaths and foetal death rates at 22 completed weeks and over of gestation by prefecture. In *Maternal and child health statistics of Japan* (p. 90).

Nagata, T., Kimura, Y., & Ishii, M. (2012). Use of a geographic information system (GIS) in the medical response to the Fukushima nuclear disaster in Japan. *Prehospital and Disaster Medicine*, *27*(2), 213–215.

Tan, C. E., Li, H. J., Zhang, X. G., Zhang, H., Han, P. Y., An, Q., et al. (2009). The impact of the Wenchuan earthquake on birth outcomes. *PLoS One*, *4*(12). https://doi.org/10.1371/journal.pone.0008200, e8200.

UNICEF. (2004). *Low birthweight, country, regional and global estimates*. World Health Organization. https://www.who.int/maternal_child_adolescent/documents/9280638327/en/.

Weissman, A., Siegler, E., Neiger, R., Jakobi, P., & Zimmer, E. Z. (1989). The influence of increased seismic activity on pregnancy outcome. *European Journal of Obstetrics, Gynecology, and Reproductive Biology*, *31*(3), 233–236.

Wertelecki, W. (2010). Malformations in a Chornobyl-impacted region. *Pediatrics*, *125*(4), e836–e843.

Wertelecki, W., Chambers, C. D., Yevtushok, L., Zymak-Zakutnya, N., Sosyniuk, Z., Lapchenko, S., et al. (2017). Chornobyl 30 years later: Radiation, pregnancies, and developmental anomalies in Rivne, Ukraine. *European Journal of Medical Genetics*, *60*(1), 2–11.

Wertelecki, W., Yevtushok, L., Kuznietsov, I., Komov, O., Lapchenko, S., Akhmedzanova, D., et al. (2018). Chornobyl, radiation, neural tube defects, and microcephaly. *European Journal of Medical Genetics*, *61*(9), 556–563.

WHO. (1977). Recommended definitions, terminology and format for statistical tables related to the perinatal period and use of a new certificate for cause of perinatal deaths. Modifications recommended by FIGO as amended October 14, 1976′. *Acta Obstetricia et Gynecologica Scandinavica*, *56*(3), 247–253.

Yasumura, S., Hosoya, M., Yamashita, S., Kamiya, K., Abe, M., Akashi, M., et al. (2012). Study protocol for the Fukushima health management survey. *Journal of Epidemiology*, *22*(5), 375–383.

Zotti, M. E., Williams, A. M., Robertson, M., Horney, J., & Hsia, J. (2013). Post-disaster reproductive health outcomes. *Maternal and Child Health Journal*, *17*(5), 783–796.

Chapter 6

Lifestyle-related diseases caused by evacuation: Results of the comprehensive health check in the Fukushima health management survey

Tetsuya Ohira[a,b], Hironori Nakano[a,b], Kanako Okazaki[a,b,c], Fumikazu Hayashi[a,b], Masanori Nagao[a,b], Akira Sakai[a,d], Mitsuaki Hosoya[a,e], Michio Shimabukuro[a,f], Seiji Yasumura[a,g], Hitoshi Ohto[a], and Kenji Kamiya[a,h], for the Fukushima Health Management Survey Group

[a]*Radiation Medical Science Center for the Fukushima Health Management Survey, Fukushima Medical University, Fukushima, Japan,* [b]*Department of Epidemiology, Fukushima Medical University School of Medicine, Fukushima, Japan,* [c]*Department of Physical Therapy, Fukushima Medical University School of Health Sciences, Fukushima, Japan,* [d]*Department of Radiation Life Sciences, Fukushima Medical University School of Medicine, Fukushima, Japan,* [e]*Department of Pediatrics, Fukushima Medical University School of Medicine, Fukushima, Japan,* [f]*Department of Diabetes, Endocrinology, and Metabolism, Fukushima Medical University School of Medicine, Fukushima, Japan,* [g]*Department of Public Health, Fukushima Medical University School of Medicine, Fukushima, Japan,* [h]*Research Institute for Radiation Biology and Medicine, Hiroshima University, Hiroshima, Japan*

Introduction

In addition to the direct damage caused by the earthquake and tsunami, the Great East Japan Earthquake that occurred on March 11, 2021, forced many residents to evacuate their homes due to the accident at the Fukushima Daiichi Nuclear Power Plant (NPP) caused by the tsunami. In Fukushima Prefecture, more than 160,000 residents were evacuated. Moreover, even now, 10 years after the earthquake, more than 35,000 residents continue to be evacuated (http://www.pref.fukushima.lg.jp/site/portal/shinsai-higaijokyo.html). In addition, direct mortalities from the Great East Japan Earthquake are reported to be 9543, 4675, and 1614 in the prefectures of Miyagi, Iwate, and Fukushima (https://www.npa.go.jp/news/other/earthquake2011/pdf/higaijokyo.pdf), respectively, while

Health Effects of the Fukushima Nuclear Disaster. https://doi.org/10.1016/B978-0-12-824098-4.00003-9

disaster-related mortalities are 929, 469, and 2313, respectively (https://www.reconstruction.go.jp/topics/main-cat2/sub-cat2-6/20201225_kanrenshi. pdf). Therefore, the high number of disaster-related mortalities in Fukushima Prefecture may be attributed to the effects of evacuation on physical and mental health.

The Fukushima Health Management Survey (FHMS) was launched to estimate external exposure doses of the people in Fukushima Prefecture and ascertain their health status, which is essential for prevention, early detection, and disease treatment following the release of radioactive materials and the evacuation of residents after the Fukushima NPP accident (Yasumura et al., 2012). The Comprehensive Health Check is one of the detailed surveys of the FHMS. Due to the earthquake and NPP accident, many residents were forced to live as evacuees, and their lifestyles (e.g., eating and exercise habits) drastically changed. Thus, this survey aims to evaluate the health status of the residents in the evacuation area, to prevent lifestyle-related diseases, and detect and treat diseases at an early stage (Radiation Medical Science Center for the Fukushima Health Management Survey, 2020).

The present study reviewed the results of longitudinal trends in lifestyle-related diseases and their risk factors among residents in the evacuation area based on the Comprehensive Health Check in the FHMS and discussed future strategies to prevent lifestyle-related diseases in the residents of Fukushima Prefecture, especially the residents in the evacuation area. We also believe that the results of this survey will be useful for the health management of evacuees due to disasters that may occur around the world.

Methods

Study population

The subjects comprised almost 210,000 men and women, except for foreign nationals, who were registered as residents in the evacuation area in Fukushima Prefecture between March 11, 2011, and April 1, 2012 (Radiation Medical Science Center for the Fukushima Health Management Survey, 2020; Yasumura et al., 2012). This area includes 13 municipalities (i.e., Hirono-machi, Naraha-machi, Tomioka-machi, Kawauchi-mura, Okuma-machi, Futaba-machi, Namie-machi, Katsurao-mura, Iitate-mura, Kawamata-machi, Tamura City, Minami-Soma City, and part of Date City, which is the area in which a specific evacuation recommendation point belongs). Following the disaster, the government ordered all residents of Hirono-machi, Naraha-machi, Tomioka-machi, Kawauchi-mura, Okuma-machi, Futaba-machi, Namie-machi, Katsurao-mura and Iitate-mura—and some residents of Tamura City, Minami-Soma City, Kawamata-machi, and Date City—to evacuate their homes.

According to the Comprehensive Health Check in the FHMS, annual health check-ups have been conducted among residents of all ages in these communities since 2011 (Radiation Medical Science Center for the Fukushima Health

Management Survey, 2020; Yasumura et al., 2012). Those <15 years old received health check-ups at designated medical institutions in or outside Fukushima Prefecture. Those ≥16 years old who resided in Fukushima Prefecture received (1) health check-ups with additional items in the specified health and comprehensive health check-ups conducted by each municipality, (2) group health check-ups conducted by Fukushima Medical University, or (3) individual health check-ups at designated medical institutions in Fukushima Prefecture. Out-of-prefecture residents received (1) health check-ups with additional items in the specified health and comprehensive health check-ups conducted by each municipality or (2) individual health check-ups at designated medical institutions in the prefecture. This study was approved by the Ethics Committee of Fukushima Medical University (#1319 and #1916).

Measures and definitions

Table 1 shows that the measurements of health check-ups vary according to age (Radiation Medical Science Center for the Fukushima Health Management Survey, 2020). Height in stocking feet and weight in light clothing were measured, and body mass index (BMI) was calculated as weight (kg)/height (m^2).

TABLE 1 Measurements in the comprehensive health check.

Age group	Measurements
0–6 years old	• Height and weight • Blood count (red blood cell count, hematocrit, hemoglobin, platelet count, white blood cell count, white blood cell differentiation)
7–15 years old	• Height, weight, and blood pressure • Blood count (red blood cell count, hematocrit, hemoglobin, platelet count, white blood cell count, white blood cell differentiation) Additional items by request • Biochemical tests (AST, ALT, γ-GT, triglycerides, HDL-cholesterol, LDL-cholesterol, HbA1c, blood glucose, serum creatinine, eGFR, and uric acid)
≥16 years old	• Height, weight, waist circumference, and blood pressure • Urine tests (urine protein, urine sugar, and urine occult blood) • Blood count (red blood cell count, hematocrit, hemoglobin, platelet count, white blood cell count, white blood cell differentiation) • Biochemical tests (AST, ALT, γ-GT, triglycerides, HDL-cholesterol, LDL-cholesterol, HbA1c, blood glucose, serum creatinine, eGFR, and uric acid)

ALT alanine aminotransferase; *AST* aspartate aminotransferase; *eGFR* estimated glomerular filtration rate; *HbA1c* hemoglobinA1c; *γ-GT* γ-glutamyl transpeptidase.

Overweight was defined as a BMI $\geq$25 kg/m^2. Systolic blood pressure (SBP) and diastolic blood pressure (DBP) were measured by trained technicians on the right arm of seated participants. Hypertension was defined as SBP $\geq$140 mmHg, a DBP $\geq$90 mmHg, or the use of antihypertensive medication.

Venous blood was collected from the subjects using ethylenediaminetetraacetic acid di-potassium as an anticoagulant for measuring the blood cell count. The counts of total white blood cells, lymphocytes, monocytes, neutrophils, eosinophils, and basophils were determined using an automated cell counter. Venous blood was also collected from the subjects using a serum-separating tube to measure aspartate aminotransferase (AST), alanine aminotransferase (ALT), γ-glutamyl transpeptidase (γ-GT), triglycerides (TG), high-density lipoprotein (HDL) cholesterol, low-density lipoprotein (LDL) cholesterol, hemoglobinA1c (HbA1c), creatinine, estimated glomerular filtration rate (eGFR), and uric acid (UA) levels. Additional measurements included fasting plasma glucose levels and urine protein, sugar, and occult blood levels. Glucose intolerance was defined as an HbA1c (NGSP) level of $\geq$5.8% or the use of an antihyperglycemic medication. Hypo-HDL and hyper-LDL cholesterolemia were defined as HDL cholesterol <40 mg/dL and LDL cholesterol $\geq$140 mg/dL, respectively. Moreover, hypertriglyceridemia was defined as fasting serum TG $\geq$150 mg/dL or being treated for dyslipidemia. The diagnosis of metabolic syndrome (MetS) was based on the definition of the Japanese Committee for Establishing Diagnostic Criteria for MetS as follows: visceral obesity (waist circumference, $\geq$85 cm in men and $\geq$90 cm in women) and the presence of at least two of the three abnormalities (fasting serum TG $\geq$150 mg/dL and/or HDL cholesterol <40 mg/dL or being treated for dyslipidemia; systolic BP $\geq$130 mmHg and/or diastolic BP $\geq$85 mmHg or being treated for hypertension; and fasting plasma glucose $\geq$110 mg/dL or being treated for diabetes).

The history of cigarette smoking and weekly alcohol intake was obtained by a trained interviewer from the participants $\geq$20 years old. Participants who consumed $\geq$44 g ethanol per day were classified as current excessive drinkers. An interviewer also obtained information on lifestyle factors (e.g., physical activity, dietary behaviors, sleeping conditions, medical history). Physical activity included regular exercise and daily walking or physical activity. Dietary behaviors included eating speed, frequency of skipping breakfast, eating dinner late, and eating between meals or bedtime snacks. Sleep conditions were evaluated by subjective sufficient sleep.

Results

Trends in the prevalence of lifestyle-related diseases after the disaster (FY 2011–18)

Table 2 shows the number of participants and participation rate in the Comprehensive Health Check from fiscal year (FY) 2011–18. In FY 2011, 64.5% and

TABLE 2 Number of participants and participation rate in the comprehensive health check from FY 2011 to 2018.

	FY 2011	FY 2012	FY 2013	FY 2014	FY 2015	FY 2016	FY 2017	FY 2018
≤15 years old								
Target subjects	27,819	27,077	26,474	25,883	25,296	24,600	23,660	22,744
Number of participants	17,934	11,780	10,248	9216	7603	6413	5403	4479
Participation rate (%)	64.5	43.5	38.7	35.6	30.1	26.1	22.8	19.7
≥16 years old								
Target subjects	182,370	184,910	186,970	188,328	190,019	191,101	191,636	191,974
Number of participants	56,399	47,009	43,040	41,874	41,211	39,990	39,367	38,815
Participation rate (%)	30.9	25.4	23.0	22.2	21.7	20.9	20.5	20.2

FY, fiscal year.

30.9% of those <15 years old and >16 years old participated in the health check-up, respectively. However, the participation rate has been continuously decreasing annually. Moreover, it was 19.7% and 20.2% for those <15 and >16 years old, respectively, in FY 2018 (Radiation Medical Science Center for Fukushima Health Management Survey, 2020).

Trends in body weight and overweight/obese

Table 3 shows changes in body weight by age for boys and girls <15 years old. In the national data, no change in the body weight of both boys and girls was noted between FY 2011 and 2018. Meanwhile, a downward trend in body weight, especially among girls, was noted in the evacuation areas of Fukushima Prefecture (Radiation Medical Science Center for Fukushima Health Management Survey, 2020). However, the body weight of boys and girls in the evacuation areas tended to be higher than that of the boys and girls in the national data when the national data were compared with the evacuation area data in FY 2018.

Table 4 shows that the mean BMI levels in all age groups remained practically unchanged among men and women between FY 2011 and 2018 (Radiation Medical Science Center for Fukushima Health Management Survey, 2020). The proportion of overweight people in all age groups among men showed a decreasing trend between FY 2011 and 2014, but an increasing trend between FY 2014 and 2018. The age-stratified proportion of overweight people in FY 2011, 2012, 2013, and 2014 was 29.8%, 30.7%, 30.0%, and 29.0%, respectively, for men 16–39 years old; 41.6%, 40.3%, 40.9%, and 39.3%, respectively, for men 40–64 years old; and 39.1%, 36.4%, 36.3%, and 35.8%, respectively, for men ≥65 years old. Moreover, the age-stratified proportion of overweight people in FY 2015, 2016, 2017, and 2018 was 29.3%, 29.4%, 33.4%, and 31.4%, respectively, for men 16–39 years old; 40.6%, 40.9%, 42.5%, and 42.2%, respectively, for men 40–64 years old; and 36.3%, 36.8%, 38.0%, and 38.6%, respectively, for men ≥65 years old. For women, the proportion of overweight people in all age groups remained practically unchanged between FY 2011 and 2018.

The National Database (NDB) showed that the proportion of obese people 40–64 years old in Japan and Fukushima Prefecture in 2017 was 35.4% and 38.2% for men and 19.3% and 24.4% for women, respectively (Ministry of Health & Labour, 2021), compared with 42.5% and 30.2% for men and women, respectively, in the Comprehensive Health Check.

Trends in lifestyle-related factors between FY 2011 and 2018 for boys and girls

Table 5 shows that the prevalence of hypertension was unchanged from FY 2011 to 2018 among boys and girls 7–15 years old. Moreover, the prevalence of hypertension (DBP ≥90 mmHg) in FY 2011 and 2018 was 0.8% and 0.4% for boys, respectively, and 0.4% and 0.3% for girls, respectively

TABLE 3 Changes in body weight between FY 2011 and 2018 in Japan and CHC for boys and girls, stratified by age.

Age	Mean values in Japan[a] in FY 2011	Mean values in Japan[a] in FY 2018	Differences between FY 2011 and 2018	Mean values in CHC in FY 2011	Mean values in CHC in FY 2018	Differences between FY 2011 and 2018	Differences between Japan and CHC in FY 2018
Boys							
6 years old	21.3	21.4	0.1	22.1	21.8	−0.3	0.4
7 years old	24.0	24.1	0.1	24.8	25.2	0.4	1.1
8 years old	27.0	27.2	0.2	28.4	28.3	−0.1	1.1
9 years old	30.3	30.7	0.4	32.6	32.8	0.2	2.1
10 years old	33.8	34.1	0.3	36.0	36.0	0.0	1.9
11 years old	38.0	38.4	0.4	40.5	40.6	0.1	2.2
12 years old	43.8	44.0	0.2	46.9	44.7	−2.2	0.7
13 years old	49.0	48.8	−0.2	51.2	49.6	−1.6	0.8
14 years old	54.2	54.0	−0.2	56.1	56.2	0.1	2.2

Continued

TABLE 3 Changes in body weight between FY 2011 and 2018 in Japan and CHC for boys and girls, stratified by age—cont'd

Age	Mean values in Japan in FY 2011	Mean values in Japan in FY 2018	Differences between FY 2011 and 2018	Mean values in CHC in FY 2011	Mean values in CHC in FY 2018	Differences between FY 2011 and 2018	Differences between Japan and CHC in FY 2018
15 years old	59.4	58.6	−0.8	60.0	58.9	−1.1	0.3
Girls							
6 years old	20.8	20.9	0.1	21.7	22.0	0.3	1.1
7 years old	23.4	23.5	0.1	24.1	24.1	0.0	0.6
8 years old	26.4	26.4	0.0	27.4	26.8	−0.6	0.4
9 years old	29.8	30.0	0.2	31.0	30.1	−0.9	0.1
10 years old	34.0	34.1	0.1	35.7	35.0	−0.7	0.9
11 years old	38.8	39.1	0.3	40.5	40.2	−0.3	1.1

12 years old	43.6	43.7	0.1	45.8	42.9	−2.9	−0.8
13 years old	47.1	47.2	0.1	48.5	47.7	−0.8	0.5
14 years old	49.9	49.9	0.0	51.8	50.3	−1.5	0.4
15 years old	51.4	51.6	0.2	53.5	52.5	−1.0	0.9

CHC, the comprehensive health check; *FY*, fiscal year.
[a]*Ministry of Education, Culture, Sports, Science and Technology School Health Survey.*

TABLE 4 Trends in mean BMI levels and overweight between FY 2011 and 2018 for men and women in CHC, stratified by age.

	FY 2011	FY 2012	FY 2013	FY 2014	FY 2015	FY 2016	FY 2017	FY 2018
Men								
BMI (kg/m^2)								
16–39 years old	23.5	23.6	23.5	23.5	23.5	23.5	23.8	23.6
40–64 years old	24.6	24.5	24.5	24.4	24.5	24.6	24.7	24.8
≥65 years old	24.2	24.0	24.0	24.0	24.0	24.1	24.2	24.2
Overweight (%)								
16–39 years old	29.8	30.7	30.0	29.0	29.3	29.4	33.4	31.4
40–64 years old	41.6	40.3	40.9	39.3	40.6	40.9	42.5	42.2
≥65 years old	39.1	36.4	36.3	35.8	36.3	36.8	38.0	38.6
Women								
BMI (kg/m^2)								
16–39 years old	21.9	21.9	21.9	22.0	22.0	22.1	22.2	22.1
40–64 years old	23.3	23.4	23.3	23.2	23.3	23.3	23.5	23.4
≥65 years old	23.9	23.8	23.8	23.7	23.6	23.7	23.7	23.7
Overweight (%)								
16–39 years old	17.2	17.1	17.3	17.6	17.6	18.8	19.1	19.1
40–64 years old	28.4	29.2	28.9	27.9	28.4	29.0	30.2	29.8
≥65 years old	35.4	34.3	34.4	33.0	32.2	32.8	33.1	32.6

CHC, the comprehensive health check; *FY*, fiscal year.

TABLE 5 Trends in lifestyle-related factors between FY 2011 and 2018 for boys and girls in CHC.

	FY 2011	FY 2012	FY 2013	FY 2014	FY 2015	FY 2016	FY 2017	FY 2018
Boys (7–15 years old)								
SBP ≥140 mmHg (%)	0.9	0.4	0.3	0.4	0.3	0.3	0.4	0.2
DBP ≥90 mmHg (%)	0.8	0.4	0.6	0.7	0.4	0.4	0.4	0.4
Hemoglobin ≤13.0 g/dL (%)	24.8	21.9	23.4	24.7	25.0	24.8	23.1	21.7
AST ≥31 U/L (%)	12.8	14.1	14.4	14.8	15.0	15.3	14.3	14.7
ALT ≥31 U/L (%)	7.0	7.4	7.5	6.7	7.2	6.3	7.6	9.5
LDL-cholesterol ≥140 mg/dL (%)	3.3	3.2	4.2	3.4	2.8	2.8	1.9	2.8
Triglycerides ≥150 mg/dL (%)	7.7	7.7	7.6	8.7	7.6	8.0	9.7	10.4
HDL-cholesterol <40 mg/dL (%)	3.1	3.1	3.1	2.7	2.9	3.3	3.4	2.8
HbA1c ≥5.8% (%)	4.6	2.6	2.6	4.3	3.2	5.0	4.2	3.9
Uric acid ≥7.1 mg/dL (%)	4.7	6.1	4.9	5.0	6.0	5.9	5.3	5.4
Girls (7–15 years old)								
SBP ≥140 mmHg (%)	0.2	0.1	0.1	0.1	0.1	0.2	0.2	–
DBP ≥90 mmHg (%)	0.4	0.3	0.4	0.2	0.3	0.5	0.3	0.3
Hemoglobin ≤12.0 g/dL (%)	7.6	6.2	5.9	6.1	6.5	7.8	8.0	5.8
AST ≥31 U/L (%)	6.4	7.0	6.4	7.4	7.2	7.4	6.9	6.7
ALT ≥31 U/L (%)	2.0	2.0	1.8	2.1	2.6	1.6	2.1	2.5

Continued

TABLE 5 Trends in lifestyle-related factors between FY 2011 and 2018 for boys and girls in CHC—cont'd

	FY 2011	FY 2012	FY 2013	FY 2014	FY 2015	FY 2016	FY 2017	FY 2018
LDL-cholesterol ≥140 mg/dL (%)	3.6	3.6	3.5	3.5	2.8	2.8	3.0	2.9
Triglycerides ≥150 mg/dL (%)	6.3	6.5	7.0	7.7	7.7	7.6	8.0	8.3
HDL-cholesterol <40 mg/dL (%)	2.8	2.3	2.7	2.0	2.4	3.4	2.9	2.2
HbA1c ≥5.8% (%)	3.6	2.2	1.9	2.7	2.0	3.1	2.6	2.2
Uric acid ≥7.1 mg/dL (%)	0.3	0.6	0.2	0.4	0.3	0.3	0.2	0.5

CHC, the comprehensive health check; *FY*, fiscal year; *HbA1c*, hemoglobinA1c.

(Radiation Medical Science Center for Fukushima Health Management Survey, 2020). The prevalence of anemia (boys: hemoglobin $\leq$13.0 g/dL; girls: hemoglobin $\leq$12.0 g/dL) was also unchanged or somewhat improved between FY 2011 and 2018. In addition, the prevalence of anemia in FY 2011 and 2018 was 24.8% and 21.7% for boys and 7.6% and 5.8% for girls, respectively. The prevalence of liver dysfunction somewhat increased between FY 2011 and 2018 for boys and girls. The prevalence of hyper-LDL cholesterolemia and hypo-HDL cholesterolemia tended to decrease between FY 2011 and 2018 for boys and girls, although the prevalence of hypertriglyceridemia increased between FY 2011 and 2018.

Trends in lifestyle-related factors between FY 2011 and 2018 for men and women

The age-stratified proportions of people with high systolic and diastolic blood pressures decreased between FY 2011 and 2018 in both men and women (Table 6; Radiation Medical Science Center for Fukushima Health Management Survey, 2020). The age-stratified prevalence of anemia (men: hemoglobin $\leq$13.0 g/dL; women: hemoglobin $\leq$12.0 g/dL) unchanged or somewhat improved between FY 2011 and 2018 in both men and women, although the prevalence of renal dysfunction (men, creatinine $\geq$1.15 mg/dL; women, creatinine $\geq$0.95 mg/dL) and hyperuricemia (uric acid $\geq$7.1 mg/dL) increased between FY 2011 and 2018.

As shown in Table 7, the age-stratified prevalence of men and women with liver dysfunction (ALT $\geq$31 U/L and/or AST $\geq$31 U/L) increased between FY 2011 and 2018 in the 16–39-year-old group, but not in the 40–64-year old and 65-year-old groups. The age-stratified proportions of men and women with hyper-LDL cholesterolemia, hypertriglyceridemia, and hypo-HDL cholesterolemia tended to decrease in all age groups between FY2011 and 2018. Moreover, the prevalence of glucose intolerance (HbA1c $\geq$5.8% and/or medication use) tended to increase in all age groups between FY 2011 and 2018 (Table 7).

Effects of evacuation on the incidence of lifestyle-related diseases after the Great East Japan earthquake

The prevalence of renal dysfunction and glucose intolerance continuously increased after the earthquake as aforementioned. Therefore, the current study investigated the factors that contributed to the development of renal dysfunction and diabetes mellitus after the earthquake (Hayashi et al., 2017; Satoh et al., 2019). Participants in the Comprehensive Health Check who were 40–74 years old and did not have chronic kidney diseases (CKD; estimated glomerular filtration rate, <60 mL/min/1.73 m^2 or proteinuria) in FY 2011 were used as the baseline (nonevacuees, $n=9780$; evacuees, $n=4712$) and followed until FY 2014 (Hayashi et al., 2017). Moreover, the association between evacuation

TABLE 6 Trends in lifestyle-related factors between FY 2011 and 2018 for men and women in CHC, stratified by age.

	FY 2011	FY 2012	FY 2013	FY 2014	FY 2015	FY 2016	FY 2017	FY 2018
Men								
SBP ≥140 mmHg (%)								
16–39 years old	5.8	4.9	4.2	4.1	3.5	3.9	4.5	4.1
40–64 years old	27.5	21.5	19.0	17.4	17.4	20.5	19.4	17.3
≥65 years old	43.1	34.2	29.4	25.1	26.5	31.6	29.2	27.3
DBP ≥90 mmHg (%)								
16–39 years old	6.6	4.8	4.7	4.3	4.1	3.5	3.9	3.9
40–64 years old	24.1	18.5	16.4	14.4	15.9	15.7	16.4	14.8
≥65 years old	17.9	12.5	9.6	7.6	8.7	9.2	8.7	7.6
Hemoglobin ≤13.0 g/dL (%)								
16–39 years old	0.6	0.5	0.6	0.9	1.2	0.8	0.7	0.6
40–64 years old	2.4	3.5	2.8	2.6	3.1	3.0	3.3	3.1
≥65 years old	8.8	12.8	11.3	10.7	12.0	12.5	13.4	12.3
Creatinine ≥1.15 mg/dL (%)								
16–39 years old	0.4	0.4	0.6	0.5	0.4	1.1	0.9	0.8
40–64 years old	2.4	2.7	2.4	3.1	3.2	3.4	3.4	3.9
≥65 years old	7.6	8.3	9.0	9.9	9.0	9.7	10.8	10.4

Uric acid $\geq$7.1 mg/dL (%)								
16–39 years old	18.5	20.3	21.8	22.0	20.9	22.6	23.8	24.2
40–64 years old	18.1	20.9	19.7	20.3	20.9	21.2	21.2	22.9
$\geq$65 years old	14.4	16.0	16.3	16.1	16.4	16.3	15.7	16.9
Women								
SBP $\geq$140 mmHg (%)								
16–39 years old	1.6	1.3	1.0	1.3	1.4	1.9	1.7	1.3
40–64 years old	19.1	14.9	12.9	11.5	12.3	14.5	13.6	12.2
$\geq$65 years old	40.4	31.8	27.0	22.7	23.4	28.6	26.3	23.7
DBP $\geq$90 mmHg (%)								
16–39 years old	1.7	1.5	1.2	1.4	2.0	2.2	2.3	1.5
40–64 years old	12.2	9.6	7.7	6.6	8.0	7.5	7.9	7.3
$\geq$65 years old	12.6	8.7	6.7	5.1	6.2	7.0	6.1	5.3
Hemoglobin $\leq$12.0 g/dL (%)								
16–39 years old	13.2	15.3	13.6	14.6	15.8	14.7	13.4	14.0
40–64 years old	11.4	12.5	11.0	9.9	11.4	11.7	11.6	10.6
$\geq$65 years old	10.5	15.4	13.4	11.8	13.5	13.9	14.5	12.7
Creatinine $\geq$0.95 mg/dL (%)								
16–39 years old	0.2	0.1	0.1	0.1	0.1	0.0	0.2	0.1
40–64 years old	0.8	0.8	0.9	0.9	1.0	1.0	1.1	0.9

Continued

TABLE 6 Trends in lifestyle-related factors between FY 2011 and 2018 for men and women in CHC, stratified by age—cont'd

	FY 2011	FY 2012	FY 2013	FY 2014	FY 2015	FY 2016	FY 2017	FY 2018
≥65 years old	4.4	4.5	5.1	5.0	5.1	5.0	5.4	5.3
Uric acid ≥7.1 mg/dL (%)								
16–39 years old	0.7	0.7	1.0	1.0	1.3	1.2	1.1	0.9
40–64 years old	1.1	1.5	1.4	1.6	1.8	1.9	1.7	2.0
≥65 years old	2.1	2.6	3.0	3.0	3.1	2.9	3.2	3.2

CHC, the comprehensive health check; *FY,* fiscal year.

TABLE 7 Trends in lifestyle-related factors between FY 2011 and 2018 in CHC, stratified by age.

	FY 2011	FY 2012	FY 2013	FY 2014	FY 2015	FY 2016	FY 2017	FY 2018
ALT ≥31 U/L (%)								
16–39 years old	8.2	8.7	8.4	8.3	8.8	9.3	9.8	9.1
40–64 years old	14.5	15.8	14.1	13.7	13.4	14.1	15.3	14.5
≥65 years old	17.7	19.5	16.8	16.8	16.8	17.0	17.6	16.4
AST ≥31 U/L (%)								
16–39 years old	15.9	16.9	16.1	15.5	16.0	16.7	17.5	17.7
40–64 years old	20.8	21.4	20.0	19.5	19.7	19.3	20.2	21.2
≥65 years old	13.6	14.2	13.0	13.3	13.1	12.7	12.2	13.4
LDL-cholesterol ≥140 mg/dL[a] (%)								
16 ≥39 years old	15.9	15.7	15.6	16.6	15.8	15.7	15.6	14.9
40 ≥64 years old	35.8	31.6	32.6	33.8	32.9	32.3	31.4	29.4
≥65 years old	28.6	22.3	23.2	23.5	24.1	23.1	22.6	20.3
Triglycerides ≥150 mg/dL[a] (%)								
16–39 years old	11.4	11.7	11.8	11.3	11.0	11.3	11.6	11.1
40–64 years old	21.3	21.5	21.5	20.5	20.9	20.9	21.7	20.8
≥65 years old	20.3	17.9	18.4	18.5	18.3	19.4	18.6	18.7

Continued

TABLE 7 Trends in lifestyle-related factors between FY 2011 and 2018 in CHC, stratified by age—cont'd

	FY 2011	FY 2012	FY 2013	FY 2014	FY 2015	FY 2016	FY 2017	FY 2018
HDL-cholesterol <40 mg/dL[a] (%)								
16–39 years old	4.0	4.3	4.1	4.2	4.3	3.6	3.9	3.9
40–64 years old	5.8	6.4	5.5	4.8	5.0	4.2	3.9	4.4
≥65 years old	8.5	8.7	7.6	7.0	7.0	5.9	5.2	5.4
HbA1c ≥5.8%[a] (%)								
16–39 years old	2.8	4.3	3.9	4.6	4.8	4.1	5.7	4.7
40–64 years old	17.7	21.8	24.6	24.2	25.4	24.9	28.6	26.5
≥65 years old	27.8	31.5	37.7	35.6	38.9	39.2	44.1	42.9

CHC, the comprehensive health check; *FY*, fiscal year.
[a]*Including medication use.*

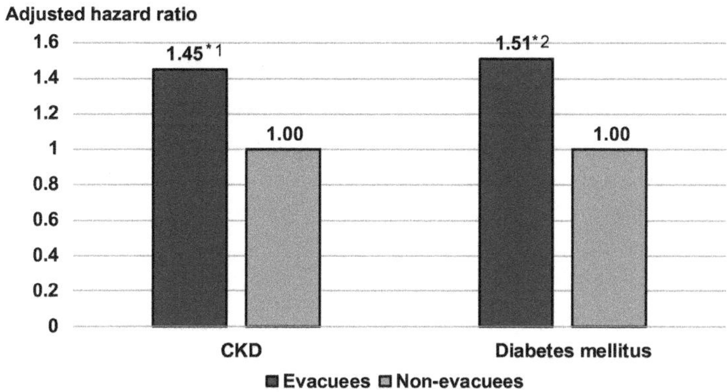

*1 Adjusted for age, sex, baseline eGFR, obesity, hypertension, diabetes, dyslipidemia, and smoking status.
*2 Adjusted for age, sex, BMI scores, current smoker, excess alcohol consumption, weight change (+10 kg) from age 20 years, weight change (+3 kg) within 1 year, sleep quality, and physical activity. CKD: chronic kidney disease.

FIG. 1 Multivariable-adjusted hazard ratios of incidence of chronic kidney disease (CKD) and diabetes mellitus for evacuees compared with nonevacuees; The Comprehensive Health Check of the Fukushima Health Management Survey (Hayashi et al., 2017; Satoh et al., 2019).

and CKD incidence during the follow-up period was examined. The CKD incidence among nonevacuees and evacuees during a mean follow-up of 2.46 years was 80.8/1000 and 100.2/1000 person-years, respectively. The evacuation was a significant risk factor for CKD incidence even after adjusting for age, sex, baseline eGFR, obesity, hypertension, diabetes, dyslipidemia, and smoking, with a hazard ratio (HR) of 1.45 (95% confidence interval (CI), 1.35–1.56) for the evacuation compared with the nonevacuation (Fig. 1).

Similarly, 13,487 participants (nonevacuees, $n=9252$; evacuees, $n=4235$) who did not have diabetes mellitus in the Comprehensive Health Check conducted in FY 2011 were followed until FY 2014 (Satoh et al., 2019). The cumulative incidence of diabetes mellitus was 4.54% ($n=612$) with a mean follow-up of 2.67 years. The incidence of diabetes mellitus was 23.2/1000 person-years in the evacuees compared with 14.4/1000 person-years in the nonevacuees. The evacuees were found to have a significant risk factor for diabetes mellitus development compared with the nonevacuees in the Cox proportional hazards model. The age- and gender-adjusted HR for evacuees was 1.78 (95% CI, 1.51–2.09), and the multivariable-adjusted HR was 1.51 (95% CI, 1.28–1.79).

Discussion

The prevalence of obesity, hypertension, diabetes mellitus, dyslipidemia, hyperuricemia, liver dysfunction, MetS, polycythemia, and atrial fibrillation was previously reported to increase among residents in the evacuation zone after the Fukushima Daiichi NPP accident (Hashimoto et al., 2017, 2020; Ohira et al., 2016b, 2016a; Sakai et al., 2014; Satoh et al., 2015, 2016; Suzuki et al., 2015;

Takahashi, Ohira, Yasumura, et al., 2017). Furthermore, evacuation is one of the significant risk factors for the incidence of lifestyle-related diseases after the disaster. The results of the Comprehensive Health Check revealed that the obesity and lifestyle-related diseases that increased after the disaster have not fully returned to their preearthquake state, although some of them have improved. These results indicate that residents who experienced evacuation after the disaster in Fukushima have a sustained risk of developing cardiovascular diseases (e.g., myocardial infarction and stroke).

A prospective observational study to evaluate the impact of the 2011 Great East Japan earthquake on the BMI of preschool children showed significantly increased BMIs among girls in Fukushima and among boys and girls in Iwate 1 month after the earthquake (Yokomichi et al., 2016). In addition, significantly increased BMIs were detected among boys and girls in Fukushima and significantly decreased BMIs were observed among boys and girls in Miyagi 19 months after the earthquake. Increases in the prevalence of overweight observed boys and girls in Fukushima who were 6–11 and 6–10 years old, respectively. Another study showed that children who experienced the disaster at 47–59 months old in Fukushima had a significantly higher standard deviation score (SDS) for BMI than the historical control group and the regional controls 1.5 years after the disaster (Isojima et al., 2017). In the Comprehensive Health Check, the change in BMI-SDS tended to decrease over 5 years among boys and girls 7–15 years old with the BMI-SDS in 2011 higher than those in 2012, 2013, 2014, and 2015 (Kawasaki et al., 2020). Furthermore, the frequency of boys and girls with a BMI score of >2 SD in 2011 was higher than that in 2012, 2013, 2014, and 2015, with the frequency of those with a BMI score >2 SD decreasing with time over the 5 years. However, the body weight of boys and girls in the evacuation areas tended to be higher than that of boys and girls in the national data when the national and evacuation area data for FY 2018 were compared. Moreover, the improvement in lipid abnormalities was delayed compared to that in obesity. Therefore, continuous follow-up on childhood obesity in the evacuation areas is necessary.

For adults, the current study reported the increase in mean body weight and the proportion of people being overweight/obese after the Great East Japan Earthquake among residents living near the Fukushima Daiichi NPP in the Fukushima prefecture and the significant association between evacuation and an increased risk of becoming overweight/obese after adjusting for confounding variables (Ohira et al., 2016a). Furthermore, the results of the Comprehensive Health Check showed that the weight gained after the earthquake has not yet returned to the preearthquake state even after 7 years. Loss of and changes in jobs, changes in eating habits, decrease in regular exercise, and increase in insomnia were reported after the disaster (Kunii et al., 2016; Yabe et al., 2014; Zhang et al., 2017). These lifestyle changes are conceivable to still have a possible impact more than 7 years later. Obesity has been reported to continuously increase in the areas that were evacuated compared to the areas that were not evacuated as a result of looking at the changes in obesity immediately after the earthquake and 5 years later (Radiation Medical Science Center for Fukushima Health Management Survey, 2020).

The body weight increase that was observed in evacuees may have played a role, although the mechanisms by which evacuation increases cardiovascular risk factors are not yet fully understood. The postdisaster increase in body weight in the FHMS was significantly associated with an increased risk of hypertension, diabetes mellitus, dyslipidemia, liver dysfunction, polycythemia, renal dysfunction, and atrial fibrillation. Furthermore, decreases in physical activity may also be associated with a higher incidence of cardiovascular risk factors in the evacuation areas. Given that most evacuees who worked before the disaster lost their jobs, the consequent decrease in physical activity may have contributed to an increase in cardiovascular risk factors via an increase in body weight. Thus, health-related behaviors could be an important factor, mediating the influence of evacuation on increased cardiovascular risk factors. In addition, the Great East Japan Earthquake in Fukushima was a complex disaster consisting of three events (i.e., a major earthquake, a tsunami, and an NPP accident) and may have increased the psychological burden. The percentage of high psychological distress as assessed by the Kessler Psychological Distress Scale (K6) was higher in Fukushima than in Miyagi (Goodwin, Takahashi, Sun, & Ben-Ezra, 2015). Furthermore, the proportion was still nearly twice as high as in other areas in 2018 although the proportion of people with high levels of psychological distress in Fukushima gradually decreased from 2011 to 2018. This may lead to longer-lasting psychological stress and an increase in stress-related diseases (e.g., MetS and diabetes). Further studies are needed to confirm the association between lifestyle and psychological factors with future incidence of cardiovascular disease (CVD), including stroke and myocardial infarction, in Fukushima residents.

This study has some limitations. First, the participation rate of the Comprehensive Health Check was not high and gradually declined, which may imply that the sample is not representative of the population. Second, this study did not evaluate socioeconomic factors other than evacuation, which may have influenced the association between evacuation and CVD risk factors. Finally, although the mean follow-up time was over 7 years, this may have been insufficient to examine the long-term effects of the evacuation on the incidence of CVDs (e.g., stroke and myocardial infarction). Therefore, evaluating these effects through a further long-term follow-up is necessary.

Following the Great East Japan earthquake, Fukushima's residents have experienced a higher incidence of both physical and psychological problems, including lifestyle-related diseases and anxiety and psychological distress, than residents of other areas. These physical and psychological issues may lead to a future CVD incidence. Therefore, continuous support in terms of health status management is required from researchers working together with local governments and communities.

Conclusions

After the Great East Japan earthquake, the proportion of people with cardiovascular risk factors such as obesity, hypertension, diabetes, and dyslipidemia, increased among residents of the evacuation zone in Fukushima Prefecture, and even more so among actual evacuees. The effects of the evacuation persisted more than 7 years after the disaster and were associated with increased cardiovascular risk factors. As of 2021, more than 35,000 Fukushima residents remain displaced. Thus, continuous prevention programs against CVD risk factors should be implemented in cooperation with local governments and communities to prevent future CVD in postdisaster evacuees.

Acknowledgments

The authors thank the staff of the Fukushima Health Management Survey for their important contributions. The findings and conclusions of this article are solely the responsibility of the authors and do not represent the official views of the Fukushima prefecture government.

References

Goodwin, R., Takahashi, M., Sun, S., & Ben-Ezra, M. (2015). Psychological distress among tsunami refugees from the great East Japan earthquake. *BJPsych Open*, *1*(1), 92–97. https://doi.org/10.1192/bjpo.bp.115.000422.

Hashimoto, S., Nagai, M., Fukuma, S., Ohira, T., Hosoya, M., Yasumura, S., et al. (2017). Influence of post-disaster evacuation on incidence of metabolic syndrome. *Journal of Atherosclerosis and Thrombosis*, *24*(3), 327–337. https://doi.org/10.5551/jat.35824.

Hashimoto, S., Nagai, M., Ohira, T., Fukuma, S., Hosoya, M., Yasumura, S., et al. (2020). Influence of post-disaster evacuation on incidence of hyperuricemia in residents of Fukushima prefecture: The Fukushima health management survey. *Clinical and Experimental Nephrology*, *24*, 1025–1032. https://doi.org/10.1007/s10157-020-01924-6.

Hayashi, Y., Nagai, M., Ohira, T., Satoh, H., Sakai, A., Ohtsuru, A., et al. (2017). The impact of evacuation on the incidence of chronic kidney disease after the great East Japan earthquake: The Fukushima health management survey. *Clinical and Experimental Nephrology*, *21*(6), 995–1002. https://doi.org/10.1007/s10157-017-1395-8.

Isojima, T., Yokoya, S., Ono, A., Kato, N., Tanaka, T., Yokomichi, H., et al. (2017). Prolonged elevated body mass index in preschool children after the great East Japan earthquake. *Pediatrics International: Official Journal of the Japan Pediatric Society*, *59*(9), 1002–1009. https://doi.org/10.1111/ped.13340.

Kawasaki, Y., Nakano, H., Hosoya, M., Yasumura, S., Ohira, T., Satoh, H., et al. (2020). Influence of post-disaster evacuation on childhood obesity and hyperlipidemia. *Pediatrics International: Official Journal of the Japan Pediatric Society*, *62*(6), 669–676. https://doi.org/10.1111/ped.14162.

Kunii, Y., Suzuki, Y., Shiga, T., Yabe, H., Yasumura, S., Maeda, M., et al. (2016). Severe psychological distress of evacuees in evacuation zone caused by the Fukushima Daiichi nuclear power plant accident: The Fukushima health management survey. *PLoS One*, *11*(7). https://doi.org/10.1371/journal.pone.0158821, e0158821.

Ministry of Health, Labour and Welfare. (2021). 5th NDB open data. Retrieved April 10, 2021, from https://www.mhlw.go.jp/stf/seisakunitsuite/bunya/0000177221_00008.html.

Ohira, T., Hosoya, M., Yasumura, S., Satoh, H., Suzuki, H., Sakai, A., et al. (2016a). Evacuation and risk of hypertension after the great East Japan earthquake: The Fukushima health management survey. *Hypertension (Dallas, Tex.: 1979)*, *68*(3), 558–564. https://doi.org/10.1161/HYPERTENSIONAHA.116.07499.

Ohira, T., Hosoya, M., Yasumura, S., Satoh, H., Suzuki, H., Sakai, A., et al. (2016b). Effect of evacuation on body weight after the great East Japan earthquake. *American Journal of Preventive Medicine*, *50*(5), 553–560. https://doi.org/10.1016/j.amepre.2015.10.008.

Radiation Medical Science Center for Fukushima Health Management Survey. (2020). *Proceedings of the 37th prefectural oversight committee meeting for Fukushima health management survey*. https://www.pref.fukushima.lg.jp/site/portal/kenkocyosa-kentoiinkai-37.html.

Radiation Medical Science Center for the Fukushima Health Management Survey. (2020). *Report of the Fukushima health management survey, FY2019*. http://kenko-kanri.jp/img/report_r1.pdf.

Sakai, A., Ohira, T., Hosoya, M., Ohtsuru, A., Satoh, H., Kawasaki, Y., et al. (2014). Life as an evacuee after the Fukushima Daiichi nuclear power plant accident is a cause of polycythemia: The Fukushima health management survey. *BMC Public Health*, *14*, 1318. https://doi.org/10.1186/1471-2458-14-1318.

Satoh, H., Ohira, T., Hosoya, M., Sakai, A., Watanabe, T., Ohtsuru, A., et al. (2015). Evacuation after the Fukushima Daiichi nuclear power plant accident is a cause of diabetes: Results from the Fukushima health management survey. *Journal of Diabetes Research*, *2015*. https://doi.org/10.1155/2015/627390, 627390.

Satoh, H., Ohira, T., Nagai, M., Hosoya, M., Sakai, A., Watanabe, T., et al. (2016). Hypo-high-density lipoprotein cholesterolemia caused by evacuation after the Fukushima Daiichi nuclear power plant accident: Results from the Fukushima health management survey. *Internal Medicine*, *55*(15), 1967–1976.

Satoh, H., Ohira, T., Nagai, M., Hosoya, M., Sakai, A., Yasumura, S., et al. (2019). Evacuation is a risk factor for diabetes development among evacuees of the great East Japan earthquake: A 4-year follow-up of the Fukushima health management survey. *Diabetes & Metabolism*, *45*(3), 312–315. https://doi.org/10.1016/j.diabet.2017.09.005.

Suzuki, H., Ohira, T., Takeishi, Y., Hosoya, M., Yasumura, S., Satoh, H., et al. (2015). Increased prevalence of atrial fibrillation after the great East Japan earthquake: Results from the Fukushima health management survey. *International Journal of Cardiology*, *198*, 102–105. https://doi.org/10.1016/j.ijcard.2015.06.151.

Takahashi, A., Ohira, T., Yasumura, S., et al. (2017). Effect of evacuation on liver function after the Fukushima Daiichi nuclear power plant accident. *Journal of Epidemiology*, *27*, 180–185.

Yabe, H., Suzuki, Y., Mashiko, H., Nakayama, Y., Hisata, M., Niwa, S., et al. (2014). Psychological distress after the great East Japan earthquake and Fukushima Daiichi nuclear power plant accident: Results of a mental health and lifestyle survey through the Fukushima health management survey in FY2011 and FY2012. *Fukushima Journal of Medical Science*, *60*(1), 57–67. https://doi.org/10.5387/fms.2014-1.

Yasumura, S., Hosoya, M., Yamashita, S., Kamiya, K., Abe, M., Akashi, M., et al. (2012). Study protocol for the Fukushima health management survey. *Journal of Epidemiology*, *22*(5), 375–383. https://doi.org/10.2188/jea.je20120105.

Yokomichi, H., Zheng, W., Matsubara, H., Ishikuro, M., Kikuya, M., Isojima, T., et al. (2016). Impact of the great East Japan earthquake on the body mass index of preschool children: A nationwide nursery school survey. *BMJ Open*, *6*(4). https://doi.org/10.1136/bmjopen-2015-010978, e010978.

Zhang, W., Ohira, T., Abe, M., Kamiya, K., Yamashita, S., Yasumura, S., et al. (2017). Evacuation after the great East Japan earthquake was associated with poor dietary intake: The Fukushima health management survey. *Journal of Epidemiology*, *27*(1), 14–23. https://doi.org/10.1016/j.je.2016.08.002.

Chapter 7

Health effects of atomic bomb radiation in 1945—Epidemiological investigations

Kotaro Ozasa
Department of Epidemiology, Radiation Effects Research Foundation, Hiroshima, Japan

Introduction

Nuclear weapons were dropped on Hiroshima and Nagasaki in August 1945. The blast and heat killed many people in a moment, and others suffered from physical injuries and acute radiation syndromes. The number of people who died before the end of 1945 was estimated to be approximately 140,000 in Hiroshima and 74,000 in Nagasaki. These numbers have been thought to represent approximately 40% and 30% of the population in each city at the time of the bombings, respectively (Hiroshima City Office, 2020; Nagasaki City Office, 2020). In addition, long-term health effects due to radiation exposure continue to this day.

Beginning just after the bombings, radioactivity in the bombed fields and the medical conditions of the survivors were intensively investigated by scientists in Japan and the United States. In 1947, the Atomic Bomb Casualty Commission (ABCC) was established for investigations of the health effects of atomic bomb radiation exposure. The ABCC was reorganized to form the Radiation Effects Research Foundation (RERF) in 1975, which assumed all research activities of ABCC. Until the early 1950s, physical injuries and acute radiation syndrome were the primary research interests, but later, long-term health effects became the targets of the investigations. In 1955, the Francis Committee, an ad hoc evaluation committee of ABCC research activities, recommended establishing a fixed population (i.e., cohort) for epidemiological follow-up of long-term outcomes. Thus, the cohorts of atomic bomb survivors, in utero-exposed survivors, and children of survivors (F_1) were constructed. Subcohorts of each cohort were also created for clinical investigations at ABCC-RERF.

Health Effects of the Fukushima Nuclear Disaster. https://doi.org/10.1016/B978-0-12-824098-4.00004-0

Consequently, the follow-up of those cohorts has continued to this day. The genetic effects of atomic bomb radiation have also been investigated in different settings of clinical and genetic studies on survivors and their children in addition to the aforementioned epidemiological studies. The results of the investigations have contributed to the health and welfare of atomic bomb survivors (RERF, 2017).

In this chapter, the author will summarize the methods and results of epidemiological studies conducted with the abovementioned cohorts so that readers can understand the details provided in the thorough review elsewhere (Ozasa, Cullings, Ohishi, Hida, & Grant, 2019; Ozasa, Grant, & Kodama, 2018). Some specific aspects of atomic bomb survivor studies will be discussed in the later sections. Hereditary effects of atomic bomb radiation will be referred to in another article (Nakamura, 2019).

Construction of cohorts, follow-up, and outcomes

A supplementary survey to the National Census of Japan identified approximately 284,000 atomic bomb survivors throughout Japan in 1950, and approximately 195,000 of them resided in Hiroshima or Nagasaki at that time. From the late 1940s, ABCC collected basic information about each survivor's situation at the time of the bombing (location, simple shielding information, etc.) in interviews by ABCC personnel and collected information from more than 99% of the resident survivors on the roster. The cohort of the Life Span Study (LSS) consisted of four groups: (1) all survivors who were located within 2.0 km of the hypocenter at the time of the bombings (inner proximal), (2) all survivors at 2.0 to <2.5 km (outer proximal), (3) a subset of survivors at 2.5 to <10 km who were city-, sex-, and age-matched to the inner proximal survivors (distal), and (4) people who were not in either city (i.e., >10 km of the hypocenters) at the time of the bombings and who were city-, sex-, and age-matched to the inner proximal survivors (not-in-city). Subjects in the last group were selected from other rosters provided by Hiroshima and Nagasaki offices and ABCC surveys to represent the population at the 1950 Census in both cities. Later, the LSS included all proximal survivors and some distal survivors in Nagasaki to improve comparison performance. The final cohort consists of a total of 120,321 subjects, including 34,363 in group (1), 19,959 in group (2), 39,419 in group (3), and 26,580 in group (4), according to the information used for DS02R1 individual dose (Cullings et al., 2017; RERF, 2017).

All cohort members have been followed for their vital status and cause of death beginning in 1950 using the family registry system (*koseki*) and vital statistics of Japan. As leukemia explicitly developed among the survivors a few years after the bombings, the Leukemia Registry, a special registry for leukemia, was created by local hematologists in Hiroshima and Nagasaki beginning in the late 1940s (then later merged with local cancer registry systems). Cancer incidence data have been collected through Hiroshima and Nagasaki local cancer registries since the late 1950s. Pathological tumor tissue registries, collecting detailed and precise information on the histological diagnosis of tumors, were initiated by local medical associations in Hiroshima and Nagasaki in the early 1970s. Autopsies of deceased survivors and other residents in Hiroshima and Nagasaki and pathological examinations of surgically resected tissue materials were carried out by ABCC-RERF from the late 1940s to the late 1980s, and the pathological samples are stored at RERF. Other pathological specimens resected by surgical procedures are stored at local hospitals where survivors were medically treated. These materials have been used in collaborative studies with local pathologists through the network of the aforementioned pathological tumor tissue registries (RERF, 2017).

The Adult Health Study (AHS) is a comprehensive health examination program at ABCC-RERF. The participants were asked to provide blood and urine for further biological investigations with informed consent. At the inception of the AHS in 1958, the cohort included four groups: (1 and 2) survivors who were within 2.0 km of either hypocenter and (1) exhibited acute symptoms (including all eligible subjects), (2) did not exhibit any acute symptoms; (3) survivors at 3.0 to <3.5 km from the hypocenter in Hiroshima and 3.0 to <4.0 km in Nagasaki, and (4) persons not in either city at the time of the bombings. Members of groups (2) to (4) were city-, sex-, and age-matched to those of group (1). Therefore, approximately 5000 subjects were selected in each group. Later, the cohort was expanded, and a total of 24,358 subjects were finally invited to the AHS. Health examinations of group (4) were terminated in 1977. Approximately 21,000 subjects visited the health examination at least once, and a high participation rate (75%–90%) was maintained each year (RERF, 2017).

As genetic effects of radiation exposure were a large concern among the public as early as the late 1940s, a series of genetic studies has been conducted (Nakamura, 2019). Based on the subjects of those studies and others, the first filial (F_1) cohort was constructed with 76,814 persons who were born from May 1946 to 1984 and whose father and/or mother were atomic bomb survivors, and the cohort has been followed since 1946. Another cohort consisting of 3638 persons who were exposed in utero to atomic bomb radiation was also constructed. Both cohorts have been followed in the same manner as the LSS. In utero survivors were directly exposed to radiation in their mother's womb after conception while F_1 persons were conceived after parental exposure to radiation. As it is impossible to distinguish the biological conception at the time of bombing, the separation of subjects between in utero and F_1 cohorts was based on the birthdate using the mean gestation period. Some 1021 subjects in the in utero cohort were invited to the AHS program for clinical follow-up beginning in 1978. The health examination program for F_1 subjects (Filial One Offspring Clinical Study: FOCS) was initiated in the early 2000s by methods that are mostly the same as those implemented in the AHS. A total of 11,951 subjects participated in the examination from 2002 to 2006, and they have been followed through repeated examinations every 4 years since 2010 (RERF, 2017).

Evaluation of exposure

Radiation from atomic bomb explosions is classified into initial radiation and residual radiation. Initial radiation was released directly from the explosion fireballs within several tens of seconds, from which gamma rays and neutrons reached the ground. Residual radiation was released from neutron-induced radioactive materials in the environment and fission products contained in the fallout. Such fallout often fell down with rain just after the bombings and contained dust and soot due to the fire and destruction caused by the bombings, so the rain was called "black rain" (Cullings et al., 2017; Ozasa et al., 2018; RERF, 2017).

Individual doses of initial radiation were determined by (1) the amount of radiation emitted from the bombs, (2) the distance from the fireball to persons and transfer conditions, and (3) external and internal shielding conditions of persons. The first two factors were estimated by theories and measurements based on nuclear physics, including demonstration experiments at nuclear test fields in the United States in the 1950s and 1960s and measurements of elements in materials exposed to atomic bomb radiation in Hiroshima and Nagasaki. The distance from the fireball to the person is usually surrogated by the ground-surface distance from the hypocenter to the person. The locations of survivors at the time of the bombings were intensively surveyed by ABCC personnel. Detailed external shielding conditions due to houses and other architectural and natural structures were also intensively investigated for inner proximal survivors until the 1960s. In contrast, the individual exposure dose of distal survivors was estimated using simple information such as in-house, in-open, etc., and average transparency coefficients obtained by the detailed surveys for the inner proximal survivors. Internal shielding by the body was evaluated using phantoms. Dosimetry systems and individual dose estimates have been developed and updated several times (TD57, TD65, DS86, DS02). The current system is DS02, and DS02R1 is the latest update of individual dose estimates, which is based on DS02 and the latest updated individual shielding information (Cullings et al., 2017; Young & Kerr, 2005). "Weighted absorbed organ dose," that is, a neutron dose $\times$ 10 (i.e., relative biological effectiveness) + gamma dose for 15 organs has been used for the evaluation of health effects.

DS02R1 individual doses were estimated for 86,720 members of in-city LSS subjects (groups (1) to (3)) but could not be estimated for the rest of the 7021 subjects because of insufficient information or excessively complicated shielding conditions, especially for inner proximal survivors. For the not-in-city subjects, a dose of zero was assumed. Among the 86,720 subjects, approximately 2.6% were exposed to 1 Gy or higher at the colon (colon dose is used as a proxy for evaluation of cancer risk of internal organs). Approximately 18% were exposed to 100 mGy to 1 Gy at the colon, and 45% were exposed to less than 5 mGy. The individual radiation dose is thought to distribute up to shielded kerma of approximately 4 Gy as a plausible limit of radiation exposure, which is roughly equivalent to a bone marrow dose of 2.7 Gy in the LSS, as $LD_{50/60}$

(the dose at which half of exposed people should die within 60 days) is estimated to be 2.9–3.3 Gy in DS02 (Cullings et al., 2017; Fujita, Kato, & Schull, 1991; RERF, 2017).

Residual radiation was not taken into account in individual dose estimates because full information on the spatial and temporal distribution of ground surface radiation activity and the migration of each survivor after the bombings was not obtained. Although some survivors seemed to be exposed to residual radiation at relatively high levels, exposure to residual radiation would not have a large influence on radiation risk estimates based on the initial radiation exposure. This is supported by the relatively low levels of average estimates of residual radiation by the DS86 and the findings that atomic bomb radiation risk was estimated largely from the subjects exposed to initial radiation of 1 Gy or higher (RERF, 1987, 2017).

Lifestyle risk factors for cancer and other diseases, such as smoking and alcohol consumption, which would be potentially confounded by radiation risk, were collected several times using self-administered questionnaires by mail during the follow-up period (Ozasa et al., 2018; RERF, 2017).

Major results from atomic bomb survivors

The leukemia risk due to atomic bomb radiation increased a few years after the bombings and peaked at approximately 6–8 years, especially for those exposed at young ages. The risk decreased rapidly but persisted slightly until almost 50 years after the exposure (Hsu et al., 2013; Preston et al., 1994; Richardson et al., 2009). Solid cancer risks appeared to increase approximately 10 years after the exposure and continue to be elevated today. The risk was indicated as a relative increase of approximately 40%–50% per Gy compared to the non-exposure condition, that is, excess relative risk (ERR)/Gy = 0.4–0.5, for both mortality and incidence for a person at age 70 who was exposed at age 30. The ERR dose-response shape for all solid cancers was thought to be linear, and the lower confidence interval for the threshold was consistently 0 Gy. However, the shape at the low-dose level varied among observations (discussed later). To demonstrate the uncertainty of risk at the low-dose level, the lowest dose range with significant risk and the upper confidence limits for threshold were estimated. Both indicators ranged up to approximately 0.1–0.2 Gy. This finding implies that radiation risk is significantly high at that level or higher. The risk of cancer associated with radiation varies between organs. However, it is not clear whether this variety is truly cancer-specific or statistically incidental or depends on the secular difference in baseline cancer incidence by site to detect the power of risk. The ERR of radiation for all solid cancers was higher in those exposed at young ages in general, but the tendency varies between cancer sites. For example, a strong association was observed for thyroid cancer, but a less remarkable association was found for others; a negative association was even observed for lung cancer. For female breast and uterine corpus cancer, the highest risk was observed due to exposure around the age of menarche. With attained age, the ERR decreased, and the tendency was similar in most organs (Brenner et al., 2018, 2020; Cahoon et al., 2017; Furukawa et al., 2013; Grant et al., 2017; Ozasa et al., 2012; Preston et al., 2007; Preston, Shimizu, Pierce, Suyama, & Mabuchi, 2003; Sadakane et al., 2019; Sakata et al., 2019; Sugiyama et al., 2020; Utada et al., 2018). The observation of LSS is disadvantageous for nonfatal cancers with childhood or early onset, such as thyroid cancer, because cancer registry systems were initiated in the late 1950s when 10 and several years had passed since radiation exposure.

The risk of noncancer diseases has been evaluated using mortality information in the LSS and clinical information, including nonfatal diseases in the AHS. In the LSS, circulatory diseases showed some increase in ERR at a relatively high dose (1 Gy or higher; Ozasa et al., 2012; Shimizu et al., 2010). Respiratory diseases also indicated some increased ERRs (Pham et al., 2013). Digestive diseases as a whole showed a tendency to increase in ERR in regard to mortality, but liver cirrhosis showed no association with radiation exposure (Ozasa et al., 2012). Diseases that are common in the elderly population with ill-defined diagnoses, such as heart failure, nonspecified stroke, and pneumonia, likely showed increased ERRs of radiation. Such diagnoses on death certificates might be low in preciseness and might include misclassification of the underlying cause of death, such as potentially hidden malignant diseases. This possible bias might partially explain the observed increase in ERRs (Ozasa, Takahashi, & Grant, 2016; Ozasa, Takahashi, Grant, & Kodama, 2017).

In the AHS, positive associations with radiation exposure were observed for the incidence of thyroid diseases and chronic liver diseases, including cirrhosis, uterine myoma, and cataracts, but a negative association was found for glaucoma. For circulatory diseases, some positive associations with radiation exposure were observed for hypertension and myocardial infarction among survivors exposed when they were younger than 40 (Wong et al., 1993; Yamada, Wong, Fujiwara, Akahoshi, & Suzuki, 2004). A positive association was observed for the incidence of hemorrhagic stroke, including subarachnoid hemorrhage (Takahashi et al., 2012). These findings are different from LSS mortality findings wherein associations were observed for hypertensive heart disease but not myocardial infarction, hemorrhagic stroke, or subarachnoid hemorrhage (Shimizu et al., 2010). These discrepancies may arise from differences in factors such as outcome indicator (incidence and mortality), aforementioned misclassification of mortality information, sample size, and examination periods.

Longitudinal analyses of total serum cholesterol in the AHS showed that the cholesterol levels were significantly higher in the subjects exposed to radiation than those who were unexposed (Wong, Yamada, Sasaki, Kodama, & Hosoda, 1999), and a similar tendency was observed for blood pressure among the subjects with young age at the time of the bombing (Sasaki, Wong, Yamada, & Kodama, 2002). As these clinical conditions are the main risk factors for circulatory diseases, these associations may partially explain the aforementioned association with myocardial infarction. In contrast, radiation exposure was negatively associated with BMI and appendicular lean mass (Tatsukawa, Misumi, et al., 2013).

For thyroid diseases, a positive association with radiation exposure was observed for the incidence of thyroid nodules, including malignant tumors and benign nodules, as detected by ultrasonography and subsequent cytological tests. The association was stronger among those exposed in earlier childhood. No significant association was observed for hypothyroidism, Graves' disease, or thyroid autoantibody positivity (Imaizumi et al., 2015, 2017, 2006). For liver diseases, a positive association was consistently observed with a positive test of hepatitis B virus (HBV) surface antigen (i.e., HBV carrier; Fujiwara et al., 2003; Kato, Mayumi, Nishioka, & Hamilton, 1983; Neriishi, Akiba, Amano, Ogino, & Kodama, 1995), but such association with the hepatitis C virus (HCV) antibody test was not observed (Fujiwara et al., 2000). Another study indicated that radiation exposure and HBV and HCV infection were independently associated with an increased incidence of hepatocellular carcinoma (HCC). Particularly, radiation exposure significantly increased the risk of non-B and non-C HCC without confounding by alcohol consumption, obesity, or smoking (Ohishi et al., 2011).

Cataracts, especially posterior subcapsular cataracts, were reported as one of the marked late effects of atomic bomb radiation in the early period (Cogan, Martin, & Kimura, 1949), and several studies indicated consistent evidence on posterior subcapsular cataracts in the AHS (Choshi et al., 1983; Miller, Fujino, & Nefzger, 1967; Minamoto et al., 2004). However, the findings on radiation exposure and cortical/nuclear opacities are insufficient and inconsistent, so the investigations are still underway. Although a report indicated a positive association between radiation and normal tension glaucoma (Kiuchi et al., 2013), the findings need to be interpreted cautiously because of potential selection bias due to the low participation rate in the study. No associations were observed between radiation and the prevalence of age-related macular degeneration (Itakura et al., 2015).

Only a few epidemiological studies have been performed on the psychological effects of atomic bombings. In a self-administered medical questionnaire conducted between 1962 and 1965, an elevated prevalence of anxiety symptoms and somatization symptoms was observed in survivors 17–20 years after the bombings (Yamada & Izumi, 2002). No apparent radiation effects on the prevalence or incidence of cognitive function were observed among survivors aged approximately 60 or older who were exposed at adolescence or later in a cross-sectional study conducted in the 1990s and a follow-up study of the subjects (Yamada, Landes, Mimori, Nagano, & Sasaki, 2016; Yamada et al., 2002).

Major results from in utero exposed persons

Children exposed in utero, especially those exposed at a level of 1 Gy or higher during 8–15 gestational weeks, suffered a high frequency of severe intellectual disability and small head size. Growth impairment, such as shorter height during teenage years, was also observed (Otake & Schull, 1998). In the follow-up of the in utero cohort, the risks of solid cancer mortality and incidence seemed to be comparable to those observed in survivors who were exposed at young ages (Delongchamp, Mabuchi, Yoshimoto, & Preston, 1997; Preston et al., 2008; Yoshimoto, Kato, & Schull, 1988). In clinical examinations, some positive association between radiation exposure and some types of hypertension was observed in adolescence (Nakashima, Akahoshi, Neriishi, & Fujiwara, 2007) while no association was observed for hypertension, hypercholesterolemia, myocardial infarction, or stroke in adulthood (Tatsukawa et al., 2008). No association was shown for cataracts or thyroid diseases, either (Imaizumi et al., 2008; Nakashima, Neriishi, & Minamoto, 2006). However, these uncertain results might be due to the small size of the cohort and observation at young ages (mostly up to 60 years), so continued follow-up is required.

Major results from children of survivors (F_1)

A series of studies on genetic effects of atomic bomb radiation exposure investigated birth defects (still birth, malformation, and neonatal death), the sex ratio of newborns, chromosomal aberrations, blood protein mutations, and DNA mutations by conventional methodologies among children of survivors, but no association between parental radiation exposure and abnormality among offspring was indicated (see more details elsewhere Nakamura, 2019; RERF, 2017; Schull, Otake, & Neel, 1981). The most recent report of the F_1 cohort analyzed mortality through 2009, and 23% of the members had exceeded 60 years. No significant increase in cancer or noncancer mortality associated with maternal or paternal radiation exposure levels was observed (Grant et al., 2015). The initial health examination survey of FOCS provided no evidence of increased prevalence of adult-onset circulatory or metabolic disease due to parental radiation exposure (Fujiwara et al., 2008; Tatsukawa, Cologne, et al., 2013). As radiation doses of parental exposure in the F_1 cohort were lower than the doses in the LSS and the age of F_1 cohort subjects was still young, the detection power of genetic effects on the above outcomes might be insufficient. Therefore, follow-up needs to be continued.

Some special aspects of atomic bomb survivor studies

In the LSS, the radiation risk estimates at high levels of approximately 1 Gy or higher are considered to be the most reliable. The reasons are as follows. Persons who were exposed to radiation at a level of 1 Gy or higher were distributed within relatively narrow areas between concentric circles approximately 0.7 and 1.5 km from the hypocenter both in Hiroshima and Nagasaki, so those persons are thought to be relatively similar in socioeconomic and lifestyle factors. They were exposed to radiation nonselectively, so that there should be no association between radiation exposure and other major risk factors for cancer, such as smoking. Thus, bias or confounding would not be likely introduced in the analyses of radiation risks. In addition, radiation risk estimates at high-dose levels are not strongly influenced by variations in risk estimates at low-dose levels or fluctuations in baseline rates. Actually, radiation risk estimates were almost the same between the analyses not including smoking (Preston et al., 2007) and including smoking (Grant et al., 2017).

On the other hand, many factors are thought to affect the estimated risk associated with low levels of approximately 0.1–0.2 Gy or lower. The first is low statistical power, as the effects observed at low-dose levels are very small. Next, bias and confounding factors could be introduced if the special distribution of socioeconomic and lifestyle factors was associated with distance from the hypocenter, which is strongly related to radiation dose. For example, in Hiroshima, proximal survivors were mostly urban residents while distal survivors were largely rural residents at the time of the bombing in 1945. Therefore, socioeconomic and lifestyle factors were likely different between urban and rural residents, so confounding factors could be introduced in radiation risk estimation. Another issue is the variety in the background rate between areas, so the assumption of the same background rate across the areas could affect the risk estimation. Actually, background rates of all-cause mortality and cancer incidence differed between proximal and distal survivors and not-in-city subjects (Cologne & Preston, 2000; Grant et al., 2017; Preston et al., 2007). Less precise methods for estimation of radiation exposure among distal survivors compared to proximal survivors could induce uncertain risk estimation. Exposure to residual radiation from atomic bombings and exposure to medical radiation in later life could have a relatively large influence at low-dose levels of the initial radiation among distal survivors.

Atomic bombings resulted in an extreme negative influence on the socioeconomic and lifestyle factors among the survivors. When we evaluate "the effects of atomic bombing," such bad statuses have "causal association from atomic bombings to health effects," so they should not be treated as confounding factors. In contrast, when we evaluate "the biological effects of atomic bomb radiation," nonbiological association is thought to be "noncausal association for atomic bomb radiation exposure," so they could be treated as confounding factors. If outcomes are biological conditions such as cancer and physical diseases, the latter treatment could usually be applied. However, when psychological or neurocognitive disorders are the outcomes, both social and biological causation may need to be considered.

The LSS subjects were exposed to initial radiation of up to 4 Gy or more for several tens of seconds, and radiation risk estimates in the LSS were mostly determined at a dose level of 1 Gy or higher. Therefore, exposure to atomic bomb radiation is thought to lead to typical high-dose and high-dose-rate exposure. Radiation effects are thought to be lower for low-dose or low-dose-rate exposure than for high-dose and high-dose-rate exposure, and the magnitude is traditionally thought to be approximately half. The coefficient or reduction is called the "dose and dose rate effectiveness factor (DDREF)" ("ICRP Publication 103. The 2007 Recommendations of the International Commission on Radiological Protection," 2007; *UNSCEAR Report 2017. Annex B. Epidemiological Studies of Cancer Risk Due to Low-Dose-Rate Radiation from Environmental Sources*, 2018). Although it is not clear at which level of dose and dose rate the reduction in effects would certainly occur or the coefficients would be applied, exposure at areas with high background radiation levels and environmental pollution, for example, continuous exposure to radiation of several tens of milligrays per year, would meet the criteria of low dose and low dose rate.

RERF defines its objectives in the Articles of Incorporation as "to conduct research and studies for peaceful purposes on medical effects of radiation and associated diseases in humans, with a view to contributing to maintenance of the health and welfare of the atomic bomb survivors and to enhancement of the health of all humankind" (RERF, 2012). According to these objectives, the following future activities are considered: extension of the epidemiological and clinical studies of health effects in the survivors and their offspring; analysis of biospecimens to clarify the mechanisms of radiation-related diseases; curation of all information and biospecimens for future research; dissemination of lessons learned to survivors, their offspring, and others; informing the world about the survivors' experiences and their contributions; and promoting global understanding of radiation effects and the importance of conducting advanced science for humanity (Samet & Niwa, 2020).

Conclusions

Long-term health effects of atomic bomb radiation in regard to cancer development, especially at high-dose levels, are thought to be conclusive among atomic bomb survivors, but the effects at low-dose levels are still a current focus of investigations. Radiation effects on noncancer outcomes are also a current focus, as possible effects have been observed, especially at high doses. Persons who were exposed in utero have some radiation effects as they were directly exposed to radiation as fetuses, whereas persons who were conceived after parental exposure to radiation (F_1) have shown no increased risks for malignancies or other diseases due to parental exposure to radiation at the present time. These results have been used for medical and welfare support for survivors and their children by the Japanese government and for radiological protection worldwide. We will need to continue the follow-up of the children of survivors for several decades to obtain conclusive findings.

Our mission is the clarification of the long-term health effects of atomic bomb radiation by continuing close follow-up of the survivors and their children throughout their lives. We sincerely give our thanks and pay respect to the persons who had or have been associated with our research activities and all victims of the atomic bombings in Hiroshima and Nagasaki.

Acknowledgments

The Radiation Effects Research Foundation, Hiroshima and Nagasaki, Japan is a public interest foundation funded by the Japanese Ministry of Health, Labor, and Welfare (MHLW) and the US Department of Energy. This publication was supported by RERF Research Protocols 1-75, 2-75, 2-61, 4-75, and 4-10. The views of the authors do not necessarily reflect those of the two governments.

References

Brenner, A. V., Preston, D. L., Sakata, R., Sugiyama, H., Berrington de Gonzalez, A., French, B., et al. (2018). Incidence of breast cancer in the life span study of atomic bomb survivors: 1958–2009. *Radiation Research, 190*(4), 434–444. https://doi.org/10.1667/RR15015.1.

Brenner, A. V., Sugiyama, H., Preston, D. L., Sakata, R., French, B., Sadakane, A., et al. (2020). Radiation risk of central nervous system tumors in the life span study of atomic bomb survivors, 1958–2009. *European Journal of Epidemiology, 35*(6), 591–600. https://doi.org/10.1007/s10654-019-00599-y.

Cahoon, E. K., Preston, D. L., Pierce, D. A., Grant, E., Brenner, A. V., Mabuchi, K., et al. (2017). Lung, laryngeal and other respiratory cancer incidence among Japanese atomic bomb survivors: An updated analysis from 1958 through 2009. *Radiation Research, 187*(5), 538–548. https://doi.org/10.1667/RR14583.1.

Choshi, K., Takaku, I., Mishima, H., Takase, T., Neriishi, S., Finch, S. C., et al. (1983). Ophthalmologic changes related to radiation exposure and age in adult health study sample, Hiroshima and Nagasaki. *Radiation Research, 96*(3), 560–579.

Cogan, D. G., Martin, S. F., & Kimura, S. J. (1949). Atomic bomb cataracts. *Science, 110*(2868), 654. https://doi.org/10.1126/science.110.2868.654.

Cologne, J. B., & Preston, D. L. (2000). Longevity of atomic-bomb survivors. *Lancet, 356*, 303–307. https://doi.org/10.1016/S0140-6736(00)02506-X.

Cullings, H. M., Grant, E. J., Egbert, S. D., Watanabe, T., Oda, T., Nakamura, F., et al. (2017). DS02R1: Improvements to atomic bomb survivors' input data and implementation of dosimetry system 2002 (DS02) and resulting changes in estimated doses. *Health Physics, 11*(1), 56–97. https://doi.org/10.1097/HP.0000000000000598.

Delongchamp, R. R., Mabuchi, K., Yoshimoto, Y., & Preston, D. L. (1997). Cancer mortality among atomic bomb survivors exposed *in utero* or as young children, October 1950–May 1992. *Radiation Research, 147*(3), 385–395.

Fujita, S., Kato, H., & Schull, W. (1991). The LD50 associated with exposure to the atomic bombings of Hiroshima and Nagasaki. *Journal of Radiation Research, 32*(Suppl), 154–161. https://doi.org/10.1269/jrr.32.supplement_154.

Fujiwara, S., Kusumi, S., Cologne, J., Akahoshi, M., Kodama, K., & Yoshizawa, H. (2000). Prevalence of anti-hepatitis C virus antibody and chronic liver disease among atomic bomb survivors. *Radiation Research, 154*(1), 12–19. https://doi.org/10.1667/0033-7587(2000)154[0012:poahcv]2.0.co;2.

Fujiwara, S., Sharp, G. B., Cologne, J. B., Kusumi, S., Akahoshi, M., Kodama, K., et al. (2003). Prevalence of hepatitis B virus infection among atomic bomb survivors. *Radiation Research, 159*(6), 780–786. https://doi.org/10.1667/0033-7587(2003)159[0780:pohbvi]2.0.co;2.

Fujiwara, S., Suyama, A., Cologne, J. B., Akahoshi, M., Yamada, M., Suzuki, G., et al. (2008). Prevalence of adult-onset multifactorial disease among offspring of atomic bomb survivors. *Radiation Research, 170*(4), 451–457. https://doi.org/10.1667/rr1392.1.

Furukawa, K., Preston, D., Funamoto, S., Yonehara, S., Ito, M., Tokuoka, S., et al. (2013). Long-term trend of thyroid cancer risk among Japanese atomic-bomb survivors: 60 years after exposure. *International Journal of Cancer*, *132*(5), 1222–1226. https://doi.org/10.1002/ijc.27749.

Grant, E. J., Brenner, A., Sugiyama, H., Sakata, R., Sadakane, A., Utada, M., et al. (2017). Solid cancer incidence among the life span study of atomic bomb survivors: 1958–2009. *Radiation Research*, *187*(5), 513–537. https://doi.org/10.1667/RR14492.1.

Grant, E. J., Furukawa, K., Sakata, R., Sugiyama, H., Sadakane, A., Takahashi, I., et al. (2015). Risk of death among the children of the atomic bomb survivors, an update after 62 years of follow-up: A cohort study. *The Lancet Oncology*, *16*(13), 1316–1323. https://doi.org/10.1016/S1470-2045(15)00209-0.

Hiroshima City Office. (2020). *Genbaku-hibakusha-taisakujigyo-gaiyo (summary of relief measures for atomic bomb survivors) 2020*. Hiroshima: Hiroshima City Office (in Japanese).

Hsu, W.-L., Preston, D. L., Soda, M., Sugiyama, H., Funamoto, S., Kodama, K., et al. (2013). The incidence of leukemia, lymphoma and multiple myeloma among atomic bomb survivors: 1950–2001. *Radiation Research*, *179*(3), 361–382. https://doi.org/10.1667/RR2892.1.

Imaizumi, M., Ashizawa, K., Neriishi, K., Akahoshi, M., Nakashima, E., Usa, T., et al. (2008). Thyroid diseases in atomic bomb survivors exposed *in utero*. *The Journal of Clinical Endocrinology and Metabolism*, *93*(5), 1641–1648. https://doi.org/10.1210/jc.2008-0042.

Imaizumi, M., Ohishi, W., Nakashima, E., Sera, N., Neriishi, K., Yamada, M., et al. (2015). Association of radiation dose with prevalence of thyroid nodules among atomic bomb survivors expose in childhood (2007–2011). *JAMA Internal Medicine*, *175*(2), 228–236. https://doi.org/10.1001/jamainternmed.2014.6692.

Imaizumi, M., Ohishi, W., Nakashima, E., Sera, N., Neriishi, K., Yamada, M., et al. (2017). Thyroid dysfunction and autoimmune thyroid diseases among atomic bomb survivors exposed in childhood. *The Journal of Clinical Endocrinology and Metabolism*, *102*(7), 2516–2524. https://doi.org/10.1210/jc.2017-00102.

Imaizumi, M., Usa, T., Tominaga, T., Neriishi, K., Akahoshi, M., Nakashima, E., et al. (2006). Radiation dose-response relationships for thyroid nodules and autoimmune thyroid diseases in Hiroshima and Nagasaki atomic bomb survivors 55–58 years after radiation exposure. *Journal of the American Medical Association*, *295*(9), 1011–1022. https://doi.org/10.1001/jama.295.9.1011.

Itakura, K., Takahashi, I., Nakashima, E., Yanagi, M., Kawasaki, R., Neriishi, K., et al. (2015). Exposure to atomic bomb radiation and age-related macular degeneration in later life: The Hiroshima-Nagasaki atomic bomb survivor study. *Investigative Ophthalmology & Visual Science*, *56*(9), 5401–5406. https://doi.org/10.1167/iovs.15-16680.

Kato, H., Mayumi, M., Nishioka, K., & Hamilton, H. B. (1983). The relationship of hepatitis B surface antigen and antibody to atomic bomb radiation in the adult health study sample, 1975–1977. *American Journal of Epidemiology*, *117*(5), 610–620. https://doi.org/10.1093/oxfordjournals.aje.a113584.

Kiuchi, Y., Yokoyama, T., Takamatsu, M., Tsuiki, E., Uematsu, M., Kinoshita, H., et al. (2013). Glaucoma in atomic bomb survivors. *Radiation Research*, *180*(4), 422–430. https://doi.org/10.1667/RR3273.2.

Miller, R. J., Fujino, T., & Nefzger, M. D. (1967). Lens findings in atomic bomb survivors. A review of major ophthalmic surveys at the atomic bomb casualty commission (1949–1962). *Archives of Ophthalmology*, *78*(6), 697–704. https://doi.org/10.1001/archopht.1967.00980030699002.

Minamoto, A., Taniguchi, H., Yoshitani, N., Mukai, S., Yokoyama, T., Kumagami, T., et al. (2004). Cataract in atomic bomb survivors. *International Journal of Radiation Biology*, *80*(5), 339–345. https://doi.org/10.1080/09553000410001680332.

Nagasaki City Office. (2020). *Genbaku-hibakusha-taisakujigyo-gaiyo (summary of relief measures for atomic bomb survivors) 2020*. Nagasaki: Nagasaki City Office (in Japanese).

Nakamura, N. (2019). History of radiation genetics: Light and darkness. *International Journal of Radiation Biology*, *95*(7), 999–1014. https://doi.org/10.1080/09553002.2019.1572251.

Nakashima, E., Akahoshi, M., Neriishi, K., & Fujiwara, S. (2007). Systolic blood pressure and systolic hypertension in adolescence of atomic bomb survivors exposed *in utero*. *Radiation Research*, *168*(5), 593–599. https://doi.org/10.1667/RR1060.1.

Nakashima, E., Neriishi, K., & Minamoto, A. (2006). A reanalysis of atomic-bomb cataract data, 2000–2002: A threshold analysis. *Health Physics*, *90*(2), 154–160. https://doi.org/10.1097/01.hp.0000175442.03596.63.

Neriishi, K., Akiba, S., Amano, T., Ogino, T., & Kodama, K. (1995). Prevalence of hepatitis B surface antigen, hepatitis B e antigen and antibody, and antigen subtypes in atomic bomb survivors. *Radiation Research*, *144*(2), 215–221.

Ohishi, W., Fujiwara, S., Cologne, J. B., Suzuki, G., Akahoshi, M., Nishi, N., et al. (2011). Impact of radiation and hepatitis virus infection on risk of hepatocellular carcinoma. *Hepatology (Baltimore, MD)*, *53*(4), 1237–1245. https://doi.org/10.1002/hep.24207.

Otake, M., & Schull, W. J. (1998). Radiation-related brain damage and growth retardation among the prenatally exposed atomic bomb survivors. *International Journal of Radiation Biology*, *74*(2), 159–171. https://doi.org/10.1080/095530098141555.

Ozasa, K., Cullings, H. M., Ohishi, W., Hida, A., & Grant, E. J. (2019). Epidemiological studies of atomic bomb radiation at the radiation effects research foundation. *International Journal of Radiation Biology*, *95*(7), 879–891. https://doi.org/10.1080/09553002.2019.1569778.

Ozasa, K., Grant, E. J., & Kodama, K. (2018). Japanese legacy cohorts: The life span study atomic bomb survivor cohort and survivors' offspring. *Journal of Epidemiology*, *28*(4), 162–169. https://doi.org/10.2188/jea.JE20170321.

Ozasa, K., Shimizu, Y., Suyama, A., Kasagi, F., Soda, M., Grant, E. J., et al. (2012). Studies of the mortality of atomic bomb survivors, report 14, 1950–2003: An overview of cancer and noncancer diseases. *Radiation Research*, *177*(3), 229–243. https://doi.org/10.1667/rr2629.1. Published erratum appears in Radiation Research, (2013), 179, e40-41.

Ozasa, K., Takahashi, I., & Grant, E. J. (2016). Radiation-related risks of non-cancer outcomes in the atomic bomb survivors. *Annals of the International Commission on Radiological Protection*, *45*(Suppl. 1), 253–261. https://doi.org/10.1177/0146645316629318.

Ozasa, K., Takahashi, I., Grant, E. J., & Kodama, K. (2017). Cardiovascular disease among atomic bomb survivors. *International Journal of Radiation Biology*, *93*(10), 1145–1150. https://doi.org/10.1080/09553002.2017.1290849.

Pham, T. M., Sakata, R., Grant, E. J., Shimizu, Y., Furukawa, K., Takahashi, I., et al. (2013). Radiation exposure and the risk of mortality from noncancer respiratory diseases in the life span study, 1950–2005. *Radiation Research*, *180*(5), 539–545. https://doi.org/10.1667/RR13421.1.

Preston, D. L., Cullings, H., Suyama, A., Funamoto, S., Nishi, N., Soda, M., et al. (2008). Solid cancer incidence in atomic bomb survivors exposed *in utero* or as young children. *Journal of the National Cancer Institute*, *100*(6), 428–436. https://doi.org/10.1093/jnci/djn045.

Preston, D. L., Kusumi, S., Tomonaga, M., Izumi, S., Ron, E., Kuramoto, A., et al. (1994). Cancer incidence in atomic bomb survivors. Part III. Leukemia, lymphoma and multiple myeloma, 1950–1987. *Radiation Research*, *137*(Suppl. 2), S68–S97.

Preston, D. L., Ron, E., Tokuoka, S., Funamoto, S., Nishi, N., Soda, M., et al. (2007). Solid cancer incidence in atomic bomb survivors: 1958–1998. *Radiation Research*, *168*(1), 1–64. https://doi.org/10.1667/RR0763.1.

Preston, D. L., Shimizu, Y., Pierce, D. A., Suyama, A., & Mabuchi, K. (2003). Studies of mortality of atomic bomb survivors. Report 13: Solid cancer and noncancer disease mortality: 1950–1997. *Radiation Research, 160*(4), 381–407. https://doi.org/10.1667/rr3049.

Radiation Effects Research Foundation (RERF). (1987). *US-Japan joint reassessment of atomic bomb radiation dosimetry in Hiroshima and Nagasaki (DS86).* Radiation Effects Research Foundation. https://www.rerf.or.jp/en/library/list-e/scids/ds86-en/.

Radiation Effects Research Foundation (RERF). (2012). *Articles of incorporation.* RERF. https://www.rerf.or.jp/en/about/report_e/.

Radiation Effects Research Foundation (RERF). (2017). *A brief description.* RERF. https://www.rerf.or.jp/uploads/2017/09/briefdescript_e.pdf.

Richardson, D. B., Sugiyama, H., Nishi, N., Sakata, R., Shimizu, Y., Grant, E. J., et al. (2009). Ionizing radiation and leukemia mortality among Japanese atomic bomb survivors, 1950–2000. *Radiation Research, 172*(3), 368–382. https://doi.org/10.1667/RR1801.1.

Sadakane, A., French, B., Brenner, A. V., Preston, D. L., Sugiyama, H., Grant, E. J., et al. (2019). Radiation and risk of liver, biliary tract, and pancreatic cancers among atomic bomb survivors in Hiroshima and Nagasaki: 1958–2009. *Radiation Research, 192*(3), 299–310. https://doi.org/10.1667/RR15341.1.

Sakata, R., Preston, D. L., Brenner, A. V., Sugiyama, H., Grant, E. J., Rajaraman, P., et al. (2019). Radiation-related risk of cancers of the upper digestive tract among Japanese atomic bomb survivors. *Radiation Research, 192*(3), 331–344. https://doi.org/10.1667/RR15386.1.

Samet, J. M., & Niwa, O. (2020). At the 75th anniversary of the bombings of Hiroshima and Nagasaki, the radiation effects research foundation continues studies of the atomic bomb survivors and their children. *Carcinogenesis, 41*(11), 1471–1472. https://doi.org/10.1093/carcin/bgaa104.

Sasaki, H., Wong, F. L., Yamada, M., & Kodama, K. (2002). The effects of aging and radiation exposure on blood pressure levels of atomic bomb survivors. *Journal of Clinical Epidemiology, 55*(10), 974–981. https://doi.org/10.1016/s0895-4356(02)00439-0.

Schull, W. J., Otake, M., & Neel, J. V. (1981). Genetic effects of the atomic bombs: A reappraisal. *Science, 213*(4513), 1220–1227. https://doi.org/10.1126/science.7268429.

Shimizu, Y., Kodama, K., Nishi, N., Kasagi, F., Suyama, A., Soda, M., et al. (2010). Radiation exposure and circulatory disease risk: Hiroshima and Nagasaki atomic bomb survivor data, 1950–2003. *BMJ (Clinical Research Ed.), 340.* https://doi.org/10.1136/bmj.b5349, b5349.

Sugiyama, H., Misumi, M., Brenner, A. V., Grant, E. J., Sakata, R., Sadakane, A., et al. (2020). Radiation risk of incident colorectal cancer by anatomical site among atomic bomb survivors: 1958–2009. *International Journal of Cancer, 146*(3), 635–645. https://doi.org/10.1002/ijc.32275.

Takahashi, I., Abott, R. D., Ohshita, T., Takahashi, T., Ozasa, K., Akahoshi, M., et al. (2012). A prospective follow-up study of the association of radiation exposure with fatal and non-fatal stroke among atomic bomb survivors in Hiroshima and Nagasaki (1980–2003). *BMJ Open, 2*(1). https://doi.org/10.1136/bmjopen-2011-000654, e000654.

Tatsukawa, Y., Cologne, J. B., Hsu, W.-L., Yamada, M., Ohishi, W., Hida, A., et al. (2013). Radiation risk of individual multifactorial diseases in offspring of the atomic-bomb survivors: A clinical health study. *Journal of Radiological Protection, 33*(2), 281–293. https://doi.org/10.1088/0952-4746/33/2/281.

Tatsukawa, Y., Misumi, M., Yamada, M., Masunari, N., Oyama, H., Nakanishi, S., et al. (2013). Alteration of body mass index and body composition in atomic bomb survivors. *International Journal of Obesity, 37*(8), 1123–1128. https://doi.org/10.1038/ijo.2012.193.

Tatsukawa, Y., Nakashima, E., Yamada, M., Funamoto, S., Hida, A., Akahoshi, M., et al. (2008). Cardiovascular disease risk among atomic bomb survivors exposed *in utero*, 1978–2003. *Radiation Research, 170*(3), 269–274. https://doi.org/10.1667/RR1434.1.

The International Commission on Radiological Protection (ICRP). (2007). ICRP publication 103. The 2007 recommendations of the international commission on radiological protection. *Annals of the ICRP*, *37*(2–4), 1–332. https://doi.org/10.1016/j.icrp.2007.10.003.

The United Nations Scientific Committee on the Effects of Atomic Radiation (UNSCEAR). (2018). *UNSCEAR report 2017. Annex B. Epidemiological studies of cancer risk due to low-dose-rate radiation from environmental sources*. UNSCEAR. https://www.unscear.org/unscear/en/publications/2017.html.

Utada, M., Brenner, A. V., Preston, D. L., Cologne, J. B., Sakata, R., Sugiyama, H., et al. (2018). Radiation risks of uterine cancer in atomic bomb survivors: 1958–2009. *JNCI Cancer Spectrum*, *2*(4), pky081. https://doi.org/10.1093/jncics/pky081.

Wong, F. L., Yamada, M., Sasaki, H., Kodama, K., Akiba, S., Shimaoka, K., et al. (1993). Noncancer disease incidence in the atomic bomb survivors: 1958–1986. *Radiation Research*, *135*(3), 418–430.

Wong, F. L., Yamada, M., Sasaki, H., Kodama, K., & Hosoda, Y. (1999). Effects of radiation on the longitudinal trends of total serum cholesterol levels in the atomic bomb survivors. *Radiation Research*, *151*(6), 736–746.

Yamada, M., & Izumi, S. (2002). Psychiatric sequelae in atomic bomb survivors in Hiroshima and Nagasaki two decades after the explosions. *Social Psychiatry and Psychiatric Epidemiology*, *37*(9), 409–415. https://doi.org/10.1007/s00127-002-0572-5.

Yamada, M., Landes, R. D., Mimori, Y., Nagano, Y., & Sasaki, H. (2016). Radiation effects on cognitive function among atomic bomb survivors exposed at or after adolescence. *The American Journal of Medicine*, *129*(6), 586–591. https://doi.org/10.1016/j.amjmed.2015.09.002.

Yamada, M., Sasaki, H., Kasagi, F., Akahoshi, M., Mimori, Y., Kodama, K., et al. (2002). Study of cognitive function among the adult health study (AHS) population in Hiroshima and Nagasaki. *Radiation Research*, *158*(2), 236–240. https://doi.org/10.1667/0033-7587(2002)158[0236:socfat]2.0.co;2.

Yamada, M., Wong, F. L., Fujiwara, S., Akahoshi, M., & Suzuki, G. (2004). Noncancer disease incidence in atomic bomb survivors, 1958–1998. *Radiation Research*, *161*(6), 622–662. https://doi.org/10.1667/rr3183.

Yoshimoto, Y., Kato, H., & Schull, W. J. (1988). Risk of cancer among children exposed *in utero* to A-bomb radiations, 1950–84. *Lancet*, *2*(8612), 665–669. https://doi.org/10.1016/s0140-6736(88)90477-1.

Young, W. Y., & Kerr, G. D. (2005). *Reassessment of the atomic bomb radiation dosimetry for Hiroshima and Nagasaki-dosimetry system 2002 (DS02)*. Radiation Effects Research Foundation. https://www.rerf.or.jp/en/library/list-e/scids/ds02-en/.

Section III

Mental health and psychosocial issues

Chapter 8

Posttraumatic stress responses and related issues: A chronological view

Masaharu Maeda, Yui Takebayashi, and Hideki Sato
Department of Disaster Psychiatry, Fukushima Medical University School of Medicine, Fukushima, Japan

Introduction

The Great East Japan Earthquake and tsunami that occurred on March 11, 2011, caused devastating damage in the Tohoku area and led to subsequent explosions at the Fukushima Daiichi Nuclear Power Plant (FDNPP), which resulted in long-term effects on the residents of Fukushima Prefecture. In particular, the coastal area where the FDNPP is located (*Hama-Dori*) was seriously affected, and as a result, more than 160,000 people were estimated to have been evacuated from this area within 1 year after the accident (Fukushima Prefecture, 2019). In contrast to Miyagi and Iwate Prefectures, which were mainly affected by the tsunami, the evacuation zone in Fukushima Prefecture was extended by governmental order day-by-day, and numerous people living in the evacuation zone were evacuated to safer remote areas, from Okinawa to Hokkaido. However, these evacuations were accompanied by uncertain and often conflicting information. During this initial period, many evacuees, who were possibly suffering posttraumatic stress disorder (PTSD) from around the time of the accident, were moving from place to place in an attempt to find safe haven, resulting in a wide dispersal.

Although it will soon be 10 years since the disaster, nearly 40,000 people are still evacuated (as of August 2020) (Fukushima Prefecture, 2019), a huge proportion of whom relocated to remote places outside Fukushima Prefecture. In this chapter, we review the long-term evacuation process, one that had never been seen after a natural disaster in Japan, and demonstrate how the Fukushima accident affected the mental health of residents living in Fukushima prefecture in a chronological manner. The evacuation process, however, is extremely complicated. For example, to our knowledge, as many as 26 official evacuation

Health Effects of the Fukushima Nuclear Disaster. https://doi.org/10.1016/B978-0-12-824098-4.00006-4

maps have been issued to date. In addition, the evacuation areas have been divided into many parts, given many different names, and changed many times. To gain a better understanding of and provide readers with a comprehensive view of this complicated long-term evacuation process, the time since the disaster is divided into four phases mainly according to major changes in evacuation orders and the actual movement of evacuees after the accident (Table 1).

TABLE 1 **Evacuation process and mental health issues after the 2011 Fukushima disaster.**

	Evacuation	Mental health issues[a]
Phase 1: Evacuation and dispersal phase (March 11, 2011, to August 2011; 8 months)	Rapid expansion of the evacuation zone and confusion Surge of evacuees and dispersal	Terror of radiation exposure, traumatization, confusion, self-condemnation, distrust Poor epidemiological data
Phase 2: Confusion phase (September 2011 to August 2015; 48 months)	Expansion of the evacuation zone ceased Decontamination and improvement of infrastructure began Some evacuees began to return	Major mental health care facilities established Improvements in mental health status and radiation risk perception Increase in number of disaster-related suicides
Phase 3: Great return phase (September 2015 to April 2017; 20 months)	Relocated local offices began to return Rapid increase of returnees Reversal of numbers of evacuees inside and outside Fukushima; the number of evacuees staying outside Fukushima exceeded the numbers inside	Little change in general mental health status and radiation risk perception Exhaustion and burnout among public employees, and increase in early retirement and suicides
Phase 4: Stagnation phase (May 2017 to present)	Return of evacuees came to standstill Many returned municipalities faced low return and high aging rates	Worsening of mental health status among evacuees staying outside of Fukushima

[a] *This description is based on the Mental Health and Lifestyle Survey conducted by Fukushima Medical University (Fukushima Prefecture, 2020).*

The mental health data referred to in this chapter are primarily based on the Mental Health and Lifestyle Survey (MHLS), which has been conducted by Fukushima Medical University as part of the Fukushima Health Management Survey. This examines possible health issues after a nuclear accident, such as thyroid cancer, leukemia, and pregnancy-related issues (Yasumura et al., 2012). The MHLS, which is conducted as a mail survey, has several purposes. The first purpose is to clarify current mental health problems and lifestyle-related issues among people who were living in the area under the evacuation order at the time of the disaster using several questionnaires, as described below. The second purpose is to provide brief interventions, including psychoeducation and advice by telephone or mail, mainly to people at risk of PTSD, depression, and other life-related issues based on the results of the MHLS. The third purpose is to share information as needed with available resources both inside and outside Fukushima Prefecture, including psychiatric clinics and local health centers.

The target population of the MHLS includes people who were registered as original residents of municipalities designated as evacuation areas after the disaster, namely Hirono, Naraha, Tomioka, Okuma, Futaba, Namie, and Kawamata town, Kawauchi, Katsurao, and Iitate village, and Minamisōma, Tamura, and part of Date city. The total number of target participants reached 208,044 on October 31, 2017. Participants are divided into the following five age groups: 0–3, 4–6, 7–12 (primary school age group), 13–15 (middle school age group), and 16 and older (adult group). The average response rate obtained every year has been approximately 20%.

Reviewing the data collected by the MHLS as well as other epidemiological studies, we clarify the relief process and changes in mental health status among those affected after the disaster.

Chronological changes in mental health status

Phase 1: Evacuation and dispersal phase (March 11, 2011 to August 2011; 8 months)

Surge and dispersal of evacuees

Very soon after the blackout at the FDNPP due to the massive tsunami and first explosion on March 11, 2011, a "nuclear emergency situation" declaration was issued amid terrible confusion by the Japanese government based on a special measure of the Nuclear Disaster Act. Hardly any of the people living near the FDNPP had ever expected the combination of a tsunami and a severe nuclear accident (Ota, Miyazaki, & Nakayama, 2017). The areas within a 3- and 10-km radius of the FDNPP were quickly designated as an "evacuation order zone" and an "indoor evacuation order zone", respectively. However, explosions in the buildings housing the nuclear reactors occurred in rapid succession, so these evacuation zones were extended to 20 and 30 km, respectively, within several days. Therefore, people were trying to escape from a critical situation amid substantial panic.

Exposure to this catastrophe likely produced long-term traumatic memories in the minds of many evacuees. Many residents should start preparing for death, as they had received unclear and gruesome information regarding the occurrence of the worst scenario: a meltdown. Running out of gasoline and, sometimes, ignorance or denial of the crisis often interfered with smooth evacuation plans and spurred confusion (such initial turmoil is described in detail in Chapter 1). In particular, hospital staff working in the evacuation zone were forced into a more serious situation. Many hospitals in the area had numerous inpatients with different preexisting physical and/or mental difficulties; these patients were generally aged, had a long duration of hospital stay, and could not perform activities independently (Hasegawa et al., 2015) (see Chapter 2). For example, Futaba Hospital, which is located only 4.5 km from the FDNPP, was in a desperate situation because of delayed rescue, with many hospital staff and patients forced to stay behind at the hospital (Mori, 2012). Although they finally managed to escape by themselves, regrettably, 50 of 436 patients died while evacuating because of their poor physical condition and inability to be transported (Mori, 2012). Similar to Futaba Hospital, many hospitals and medical facilities in the evacuation area with a great number of patients fell into similar crises during the first 7 days after the accident (Maeda, 2012). Those who barely escaped from these situations still believed that the evacuation order would be lifted in a few weeks, which would allow them to return to their homes (Maeda, 2012). Unfortunately, nuclear disaster preparedness had been hardly considered by the residents until the crisis had actually occurred.

Dispersal and traumatic experiences

To our knowledge, no epidemiological studies have been conducted to clarify mental health status among evacuees during this period. On the other hand, mental health surveys conducted later, such as the MHLS, showed intense symptoms of PTSD among residents who had lived in the evacuation zone as a result of overwhelming, unexpected experiences described as being forced to live at the gate of death (Maeda, Oe, & Suzuki, 2018). In addition to such posttraumatic stress responses related to fear, intense feelings of guilt about their own behavior after the accident also seemed to have tormented people who had taken on some social roles to protect others from threats associated with the disaster, including caregivers, medical staff, and public employees. For example, mothers often blamed themselves for their children's exposure to radiation risks because they had temporarily stayed in dangerous places during the evacuation process, although they had not realized it at that time (Maeda & Oe, 2015; Maeda, Ueda, Nagai, Fujii, & Oe, 2016). In addition, the people in charge of taking care of affected people, such as medical staff and public employees, also faced serious role conflicts between helping others as a duty and escaping as a victim (Maeda, Ueda, et al., 2016).

The number of evacuees was eventually estimated to reach at least 160,000 dispersed across Japan, from Okinawa to Hokkaido (Fukushima Prefecture, 2019). This was in great contrast to evacuees from natural disasters, who generally tend to evacuate to as near a place as possible from their residence. While the Fukushima evacuees tended to relocate far from the disaster area based on a fear of radiation exposure, but they did not always find a familiar or appropriate place. Because of poor preparation and information, they were forced to move frequently from area to area with—and sometimes without—their family members. In fact, the results of the MHLS indicated that the evacuees changed their living place as many as four times on average during the first year after the disaster (Yabe et al., 2014). Indeed, many evacuees, swayed by unclear and often conflicting rumors and information, continued wandering to seek a safer, more familiar place to live.

Phase 2: Confusion phase (September 2011 to August 2015; 48 months)

Prolonged evacuation

The rapid expansion of the evacuation zone finally ceased at around 6 months after the disaster, and the declaration of the emergency evacuation preparedness zone (Hirono town and part of Minamisōma city) was lifted in September 2011. This was the first step toward having the evacuees return. In addition, areas with lower radiation doses, including Odaka town in Minamisōma city, part of Tamura city, and part of Kawauchi village, were designated as "zones in preparation for the lifting of the evacuation order," indicating that these areas would become habitable again in the near future and that people could temporarily visit but not yet return to live. Furthermore, along with progress in reconstruction, the main road and highway running north-south near the FDNPP were reopened during this phase.

The large-scale decontamination effort initiated during this phase, considered the first such challenge in human history, was considered a significant breakthrough in providing relief to Fukushima, even though it was projected to cost upwards of hundreds of billions of dollars. The decontamination effort substantially ended in 2018, and as a result, contributed greatly to the return of evacuees. Meanwhile, several other municipalities, including Iitate village and Okuma town, which had been designated as "zones in preparation for the lifting of the evacuation order," required several more years until the evacuation orders could finally be lifted. For example, the evacuation order for Odaka town, described above, was lifted in the summer of 2016; that is, the "preparation period for returning" for the original residents was as long as 5 years. This process was too long for many evacuees who had expected an early return, probably resulting in great disappointment and hesitation in returning as described below.

In April 2014, the evacuation order for part of Tamura city was lifted. This was the first case among the evacuation zones after the accident for which the restriction was completely lifted. Some of the evacuees began returning to their hometown because of either the end of the restrictions or the exhaustion associated with living uncomfortable lives as evacuees over the long term.

Establishment of new mental health care facilities

Notably, in this phase, several organizations and facilities employing mental health professionals (e.g., social workers, clinical psychologists, community nurses) were newly established in Fukushima Prefecture. For example, the Radiation Medical Science Center was established at Fukushima Medical University in September 2011 to launch major community-based surveys (collectively called "the Fukushima Health Management Survey") that could identify different radiation-related health problems (Yasumura et al., 2012). The MHLS has been conducted as a part of the Fukushima Health Management Survey, and based on the results, telephone interventions have been performed since February 2012 for respondents at risk of mental and/or physical health problems (Horikoshi et al., 2020).

Another large-scale mental health care facility, the Fukushima Center for Disaster Mental Health, was established in November 2011 to deal with mental health issues among affected residents living in Fukushima Prefecture. This facility began providing various types of psychological interventions, such as outreach services (visit services), consultation services, group activities including psychoeducation, and workshops for relief workers or other care providers. This multidisciplinary center employs about 40 mental health professionals and has five branches in Fukushima Prefecture.

These facilities are funded by the national government and were organized by numerous professionals from different areas across Japan, as a large number of health professionals who had once worked in Fukushima Prefecture, including doctors and nurses, were evacuated after the disaster. Considering the serious shortage of mental health professionals in Fukushima Prefecture at the time of the accident, the establishment of these new facilities, which aimed to provide long-term mental health care through many newly arriving professionals, should have been a breakthrough; however, an enormous amount of time was needed for these new workers to adjust to the unclear and unfamiliar disaster situation (Maeda, Ueda, & Hiruta, 2014). In addition, they often faced public stigmas about working in a "contaminated area," especially in the initial phase after the accident.

Improvement in mental health status

The MHLS data generally showed marked improvements in mental health status among the evacuees. The results of the 6-item Kessler Psychological Distress Scale (K6) (Kessler et al., 2003), which can identify general psychological distress, especially the risk of depression, indicated that the prevalence rate of probable depression among the respondents decreased from 14.6% in 2012 to 7.7% in 2015 (Fig. 1) (Fukushima Prefecture, 2020). Similarly, in the first survey year, based on the results of the Strength and Difficulty Questionnaire (SDQ) (Matsuishi et al., 2008), which can identify risks of having some behavioral and/or emotional problems in children based on the estimations of their caregivers, the prevalence rate for children at risk in all age groups (4–6-year-old age group, elementary and junior high school group) was extremely high compared with that in the control population (Fig. 2) (Fukushima Prefecture, 2020); however, this rate rapidly decreased during this phase. For example, in the case of the 4–6-year-old age group, the prevalence rate greatly decreased, from about 24% in 2012 to 13% in 2015 (Fukushima Prefecture, 2020). With regard to individual perceptions of radiation health risks related to the Fukushima disaster (risk perception for late-onset problems, such as leukemia or thyroid cancer, and genetic effects), the proportion of respondents expressing concern about such health risks also gradually declined during this phase, whereas this had been very large in the first survey year (Fukushima Prefecture, 2020).

The MHLS also identified respondents at risk of PTSD using a version of the PTSD Checklist for Specific Stressors (PCL-S) (Suzuki et al., 2017). The findings in the first survey year showed that 21.6% of the adults surveyed scored above the cutoff value ($\geq$44) on the PCL-S at 10 months after the disaster (Yabe et al., 2014), which is almost equal to that for workers after the 9/11 World Trade Center attacks in the United States (Stellman et al., 2008). Another study (Tsujiuchi et al., 2016) conducted in Saitama Prefecture on evacuees from

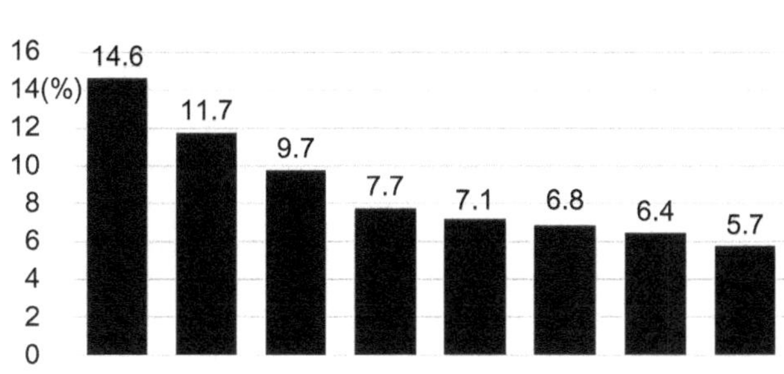

FIG. 1 Prevalence of adults at risk of probable depression. A high risk of psychological distress was considered a score $\geq$13 on the 6-item Kessler Psychological Distress Scale. The prevalence of being at high risk for psychological distress in the community sample was 3.0% (Kawakami, 2007). *(No permission required.)*

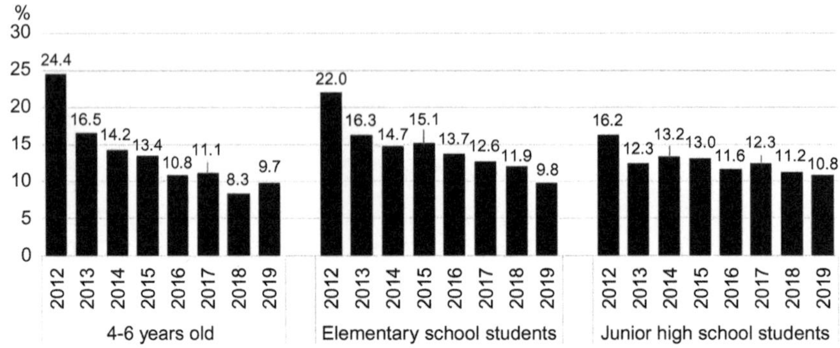

FIG. 2 Prevalence of children at risk of some behavioral and/or emotional problems. A high risk of psychopathological problems in children was considered a score ≥16 on the Strength and Difficulties Questionnaire. The prevalence of being at high risk for psychopathological problems in the community sample was 9.5% (Matsuishi et al., 2008). *(No permission required.)*

Fukushima Prefecture 1 year after the disaster showed that 59.4% of residents scored ≥30 on the Impact of Events Scale-Revised (Asukai et al., 2002), which is a 22-item self-rating scale for screening PTSD symptoms. The findings revealed that PTSD symptoms were also associated with chronic physical and mental diseases; anxiety about livelihoods, lost jobs, and social ties; and concerns about monetary compensation (Tsujiuchi et al., 2016).

These sources of evidence obtained about 1 year after the disaster clearly demonstrated that intense traumatic responses to the disaster were afflicting affected people. However, the results of the MHLS also indicated that the age-adjusted prevalence of probable PTSD based on the results of the PCL-S gradually decreased during this phase, from 19.0% at 10 months postdisaster to 17.8% at 35 months postdisaster for men, and from 25.3% at 10 months postdisaster to 23.3% at 35 months postdisaster for women (Oe et al., 2016).

This general improvement in mental health status among evacuees might reflect a recovery from an initial bout of great confusion or some previous psychological resilience. Given the marked decline in the high radiation risk perception among evacuees, they may have been able to calm down after receiving more correct information about the effects of radiation. Alternatively, the remote evacuation itself could have reduced worries about radiation risks, so mitigating psychological distress. In fact, an analysis of the results of the MHLS showed a strong association between such risk perception and mental health problems, including depression, although the causality remained unclear (Oe et al., 2016; Suzuki et al., 2015).

Despite the apparent improvement of mental health outcomes shown in the MHLS in this phase, it should be noted that these data only appeared to improve compared with the results in the first survey year; no improvements have been seen compared with those in control groups. For example, according to the K6, the prevalence of people at risk of depression among the evacuees in 2015 was still considerably higher than that in the general population in Japan (7.7% vs. 3.0%, respectively) (Fukushima Prefecture, 2020; Kawakami, 2007).

Increase in disaster-related suicide

While several outcomes shown in the MHLS indicated certain improvements in general mental health status among evacuees, the number of "disaster-related suicides" conversely increased during this phase (National Police Agency, 2019; Takebayashi, Hoshino, Kunii, Niwa, & Maeda, 2020). Suicide cases among affected people might be certified as "disaster-related suicide" by national and local governments when fulfilling some criteria involving the contents of a final note, the testimony of a bereaved family member, or the scene of the suicide. In the first year after the disaster, the numbers of disaster-related suicides in three affected prefectures were 22, 17, and 10 in Miyagi, Iwate, and Fukushima Prefectures, respectively (Fig. 3) (National Police Agency, 2019). The fact that this order was the same as that for the death tolls from the earthquake and tsunami (9540, 4673, and 1613, respectively) might reflect acute grief or loss reactions in the initial period.

Thereafter, the numbers of disaster-related suicides in Miyagi and Iwate Prefectures generally decreased yearly, never exceeding those in the first year. The number of disaster-related suicides in Fukushima Prefecture, however, increased to 23 in 2012, exceeding that in Miyagi Prefecture in the first year (Fig. 3) (National Police Agency, 2019). The total number of disaster-related suicides in Fukushima Prefecture during this phase (2012–15) eventually reached as many as 70, which was substantially higher compared with Miyagi and Iwate Prefectures (18 and 15, respectively) (National Police Agency, 2019).

The improvements in general mental health status and the increase in disaster-related suicides among evacuees during this phase seemed to be

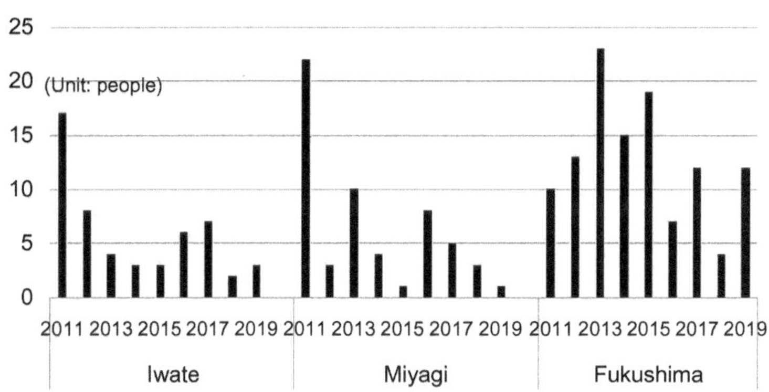

FIG. 3 Disaster-related suicides in prefectures affected by the Great East Japan Earthquake. *(From Japan Ministry of Health, Labor and Welfare. (2021). The number of suicides relating to Great East Japan Earthquake 2021. Available online at: https://www.mhlw.go.jp/content/202111-shinsai.pdf. (Accessed 21 December 2021). No permission required.)*

contrary to each other. This discrepancy could be a kind of polarization that was also seen after the Great Hanshin Earthquake in 1995, subsequently termed a "scissors-like gap" (Nishio et al., 2009). Moreover, a study analyzing disaster-related suicides in Fukushima Prefecture revealed that, in contrast to the general tendency of suicides in Japan, aged women were more likely to commit suicide (Takebayashi et al., 2020), suggesting that they were more vulnerable to community loss. Another study based on a national survey on suicide in Japan from 2010 to 2014 also demonstrated that the standardized mortality suicide rates in the three affected prefectures increased after a brief decline (a so-called "U-shaped curve") (Ohto, Maeda, Yabe, Yasumura, & Bromet, 2015). In other studies examining suicide, similar tendencies, such as a U-shaped curve for suicide rates among affected populations, were often seen after major natural disasters (Kõlves, Kõlves, & De Leo, 2013; Maeda, Oe, Bromet, Yasumura, & Ohto, 2016). Such a delayed suicide pattern after the Fukushima disaster is discussed in more detail in a later chapter.

Remarkable exhaustion or burnout among various types of relief workers was another issue that gradually emerged during this phase. In particular, public employees working in the disaster area faced more difficult situations. This is also considered to be another unique, serious feature of the current disaster, and is discussed later.

Phase 3: Great return phase (September 2015 to April 2017; 20 months)

Accelerated process of returning

The return of evacuees was accelerated in this phase while being apt to stagnate until then. As described above, little progress seemed to be made in many of the evacuated municipalities, even though they had already been designated as "zones in preparation for the lifting of the evacuation order" for several years. The evacuation order for Naraha town was removed first in September 2015, followed by many other municipalities in succession: the orders for Katsurao village, Kawauchi village, and the Odaka area (Minamisōma city) were lifted in 2016, and those for Kawamata town, Namie town, Iitate village, and Tomioka town were lifted in 2017.

The lifting of these restrictions for the evacuated municipalities signified great progress in terms of relief after the disaster, and thus became the touchstone for rebuilding communities among many municipalities. Opinion surveys conducted by these municipalities apparently showed that many evacuees, especially young people, were still hesitant to return to their hometowns for various reasons, such as concerns about radiation effects, subsequent accidents at the FDNEPS, an unclear future, financial problems, the education of their children, and poor medical and welfare resources (Maeda & Oe, 2015; NHK, 2015). These municipalities certainly had hopeful and grand conceptions, but they never had convincing measures and prospects for the near future regarding how and when they could achieve success. In other words, the decision to return was a painful one and represented a double-edged sword for the municipalities. Furthermore, during this phase, with the lifting of the evacuation orders, the assistance programs that had been put in place for evacuees to provide housing and financial support started to be reduced. Many evacuees from these municipalities were entering into a crucial phase to determine whether they should return (NHK, 2015).

Despite such difficulties, the number of returnees rapidly grew during this phase. In particular, the number of evacuees staying inside Fukushima Prefecture decreased by about 40,000, eventually falling below 20,000 (Fig. 4) (Fukushima Prefecture, 2019). On the other hand, the number of those evacuated outside Fukushima Prefecture showed only a small decrease; as of June 2017, there were still nearly 40,000 evacuees staying outside the prefecture. The order of the number of evacuees staying outside and inside Fukushima Prefecture was reversed during this phase, and thereafter, the difference grew larger (Fig. 4) (Fukushima Prefecture, 2019).

Unchanged mental health status and exhaustion among workers

Contrary to the progress in returning, the mental health status among the evacuees, which had been ameliorating rapidly in phase 2, appeared to be unchanged during this phase. For example, the prevalence of people at risk of mental health problems based on the results of the MHLS was 7.7% in 2015; this had only decreased to 6.8% by 2017 (Fig. 1) (Fukushima Prefecture, 2020). Another study (Nishi, Susukida, Usuda, & Yamanouchi, 2018) conducted during this phase using national demographic data also revealed that the psychological distress level among the affected people living in Fukushima Prefecture was still significantly higher than that among those living in other prefectures. Also, with regard to posttraumatic responses identified based on the MHLS, despite a lack

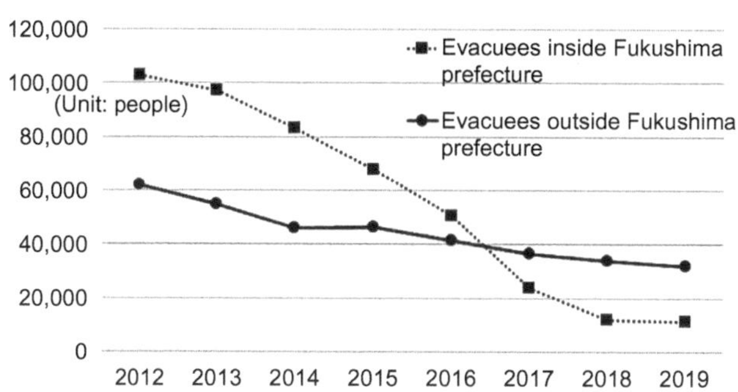

FIG. 4 Evacuees inside and outside Fukushima Prefecture. *(From Fukushima Prefecture. (2020).* Steps for reconstruction and revitalization in Fukushima Prefecture. *Available online at: https://www.pref.fukushima.lg.jp/uploaded/attachment/442270.pdf. (Accessed 21 December 2021). No permission required.)*

of data during this phase, the prevalence rate for probable PTSD in the 3 most recent years (2017–19) showed little change, and nearly 10% of the evacuees were still estimated to have substantial PTSD symptoms (Fukushima Prefecture, 2020).

Meanwhile, the number of disaster-related suicides, which greatly increased in the previous phase, showed a small decrease from the peak in 2013, but was still higher compared with other affected prefectures (Fig. 3) (National Police Agency, 2019). Another study examining the national survey data for suicides demonstrated that the number of suicide cases among people living in evacuation order zones during this phase, which had shown a great increase soon after the disaster, increased again 4 years later (Orui, Suzuki, Maeda, & Yasumura, 2018). In addition, with regard to the risk perception of radiation health effects among the evacuees, the MHLS showed few changes in this phase, and about one-third of the respondents still expressed concern about the genetic effects (Fukushima Prefecture, 2020).

Consideration should be also given to the mental health of different relief workers, including public employees working in the disaster area, in addition to the evacuees. The total number of public employees working in Fukushima Prefecture was approximately 18,000, many of whom were engaged in various types of long-term, difficult relief work. Especially, those working in the 13 evacuated municipalities faced great difficulties, such as long working hours, increased workloads, exposure to anger or complaints from affected people, and an unclear future regarding their municipalities (Maeda, Ueda, et al., 2016). Additionally, stressors involving the fact that public employees are also "evacuees" should be considered; that is to say, public employees also need to lead lives as evacuees in unfamiliar places with their family members, sometimes being separated.

When the evacuation orders for many of the evacuated municipalities were lifted during this phase, these workers were already considerably exhausted, some even falling into burnout. In fact, surprisingly, diagnostic interviews conducted with public employees working in the disaster area in 2014–15 revealed rates of depression, sleeping difficulties, and suicide risk of 17.9%, 72.6%, and 8.9%, respectively (Maeda, Ueda, et al., 2016). These results strongly suggested that many public employees working in evacuated municipalities were already seriously exhausted at the time the evacuation orders were lifted. The return to their municipalities thus appeared to add further burdens on them, sometimes leading to crisis. Unfortunately, in 2016, nine public employees in Fukushima Prefecture committed suicide (All-Japan Prefectural and Municipal Workers' Union, 2017). Furthermore, a survey conducted by the All-Japan Prefectural and Municipal Workers Union for public employees in Fukushima Prefecture revealed that the number of those taking sick leave because of psychiatric disorders increased rapidly, from 61 in 2012 to 191 in 2017, and that 171 took an early retirement in 2017 (All-Japan Prefectural and Municipal Workers' Union, 2019).

Such job-related mental health problems observed among public employees were also serious issues for other types of workers in the disaster area. Two studies conducted in 2015 also revealed that nearly 50% of medical and public health nurses working in Fukushima Prefecture had a high psychological distress level or low general mental health conditions (Nukui et al., 2017; Yoshida et al., 2016). Another study asking nurses working at Fukushima Medical University Hospital in 2013 about their intention to continue working revealed that 44% of the participants intended to leave their job after the disaster, which was strongly associated with worries or anxiety about radiation health effects (Sato et al., 2015). Moreover, workers in many newly established facilities after the disaster, such as the Fukushima Center for Disaster Mental Health, also met with similar difficulties; they often fell into burnout and, as a result, left their jobs (Maeda, Lyamzina, & Ito, 2019).

Several psychological and environmental factors were considered to influence the mental health conditions of different relief workers (Maeda, Ueda, et al., 2016; Shigemura et al., 2014). First, they were handling a massively higher workload than they had before the disaster. Second, as the relief process was prolonged, it was difficult for workers to maintain their morale or a good mental health condition. In particular, the unclear future of both their workplaces and communities made them anxious and sometimes led to early retirement. Third, they were often exposed to intense negative feelings from residents rather than respect or appreciation. For example, many individual public employees who participated in interviews in the above study (Maeda, Ueda, et al., 2016) responded that they had been traumatized by the evacuees' anger and felt more stressful about it than the explosions at the FDNEPS or their subsequent lives as evacuees. Actually, they were flooded with numerous complaints from evacuees, especially in the initial phase. Similarly, employees of the Tokyo Electric Power Company (TEPCO) working at the FDNEPS were experiencing substantial discrimination from the public (Shigemura et al., 2014). Fourth, public employees especially had many troubles in their own lives as evacuees. Many experienced role conflicts regarding whether they should place more value on their family or their job, and some had to continue working while being separated from their family. Such role conflicts drove some relief workers to make the difficult decision to stay behind for work or leave for their family (Maeda, Ueda, et al., 2016).

Phase 4: Stagnation phase (May 2017 to present)

Stagnation of returning

The great waves of returnees seen in the previous phase quickly shrank in this phase. The evacuation orders for Okuma and Futaba towns, where the FDNPP was located, were partially lifted in April 2019 and March 2020, respectively. This meant that all the evacuated municipalities were eventually cleared for return, although not necessarily at a satisfactory level. In contrast to the previous phase, however, the areas considered habitable again in this phase were too limited, and so only a small number of evacuees actually returned.

Many of these municipalities faced serious low return rates. In 10 of the municipalities, for example, as of September 2019, the total return rate was only 26.8% (Sankei Shinbun, 2019). With regard to the evacuees' reasons for hesitating to return, many expressed concerns about a variety of matters, such as poor educational and medical environments, the unclear future of their community, or changes in the atmosphere (e.g., smaller population, shuttered houses), whereas, in the initial postdisaster phase, their concerns had focused on radiation exposure or subsequent explosions at the FDNPP (Reconstruction Agency, 2020a). In addition, according to municipal opinion surveys (Reconstruction Agency, 2020a), a primary reason why the evacuees decided not to return was that they had already established a new life during their long-term evacuation. In particular, younger people had less desire to return to their hometowns because it was easier for them to resettle in and adjust to other places, whereas the aged tended to hope to live out the remainder of their lives in their hometown.

Thus, many municipalities also experienced extremely high population aging rates. For example, the mean proportion of the population aged 65 and over in 11 of the municipalities reached as high as 43% as of March 2020 (NHK, 2020), exceeding that in Fukushima Prefecture (approximately 32%) and all of Japan (approximately 27%). Considering that aged people were more likely to express concerns about radiation health effects (Suzuki et al., 2015), the prolonged relief process itself appeared to have led to the aging of the population.

Concerns about the mental health status of evacuees outside Fukushima Prefecture

Recently, the mental health status of evacuees staying outside Fukushima Prefecture worsened compared with those inside the prefecture. The latest data from the MHLS showed that the prevalence of probable depression among the respondents outside Fukushima Prefecture was higher than that among those inside (8.1% vs. 5.3%, respectively) (Fukushima Prefecture, 2020) (Fig. 5). Similarly, based on the results of the SDQ, the prevalence of young people (aged 15 and below) at risk of behavioral or emotional problems was higher among those staying outside than among those staying inside Fukushima Prefecture (12.8% vs. 9.0% among those of elementary school age, and 13.9% vs. 10.1% among those of junior high school age) (Fukushima Prefecture, 2020) (Fig. 6). Evacuees staying in remote areas might be more likely to have difficulties such as maladaptation and isolation in unfamiliar places. Given the fact that there are still nearly 30,000 evacuees outside Fukushima Prefecture,

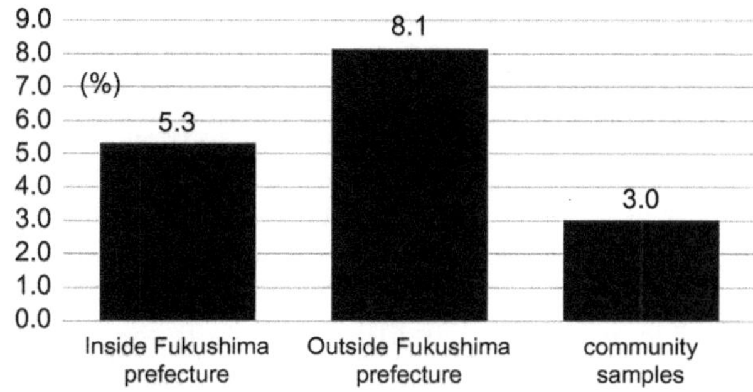

FIG. 5 Comparison of the prevalence of probable depression between adults inside and outside Fukushima Prefecture in 2019. A high risk of psychological distress was considered a score ≥13 on the 6-item Kessler Psychological Distress Scale. The prevalence of being at high risk for psychological distress in the community sample was 3.0% (Kawakami, 2007). *(No permission required.)*

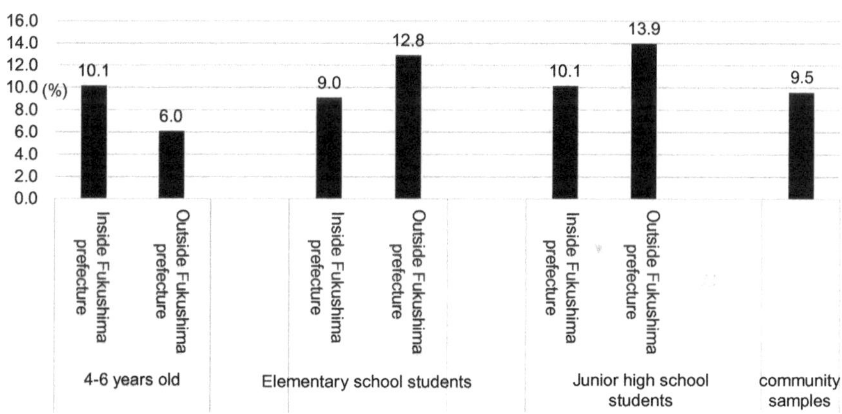

FIG. 6 Comparison of the prevalence of some behavioral problems between children inside and outside Fukushima Prefecture in 2019. A high risk of psychopathological problems in children was considered a score ≥16 on the Strength and Difficulties Questionnaire. The prevalence of being at high risk for psychopathological problems in the community sample was 9.5% (Matsuishi, 2008). *(No permission required.)*

a mental health care system that can provide more efficient, long-term interventions for these individuals is needed to help prevent their isolation.

Discussion

Mental health issues after the disaster

In this chapter, we reviewed the long-term developments in regard to recovery after the Fukushima disaster, focusing especially on the evacuation/return of affected people and their mental health issues. Evacuees have experienced substantial confusion as a result of not only the unclear aspects of relief but also the uncertain information about the accident and its radiation effects. In addition, they have suffered from various types of social reactions, such as public stigmas about marriage, reproduction, and/or even receiving compensation from TEPCO (Maeda et al., 2019). These aspects of the Fukushima disaster were thought to be greatly different from those of natural disasters. For example, in contrast to natural disasters, the impact of the Fukushima nuclear disaster was largely invisible, which made it difficult to clarify the boundary between disaster and nondisaster areas (Maeda et al., 2018). Deciding on or changing the evacuation zones, therefore, seemed to give evacuees the inevitable impression that decisions had been made arbitrarily or politically, leading to a sense of inequality and a lack of control.

These unique, long-term effects of the Fukushima disaster might produce different mental health consequences among affected people. In terms of general psychological states, based on the results of the K6 in the MHLS, an extremely high prevalence for a risk of depression (14.6% of the adult group) was seen in 2012. Thereafter, this prevalence gradually decreased during the first 3 years but remained higher than that in the general population in Japan (3.0%) according to the latest results (Kawakami, 2007). These results suggest that many evacuees continue to suffer from psychiatric symptoms, especially depressive symptoms, and subclinical psychological distress. In addition, more than 60% of the respondents indicated that they had some sleeping difficulties (Fukushima Prefecture, 2020), and the total number of disaster-related suicides during the first 5 years after the disaster was more than 100, which was substantially higher than the rates in Iwate and Miyagi Prefectures (National Police Agency, 2019). Furthermore, as mentioned above, other studies using the K6 scores in the MHLS (Oe, Maeda, et al., 2016; Suzuki et al., 2015) indicated a close association between psychological distress, including depressive symptoms, and negative risk perceptions of radiation. Considering this unique relationship, a comprehensive approach addressing both radiation risk perception and mental health might be required, instead of dealing with these concepts independently. That is to say, it is necessary for care providers to have two types of skills: risk communication and mental health care.

Furthermore, recent results from the MHLS revealed that evacuees relocating to remote areas were more likely to have mental health problems. Compared with natural disasters, the broad dispersal of evacuees is also considered to be one of the outstanding features of nuclear disasters, making adequate healthcare provision more difficult. Therefore, specific measures are required to provide seamless care and efficient interventions for evacuees staying in remote areas and for returnees. First, a broader care network should be established with local stakeholders sharing necessary information about the evacuees. Moreover, an antistigma campaign should be carried out to help prevent isolation among the evacuees. For instance, a recent study in the United Kingdom that examined the effectiveness of an antistigma campaign utilizing the mass media to dispel stigmas toward people with mental health problems showed that it substantially increased knowledge among the participants despite being only a short-term campaign (Evans-Lacko, London, Little, Henderson, & Thornicroft, 2010). In addition, remote counseling tools over the telephone and/or Internet may be effective for dispersed evacuees. Since the disaster occurred, as described above, the support team at Fukushima Medical University, consisting of mental health professionals, including psychologists, social workers, and nurses, has been providing brief telephone interventions in collaboration with other mental health care resources for evacuees at risk of various health problems since 2012. As a result, up to 30,000 evacuees have received these interventions at 8 years postdisaster (Horikoshi et al., 2020). The detailed procedures and usefulness of such telephone interventions for evacuees are described in more detail in a later chapter.

Future tasks

Soon after the accident, the evacuation order zone was extended to as many as 11 municipalities, representing an area of approximately 1150 m^2 with 840,000 residents in total. Since the first lifting of the evacuation order for the Miyakoji area of Tamura city in April 2014, those for other municipalities were lifted in succession. As of spring 2020, the evacuation order zone (officially called the "difficult-to-return zone") has been reduced to about 337 m^2, where about 25,000 residents were thought to have once lived before the disaster. In other words, this means that about 70% of the total evacuation zone has been recovered and made habitable again as a result of active relief work, including large-scale decontamination efforts. This difficult relief process, despite the criticisms, can be evaluated more positively because this was a previously unexperienced, great challenge for Japan with no clear guidelines for success.

The Japanese government designated the first 5 years after the accident as a "concentrated reconstruction period" and the subsequent 5 years as a "reconstruction and revitalization period," which will soon pass. The Japanese government has issued a new plan, called the "second reconstruction and revitalization period," that involves further relief work in the affected prefectures (Iwate, Miyagi and Fukushima) over the next 5 years (Reconstruction Agency, 2020b). This plan especially emphasizes the importance of Fukushima's recovery, which is recognized to be delayed compared with other prefectures. Actually, in the Soso area that contains the FDNPP and many evacuation zones, the population has decreased by 80,000 or more, representing a 43% decrease compared with the predisaster population (Reconstruction Agency, 2020c).

Considering the current situations in the disaster areas and the recent results of opinion surveys for evacuees, the "second reconstruction and revitalization period" plan states that the Japanese government will promote not only the return of evacuees but also the settling of newcomers from other areas through different newly launched businesses in the disaster area (Reconstruction Agency, 2020b). Given the fact that many evacuees still hesitate to return, this new policy might be realistic and acceptable for many municipalities. However, the 30,000 current evacuees must not be forgotten or abandoned. While care resources remain poor in the returned municipalities and cuts in human and financial resources are possible in the near future, we are currently facing a great task regarding how we should further maintain and develop measures to support these evacuees and returnees hereafter.

Lastly, the relief process for the Fukushima disaster has also shed light on mental health issues, including exhaustion, burnout, depression, and even suicide, among different workers that contribute to the long-term recovery efforts. The recovery phase after severe nuclear accidents takes an extremely long time compared with natural disasters, as is obvious from the accidents at Three Mile Island, Chernobyl, and Fukushima. In addition to evacuees, consideration of mental health care for relief workers is essential to promote long-term relief. Especially, care networks that can function well for a long time and provide efficient support, including monetary and mental health support, for such workers should be established. In reviewing the developments that occurred after the Fukushima disaster, establishing such a network is a core task for future recovery. Therefore, detailed information about the current situation in Fukushima should be shared with as many outside experts as possible to help avoid other similar tragedies.

References

All-Japan Prefectural and Municipal Workers' Union. (2017). http://ijimemental.web.fc2.com/sinnbunn17.3.8.pdf.

All-Japan Prefectural and Municipal Workers' Union. (2019). http://ictj-report.joho.or.jp/1904/sp10.html.

Asukai, N., Kato, H., Kawamura, N., Kim, Y., Yamamoto, K., Kishimoto, J., et al. (2002). Reliability and validity of the Japanese-language version of the impact of event scale-revised (IES-R-J): Four studies of different traumatic events. *The Journal of Nervous and Mental Disease*, *190*(3), 175–182.

Evans-Lacko, S., London, J., Little, K., Henderson, C., & Thornicroft, G. (2010). Evaluation of a brief anti-stigma campaign in Cambridge: Do short-term campaigns work? *BMC Public Health*, *10*, 339. https://doi.org/10.1186/1471-2458-10-339.

Fukushima Prefecture. (2019). https://www.pref.fukushima.lg.jp/site/portal-english/en03-08.html.

Fukushima Prefecture. (2020). https://www.pref.fukushima.lg.jp/uploaded/attachment/386028.pdf.

Hasegawa, A., Tanigawa, K., Ohtsuru, A., Yabe, H., Maeda, M., Shigemura, J., et al. (2015). Health effects of radiation and other health problems in the aftermath of nuclear accidents, with an emphasis on Fukushima. *Lancet (London, England)*, *386*(9992), 479–488. https://doi.org/10.1016/S0140-6736(15)61106-0.

Horikoshi, N., Maeda, M., Iwasa, H., Momoi, M., Oikawa, Y., Ueda, Y., et al. (2020). The usefulness of brief telephonic intervention after a nuclear crisis: Long-term community-based support for Fukushima evacuees. *Disaster Medicine and Public Health Preparedness*, 1–9. https://doi.org/10.1017/dmp.2020.161.

Kawakami, N. (2007). *National survey of mental health measured by K6 and factors affecting mental health status in research on applied use of statistics and information*. Health Labour Sciences Research Grant.

Kessler, R. C., Barker, P. R., Colpe, L. J., Epstein, J. F., Gfroerer, J. C., Hiripi, E., et al. (2003). Screening for serious mental illness in the general population. *Archives of General Psychiatry*, *60*(2), 184–189.

Kõlves, K., Kõlves, K. E., & De Leo, D. (2013). Natural disasters and suicidal behaviours: A systematic literature review. *Journal of Affective Disorders*, *146*(1), 1–14. https://doi.org/10.1016/j.jad.2012.07.037.

Maeda, M. (2012). Minamisoma-shi Hibarigaoka Byoin no kuto. *Rinsho-Seishinigaku*, *41*, 183–1191.

Maeda, M., Lyamzina, Y., & Ito, A. (2019). The psychosocial consequences of the Fukushima disaster: What are we suffering from? In G. Steinhauser, A. Koizumi, & K. Shozugawa (Eds.), *Nuclear emergencies: A holistic approach to preparedness and response* (pp. 63–75). Springer.

Maeda, M., & Oe, M. (2015). In C. Katie (Ed.), *The Great East Japan Earthquake: Tsunami and nuclear disaster* (pp. 71–90). Springer International.

Maeda, M., Oe, M., Bromet, E., Yasumura, S., & Ohto, H. (2016). Fukushima, mental health and suicide. *Journal of Epidemiology and Community Health*, *70*(9), 843–844. https://doi.org/10.1136/jech-2015-207086.

Maeda, M., Oe, M., & Suzuki, Y. (2018). Psychosocial effects of the Fukushima disaster and current tasks: Differences between natural and nuclear disasters. *Journal of the National Institute of Public Health*, *67*, 50–58. https://doi.org/10.20683/jniph.67.1_50.

Maeda, M., Ueda, Y., & Hiruta, G. (2014). Fukushima Kokoronokea center no yakuwaritoha? *Traumatic Stress*, *12*, 5–12.

Maeda, M., Ueda, Y., Nagai, M., Fujii, S., & Oe, M. (2016). Diagnostic interview study of the prevalence of depression among public employees engaged in long-term relief work in Fukushima. *Psychiatry and Clinical Neurosciences*, *70*(9), 413–420. https://doi.org/10.1111/pcn.12414.

Matsuishi, T., Nagano, M., Araki, Y., Tanaka, Y., Iwasaki, M., Yamashita, Y., et al. (2008). Scale properties of the Japanese version of the Strengths and Difficulties Questionnaire (SDQ): A study of infant and school children in community samples. *Brain & Development*, *30*(6), 410–415. https://doi.org/10.1016/j.braindev.2007.12.003.

Matsuishi, T., Nagano, M., Araki, Y., Tanaka, Y., Iwasaki, M., Yamashita, Y., et al. (2008). Scale properties of the Japanese version of the Strengths and Difficulties Questionnaire (SDQ): A study of infant and school children in community samples. *Brain and Development*, *30*(6), 410–415.

Mori, I. (2012). *Naze inncho wa tobohan ni saretanoka*. Kodansha Ltd.

National Police Agency. (2019). https://www.npa.go.jp/safetylife/seianki/jisatsu/R02/R01_jisatuno_joukyou.pdf.

NHK. (2015). http://www.nhk.or.jp/ashita/english/stories/kuro-canwe.html.

NHK. (2020). https://www3.nhk.or.jp/news/html/20200311/k10012320891000.html.

Nishi, D., Susukida, R., Usuda, K., & Yamanouchi, Y. (2018). Psychological distress among people in Fukushima prefecture before and after the Great East Japan Earthquake using a nation-wide survey. *Psychiatry and Clinical Neurosciences*, *72*(12), 878. https://doi.org/10.1111/pcn.12786.

Nishio, A., Akazawa, K., Shibuya, F., Abe, R., Nushida, H., Ueno, Y., et al. (2009). Influence on the suicide rate two years after a devastating disaster: a report from the 1995 Great Hanshin-Awaji Earthquake. *Psychiatry and Clinical Neurosciences*, *63*(2), 247–250. https://doi.org/10.1111/j.1440-1819.2009.01942.x.

Nukui, H., Murakami, M., Midorikawa, S., Suenaga, M., Rokkaku, Y., Yabe, H., et al. (2017). Mental health and related factors of hospital nurses. *Asia-Pacific Journal of Public Health, 29*(2 Suppl), 161S–170S. https://doi.org/10.1177/1010539516682589.

Oe, M., Fujii, S., Maeda, M., Nagai, M., Harigane, M., Miura, I., et al. (2016). Three-year trend survey of psychological distress, post-traumatic stress, and problem drinking among residents in the evacuation zone after the Fukushima Daiichi Nuclear Power Plant accident [The Fukushima Health Management Survey]. *Psychiatry and Clinical Neurosciences, 70*(6), 245–252. https://doi.org/10.1111/pcn.12387.

Oe, M., Maeda, M., Nagai, M., Yasumura, S., Yabe, H., Suzuki, Y., et al. (2016). Predictors of severe psychological distress trajectory after nuclear disaster: Evidence from the Fukushima Health Management Survey. *BMJ Open, 6*(10). https://doi.org/10.1136/bmjopen-2016-013400, e013400.

Ohto, H., Maeda, M., Yabe, H., Yasumura, S., & Bromet, E. E. (2015). Suicide rates in the aftermath of the 2011 earthquake in Japan. *Lancet (London, England), 385*(9979), 1727. https://doi.org/10.1016/S0140-6736(15)60890-X.

Orui, M., Suzuki, Y., Maeda, M., & Yasumura, S. (2018). Suicide rates in evacuation areas after the Fukushima Daiichi nuclear disaster. *Crisis, 39*(5), 353–363. https://doi.org/10.1027/0227-5910/a000509.

Ota, H., Miyazaki, K., & Nakayama, T. (2017). How evacuees obtained health care information after the great east Japan earthquake: A qualitative interview study. *Disaster Medicine and Public Health Preparedness, 11*(6), 729–734. https://doi.org/10.1017/dmp.2017.25.

Reconstruction Agency. (2020a). https://www.reconstruction.go.jp/topics/main-cat1/sub-cat1-4/ikoucyousa/.

Reconstruction Agency. (2020b). https://www.reconstruction.go.jp/english/topics/Laws_etc/Reconstruction_Efforts_from_FY_2021.pdf.

Reconstruction Agency. (2020c). https://www.reconstruction.go.jp/portal/chiiki/hukkoukyoku/fukusima/material/200730_fukkokasoku.pdf.

Sato, Y., Hayashida, N., Orita, M., Urata, H., Shinkawa, T., Fukushima, Y., et al. (2015). Factors associated with nurses' intention to leave their jobs after the Fukushima Daiichi Nuclear Power plant accident. *PLoS One, 10*(3). https://doi.org/10.1371/journal.pone.0122389, e0122389.

Shigemura, J., Tanigawa, T., Nishi, D., Matsuoka, Y., Nomura, S., & Yoshino, A. (2014). Associations between disaster exposures, peritraumatic distress, and posttraumatic stress responses in Fukushima nuclear plant workers following the 2011 nuclear accident: The Fukushima NEWS Project study. *PLoS One, 9*(2). https://doi.org/10.1371/journal.pone.0087516, e87516.

Stellman, J. M., Smith, R. P., Katz, C. L., Sharma, V., Charney, D. S., Herbert, R., et al. (2008). Enduring mental health morbidity and social function impairment in world trade center rescue, recovery, and cleanup workers: The psychological dimension of an environmental health disaster. *Environmental Health Perspectives, 116*(9), 1248–1253. https://doi.org/10.1289/ehp.11164.

Sankei Shinbun. (2019). https://www.sankei.com/affairs/news/190910/afr1909100034-n1.html.

Suzuki, Y., Yabe, H., Horikoshi, N., Yasumura, S., Kawakami, N., Ohtsuru, A., et al. (2017). Diagnostic accuracy of Japanese posttraumatic stress measures after a complex disaster: The Fukushima health management survey. *Asia-Pacific Psychiatry: Official Journal of the Pacific Rim College of Psychiatrists, 9*(1). https://doi.org/10.1111/appy.12248.

Suzuki, Y., Yabe, H., Yasumura, S., Ohira, T., Niwa, S.-I., Ohtsuru, A., et al. (2015). Psychological distress and the perception of radiation risks: the Fukushima Health Management Survey. *Bulletin of the World Health Organization, 93*(9), 598–605. https://doi.org/10.2471/BLT.14.146498.

Takebayashi, Y., Hoshino, H., Kunii, Y., Niwa, S.-I., & Maeda, M. (2020). Characteristics of disaster-related suicide in Fukushima prefecture after the nuclear accident. *Crisis*, 1–8. https://doi.org/10.1027/0227-5910/a000679.

Tsujiuchi, T., Yamaguchi, M., Masuda, K., Tsuchida, M., Inomata, T., Kumano, H., et al. (2016). High prevalence of post-traumatic stress symptoms in relation to social factors in affected population one year after the Fukushima nuclear disaster. *PLoS One*, *11*(3). https://doi.org/10.1371/journal.pone.0151807, e0151807.

Yabe, H., Suzuki, Y., Mashiko, H., Nakayama, Y., Hisata, M., Niwa, S.-I., et al. (2014). Psychological distress after the Great East Japan Earthquake and Fukushima Daiichi Nuclear Power Plant accident: Results of a Mental Health and Lifestyle Survey through the Fukushima Health Management Survey in FY2011 and FY2012. *Fukushima Journal of Medical Science*, *60*(1), 57–67.

Yasumura, S., Hosoya, M., Yamashita, S., Kamiya, K., Abe, M., Akashi, M., et al. (2012). Study protocol for the Fukushima Health Management Survey. *Journal of Epidemiology*, *22*(5), 375–383.

Yoshida, K., Orita, M., Goto, A., Kumagai, A., Yasui, K., Ohtsuru, A., et al. (2016). Radiation-related anxiety among public health nurses in the Fukushima Prefecture after the accident at the Fukushima Daiichi Nuclear Power Station: A cross-sectional study. *BMJ Open*, *6*(10). https://doi.org/10.1136/bmjopen-2016-013564, e013564.

Chapter 9

Suicide-related issues among affected people

Yoshitake Takebayashi
Department of Health Risk Communication, School of Medicine, Fukushima Medical University, Fukushima, Japan

Introduction

In regular times, suicide risk increases due to the complex interaction of diverse issues, such as worsened socioeconomic, physical, and mental health conditions (Van Orden et al., 2010). Large-scale natural disasters often create local vulnerabilities that increase suicide risk (Horney et al., 2020; Safarpour et al., 2020). The Great East Japan Earthquake (GEJE) in 2011 was one of the most devastating disasters in Japan, with a massive earthquake and tsunami that killed 15,899 people (as of March 2021) (National Police Agency, 2021). Fukushima was affected by a triple disaster, with the nuclear power plant accident occurring in addition to the tsunami and earthquake (Bromet, 2011). Due to the nuclear power plant accident, evacuation orders were issued for neighboring areas (Kuroda, Orui, & Hori, 2021). Many residents were forced to evacuate inside and outside the prefecture for several years (Harigane et al., 2021; Takebayashi, Hoshino, Kunii, Niwa, & Maeda, 2020). As a result, socioeconomic problems and the deterioration of physical and mental health have continued. Socioeconomic problems include reputational damage to agriculture, fisheries, forestry, job separation, job loss, job change, separation, and loss of community ties. Regarding health problems, a high prevalence of worsening lifestyle-related diseases (diabetes and hypertension) (Murakami, Tsubokura, Ono, Nomura, & Oikawa, 2017; Nomura et al., 2016; Ohira et al., 2018) and mental health problems such as depression, posttraumatic stress disorder (PTSD), and dependence or misuse of alcohol have been reported in evacuation area residents (Harigane et al., 2021; Maeda & Oe, 2017; Oe et al., 2016). Generally, each factor increases the risk of suicide (Van Orden et al., 2010). In Fukushima, there is a concern that the combination of these factors may increase the community's vulnerability to suicide risk (Maeda & Oe, 2017). This paper provides an overview of postdisaster suicide trends in Fukushima

Health Effects of the Fukushima Nuclear Disaster. https://doi.org/10.1016/B978-0-12-824098-4.00002-7

prefecture based on national statistical data. I also introduced the government's definition of disaster-related suicide and outline its characteristics. Finally, I summarized and discussed the main points of postdisaster suicide prevention based on the above discussion.

General trends in the relationship between disasters and suicide

Suicide rates generally increase in disaster-affected areas following a disaster (Horney et al., 2020; Safarpour et al., 2020). Safarpour et al. (2020) conducted a metaanalysis to synthesize the findings of 11 papers that compared suicide rates in disaster-affected areas before and after disasters, not limited to a specific country or region. The suicide rate increased from 13.61% per 100,000 population before the disaster to 16.68% per 100,000 population afterward in their metaanalysis. This trend was specific to males. Horney et al. (2020) examined the suicide rates of 281 disasters in the United States between 2003 and 2013 for 3 years before and after the disaster. The suicide rate increased remarkably after the disaster, especially in the second year of their analysis. Matsubayashi, Sawada, and Ueda (2013) examined the general association between disasters before the GEJE and suicide in Japan. They analyzed the relationship between disasters and prefectural suicide rates using panel data after natural disasters of various scales and types, such as typhoons, floods, and earthquakes in Japan from 1982 to 2010. They reported that the association between disasters and suicide varies by disaster scale, the period after the disaster, and age group. Specifically, in the year of the disaster, the suicide rate tends to increase with the number of affected persons among the elderly (65 years and older) (i.e., the larger the scale of the disaster). For elderly males, the suicide rate decreased with the number of affected persons 5 years after the disaster. Subsequently, among older women and men under 65 years of age, the suicide rate decreased with the number of victims in the year following the disaster but increased again 3 and 4 years later.

Suicide trend among residents in evacuation order area

Orui, Suzuki, Maeda, and Yasumura (2018) and Kuroda et al. (2021) reported the trend of suicide in Fukushima, which experienced the triple disaster. A more recent article by Kuroda et al. (2021) examined the monthly suicide rate (suicides per 100,000 population) in Fukushima within 10 years (from before the disaster (March 2009) to November 2018), by the evacuation ordered area (EOA) due to the nuclear accident as well as by gender and age group. For example, the suicide rate among males in the EOA was 30.4 in the year before the earthquake, which was nearly the national average at that time. However, the rate significantly increased to 47.8 after the earthquake. The decreasing trend started in the first year after the GEJE. The rate was 23.1 2.5 years after the GEJE, which was much lower than the national average. The range of suicide rates was below the national average until the fourth year after the GEJE from 2.5 years. However, 4.5 years later, the rate in the evacuation ordered areas surged again to 37.6. This was just after the evacuation orders were lifted in some areas. Since then, the rate has remained above the national average, and the suicide rate has increased markedly with each lifting of the evacuation order areas. Although there has been no apparent surge in the suicide rate among males in the non-EOA, the rate has remained above the national average since before the disaster.

The female suicide rate in evacuation ordered areas remained below the national average for 1.5 years immediately after the disaster. However, the rate increased by 23.5 2.5 years later (the national average female suicide rate was about 14 at this time). It declined again in the next 1.5 years. For males, an increase was observed immediately after lifting the evacuation order. An increase in the suicide rate was observed for females after a delay of approximately 6 months from lifting the order. Although there is no clear trend in the suicide rate for women in the non-EAO, the rate has generally remained at or above the national average.

Regarding the increase in suicide rates after lifting evacuation orders, Murakami, Takebayashi, Ono, Kubota, and Tsubokura (2020) examined the relationship between the intention to return to their hometown and well-being in EOA. They found that well-being was lower among those who had not yet decided to return compared to those who had not returned or had returned. These findings suggest the importance of psychosocial support and risk communication that considers the distress associated with the decision to return before and after lifting evacuation orders.

Furthermore, Kuroda et al. (2021) examined a 2-year moving average of suicide rates by age group. The suicide rate rose markedly among men aged 70 and over in the EOA immediately after the GEJE and declined significantly a year later. However, the rate increased again from 3 years after the GEJE (March 2014) to 6 years (February 2017) and remained high. The rate increased from 11.5 to ~17 among males under 30 in the EOA. The rate did not show a clear decreasing trend. The rate among males between 30 and 49 in the EOA decreased significantly immediately after the earthquake, and it has maintained a decreasing trend. The suicide rate among women 70 and over in the evacuation zone increased from 1 year after the disaster (March 2012) to 4 years (February 2015) and then decreased. The rate among women under 30 also increased slightly from 1 year after the disaster (March 2012) to 5 years and has maintained this level. In the non-EOA, there was no apparent increase in the suicide rate immediately after the disaster in any age category, and the rate continued to decline thereafter. These findings suggest that special attention should be paid to interventions against EOA suicide among elderly or young males immediately after and over the medium to long term, and for elderly or young females within 5 years of the disaster.

Standardized mortality ratio of suicide in EOA

The average suicide rate in Japan peaked in 2008 and maintained a downward trend until 2019. Given this trend, it is not necessarily a good situation if the suicide rate in the region stays at the same level without increasing. In this section, I examined the suicide rate in Fukushima in comparison to the national average.

Fig. 1 shows the standardized mortality ratio (SMR) for suicide in the EOA and non-EOA in Fukushima prefecture. The SMR is calculated by dividing the number of observed suicide deaths by the number of expected suicide deaths. The number of expected deaths is the sum of the product of the age-specific suicide rate in Japan and the age-specific population of a municipality in Fukushima Prefecture. Thus, SMR is equal to one when the age-adjusted suicide mortality rate is equal to the national average. The number of suicide deaths and rates by age for Japan and municipalities in Fukushima is based on the Ministry of Health, Labour and Welfare (2021a, 2021b). The age-group population of a municipality in Fukushima is based on future population estimates for 2010, 2015, and 2020, which were published by the National Institute of Population and Social Security Research (2008). The future municipal-level population estimates were updated in 2013 and 2018, excluding Fukushima Prefecture, because the population in the EOA in Fukushima is unstable. The future population estimates of municipalities in Fukushima conducted in 2008 were based on preearthquake data. Therefore, the population decline due to the evacuation order is not reflected. As the population declined due to evacuation orders, the actual expected number of deaths is assumed to be smaller. Thus, the estimated SMR may have been underestimated. The number of suicides per month in each municipality and EOA or non-EOA was estimated using a hierarchical Poisson-gamma Bayesian model (Konda, 2014). The estimated monthly SMR was averaged over the year.

Fig. 1A and B show the suicide-SMR for Fukushima prefecture as a whole. In the Fukushima prefecture, the suicide-SMR for both men and women has been higher than one. For males, the SMR has been on an upward trend since 2012, and in 2020, during the COVID-19 pandemic, SMRs increased compared to the previous year. Subsequently, there is no increase or decrease in the number of females, but the level in 2020 is lower than the last year. This trend deviates from the national average (Tanaka & Okamoto, 2021). In Japan, an increased suicide rate was observed after July 2020 among females and after October 2020 among males compared to the previous year, with an exceptionally high increase among females. This trend has been widely reported in mass media both in Japan and abroad and has attracted social attention. Generally, suicide prevention measures are implemented by local governments on a regional basis in Japan, so if I become overconfident about the national average and do not analyze regional data, there is a risk of misjudging the measures taken.

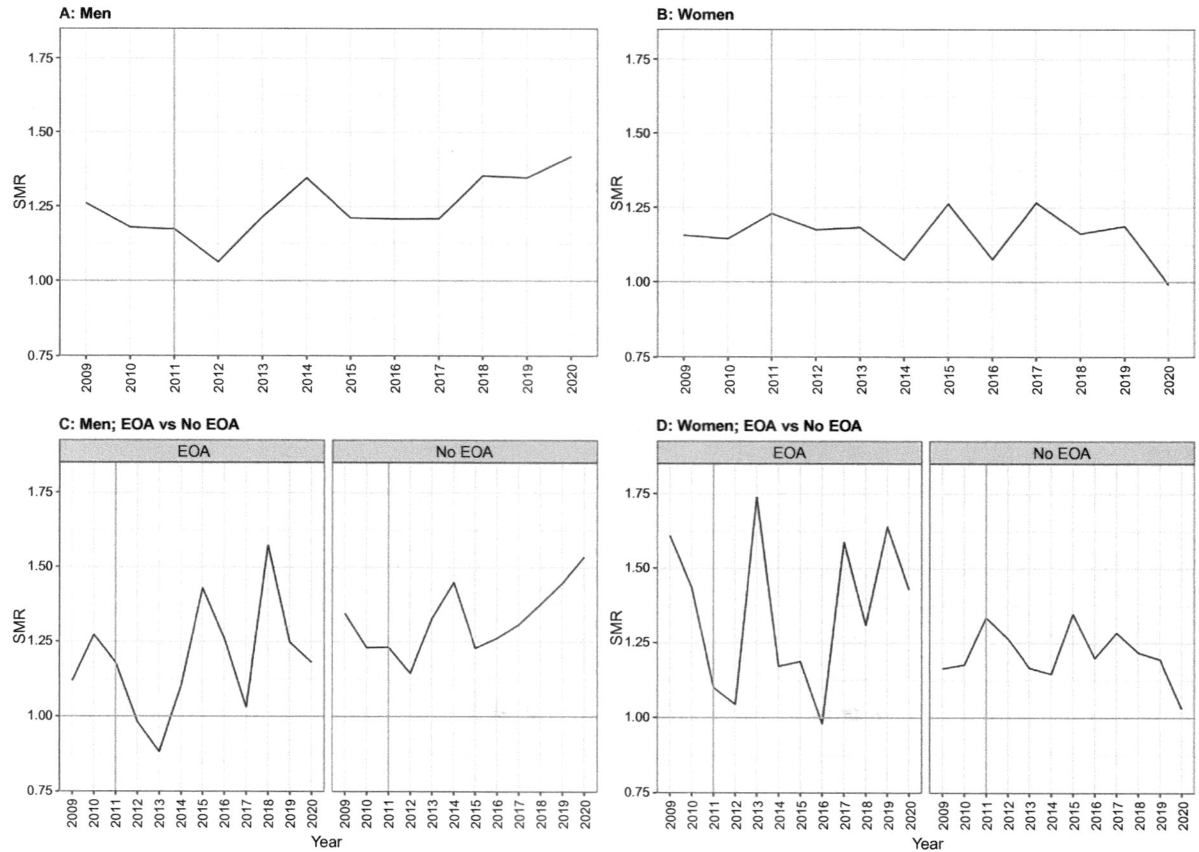

FIG. 1 Estimated annual mean of the standardized mortality ratio (SMR) for suicide by gender from a hierarchical Poisson-gamma Bayesian model.

Fig. 1C and D show the suicide-SMR for the EOA and non-EOA. Among males, there was an upward trend in both areas. The suicide SMR peaked in 2018 and then decreased from 2019 to 2020 in the EOA. Contrastingly, suicide-SMR continued to increase from 2015 to 2020 in the non-EOA group. For women, suicide-SMR in the EOA has been unstable but has remained at a high level since 2017.

Subsequently, suicide-SMR in the non-EOA among women has decreased since 2017, especially from 2019 to 2020. These findings suggest that strengthened or continued measures are needed regardless of whether the area is EOA for men. Especially during the COVID-19 pandemic, measures for men in non-EOAs may become urgent. In contrast, it is suggested that it is important to prioritize measures for women in the EOA.

Disaster-related suicide in Fukushima

The cabinet office defines "GEJE-related suicide" as suicide that meets any of the five conditions listed in Table 1 (Cabinet Office, 2011; Takebayashi et al., 2020). To specify death as a disaster-related suicide, the prefectural police collect information on the deceased from close relatives (family, friends, acquaintances, colleagues, neighbors, etc.). If the conditions in Table 1

TABLE 1 Requirements for disaster-related suicide for the Great East Japan earthquake.

1	The site of discovery of the corpse is at a shelter, temporary housing, or morgue
2	A person who died by suicide stayed in a shelter or temporary housing, confirmed by the statement of the bereaved or some other evidence
3	A person who died by suicide was evacuated from an affected area, confirmed by the statement of the bereaved or some other evidence
4	A dwelling (area of residency) or a workplace of the person who died by suicide was confirmed to be damaged by the earthquake or tsunami, confirmed by the statement of the bereaved or some other evidence
5	The cause or motivation for the suicide was found to be directly influenced by the GEJE, in ways other than mentioned above, confirmed by the statement of the bereaved or some other evidence. For example, (a) if the person who died by suicide left a suicide note that he or she ended his or her life because of the GEJE, or (b) while the person who died by suicide was alive, he or she expressed the desire to end his or her life and provided reasons related to the GEJE to his or her bereaved family

From Takebayashi, Y., Hoshino, H., Kunii, Y., Niwa, S. I., & Maeda, M. (2020). Characteristics of disaster-related suicide in Fukushima prefecture after the nuclear accident. *Crisis, 41* (6), 475–482. https://doi.org/10.1027/0227-5910/a000679.

were met based on the collected information, the death was identified as a disaster-related suicide. If it is determined that death is a disaster-related suicide, the government will provide a certain amount of support, such as disaster condolence money.

According to the number of disaster-related suicides published by the Ministry of Health, Labour and Welfare (2021a), 240 suicides were certified as disaster-related during the 10 years from 2011 to 2020. This value is equivalent to 8% of all suicides in Fukushima prefecture and 36% in evacuation ordered areas. According to gender, twice as many males compared to females (159 males, 81 females) committed disaster-related suicide. By age, suicide occurred most often among those in their 50s and 60s (50s: 23%, 60s: 22%), followed by those in their 70s (14%) and 40s (12%), 80s (10%) and 30s (9%), 20s (7%), and under 20s (2%). By cause and motive of suicide, health problems were the most common (33%), followed by family problems (15%) and economic and living problems (15%) in equal proportions. Compared to these causes, work problems (8%), gender problems (1%), and school problems (0%) were less likely cited as the cause. In terms of occupation, suicides among the unemployed were significantly more frequent (68%). Of the unemployed, 39% were living on pension or unemployment insurance (mainly the elderly), 35% were unemployed, 12% were housewives, and 11% were unemployed. The tendency for suicides to be more common among males, the middle-aged, and the unemployed, and for health problems to be the primary motivation for suicide are consistent with the general trend of suicides in Japan, not just earthquake-related suicides.

The number of disaster-related suicides by affected prefecture was 54 (23%) in Iwate prefecture, 58 (24%) in Miyagi prefecture, 118 (49%) in Fukushima prefecture, and 10 (0.04%) in other prefectures, with Fukushima prefecture accounting for about half the total. The definition of disaster-related suicide emphasizes that deaths are strongly related to evacuation caused by the disaster. In Fukushima prefecture, the number of people evacuated was higher than in other prefectures because of the evacuation order caused by the nuclear power plant accidents. As a result, the evacuation process became prolonged. Reflecting on these situations, the pattern of the number of suicides over time in Fukushima prefecture is different from that in other prefectures. In Iwate and Miyagi prefectures, the number of earthquake-related suicides was highest in the first year and then declined. However, in Fukushima prefecture, the number was highest in the second year, then this repeatedly increased and decreased (Fig. 2). The high number of disaster-related suicides in Iwate and Miyagi prefectures in the first year may reflect the intensity of the acute grief response associated with the many deaths caused by the tsunami. However, the delayed increase in Fukushima may be due to the peculiarity of the effects of the nuclear accident.

We summarized the death certificate data of 99 disaster-related suicides in Fukushima prefecture from 2011 to 2017 (Takebayashi et al., 2020). The results

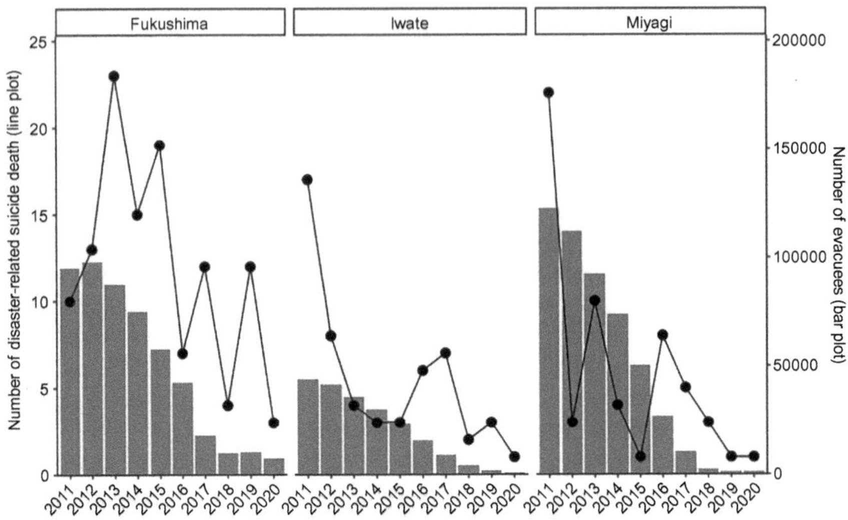

FIG. 2 Number of disaster-related suicides defined and evacuees by the affected prefecture from GEJE.

showed that the rate of signs of suicide contemplation and the presence of worries expressed to others before suicide in women were significantly higher than in men. However, more than half the men expressed worries. We also compared the development and treatment history of physical and psychiatric diseases before and after the disaster (no treatment, only physical illness treatment, only psychiatric treatment, both treatments) among disaster-related suicides (Fig. 3). After the disaster, the number of people who did not receive treatment decreased, and the number of people who received treatment for both psychiatric and physical illnesses increased. Additionally, the number of people who received treatment for only physical conditions did not increase while the number of people who received treatment for only mental illness increased. This suggests that psychiatric illnesses developed after the disaster among many individuals who eventually committed suicide. Furthermore, these results imply that the people surrounding the suicide victims could not respond appropriately to the concerns of patients even when they showed signs of suicide. These findings suggest that healthcare professionals could not adequately assess and manage suicide risk for patients who visited specialized agencies. Therefore, it would be effective to promote gatekeeper training for residents in evacuation ordered areas and improve healthcare professionals' ability to assess suicide risk. Additionally, as men are less likely than women to show signs of suicide to their close relatives, both residents and professionals should pay more careful attention to males who fall under socioeconomic health-related suicide risk factors.

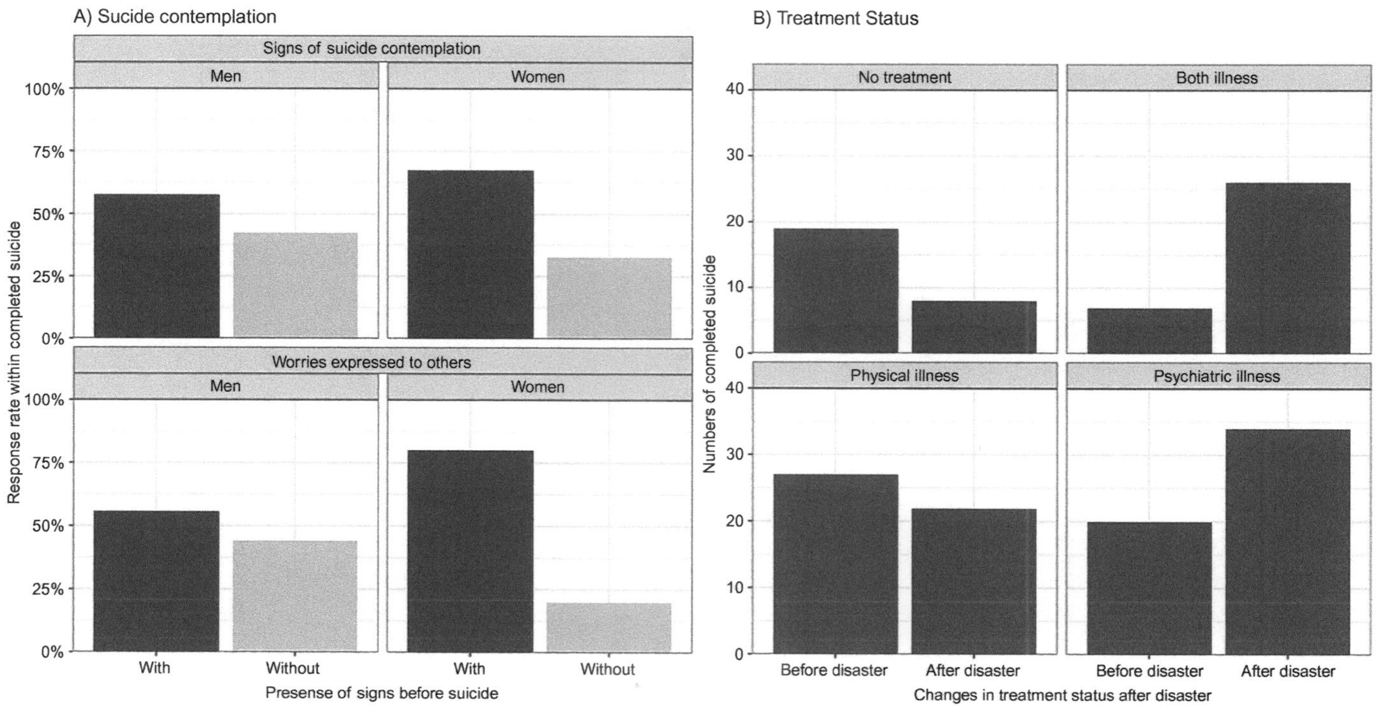

FIG. 3 (A) Prevalence of signs of suicide contemplation among individuals with disaster-related suicide by gender. (B) Treatment status among individual with disaster related suicide by treatment (no treatment, physical illness, psychiatric illness, and both illness) and period (before or after disaster).

Perspective for suicide prevention effort after the triple disaster

It is essential to understand the actual conditions of the area and the demand of residents in detail to enact effective suicide prevention measures after a disaster. Resident needs change dynamically over time after a disaster. Additionally, the suicide risk varies among individuals.

Concerning the passage of time and the changing support needs, a practical report by Orui et al. (2020) provides valuable information and perspectives useful for planning support. Their proposal is based on their experience of continuous research and support in the Fukushima Prefecture after the disaster. They divided the postdisaster period into three periods:

(1) The acute period, which is immediately after the disaster, lasts for about 6 months.
(2) The mid-term period, which lasts from 6 months to 5 years.
(3) The recovery period, which lasts more than 5 years.

Common to the acute, mid-term, and recovery periods is the need to consider that moving one's residence is a significant source of stress. For example, in the acute phase, people move from evacuation centers to temporary housing. In the mid-term and recovery periods, they move from temporary to public disaster housing. For evacuees living in evacuation centers during the acute phase, their primary needs include care for the trauma associated with the disaster and experiences of loss associated with grief. In the mid-term and recovery periods, the primary requirements are to promote social ties once formed in temporary housing that are not fragmented by migration. Furthermore, during this period, providing treatment and support is essential for individuals with mental health problems (depressive symptoms, problematic alcohol drinking, dementia, etc.). While the lifting of evacuation orders is a positive social sign that signals recovery, it can also create psychological conflicts associated with critical decision-making (Murakami et al., 2020). For example, there may be disagreements within the family about returning home, or even if they decide not to return, they may have to make decisions about what to do with the house and land of their former residence. Therefore, when such administrative orders are given or canceled, it is necessary to care for affected residents, assuming they are experiencing significant conflict.

The report by Orui et al. (2020) also suggests how to provide support to residents in the affected areas according to the severity of their mental health risks. They used a valid screening index for mental health to divide residents into three primary to tertiary prevention levels. At the primary prevention level, which includes general disaster victims, maintaining a healthy lifestyle is the focus of intervention. For example, they provided information on a healthy lifestyle through lectures and the distribution of booklets. They also offered opportunities (such as tea parties) or facilities for spontaneous gatherings and exchanges. At the secondary prevention level, which targets high-risk populations, support is provided for residents who are likely to progress to developing mental disorders, such as PTSD, depression, anxiety, alcohol abuse, and dependence. Additionally, special attention is needed for those whose lives have been disrupted by grief. The approach at this level requires disaster mental health care specialists (public health nurses, clinical psychologists, mental health workers, etc.) to collect and organize the information necessary for disaster victim support through communication, based on supportive listening and empathy. Finally, the tertiary prevention level for mental health problems at the disease level includes support for survivors with newly developed mental disorders due to the disaster experience and support for patients who have a history of mental illnesses. This level requires diagnosis, prescription, medication, and management by psychiatrists and the provision of psychotherapy by psychologists and others.

To develop a flexible suicide prevention intervention, experts should carefully listen to the demands of local stakeholders and residents to determine what support and priorities are needed at each stage and community. More importantly, I would like to emphasize that the government's financial support for survivors is essential for mental health support. Orui, Nakajima, et al. (2018) and Orui, Suzuki, et al. (2018) showed that prolonged worsened mental health conditions after a disaster are strongly associated with disaster-related unemployment and economic hardship, even after controlling other disaster-related, health-related, and social factors. Therefore, it is essential to provide financial support to disaster victims and provide detailed mental health support that meets individual demands.

Conclusion

This paper aimed to provide an overview of the trends and characteristics of suicide in the Fukushima prefecture after disasters. Postdisaster suicides in Fukushima have increased compared to the national average, especially among males. Among women, the suicide rate was higher than the national average, especially in EOAs. The analysis based on the death certificate data of disaster-related suicides as defined by the government showed that many victims gave signs of suicide to those around them before their deaths or had said something suggestive of it. Furthermore, many of these people visited specialized institutions for physical or mental illness before death. These findings indicate the importance of suicide prevention measures after a complex disaster. To make suicide prevention measures effective, it is essential to provide flexible support according to changes in demand over time and individual differences in risk.

Finally, mental health problems among support providers (e.g., medical professionals and administrative staff) concerning large-scale disasters require a long recovery period (Hori, Takebayashi, Tsubokura, & Kim, 2020; Maeda, Ueda, Nagai, Fujii, & Oe, 2016). For instance, Hori et al. (2020) reported the case of a survivor who entered a manic state immediately after the disaster but was able to function as a local support provider. Several years later, the survivor developed significant physical and mental health problems and responded well to specialized psychotherapy for PTSD. When a disaster occurs on a large scale, it is necessary to consider the risk of exhaustion and suicide among support providers during the recovery period. A sustainable support system should be developed that does not concentrate the burden on supporters in the early stages of a disaster.

References

Bromet, E. (2011). Lessons learned from radiation disasters. *World Psychiatry*, *10*(2), 83–84.

Cabinet Office. (2011). *Understanding the actual situation of suicide related to the great East Japan earthquake*.

Harigane, M., Takebayashi, Y., Murakami, M., Maeda, M., Mizuki, R., Oikawa, Y., et al. (2021). Higher psychological distress experienced by evacuees relocating outside Fukushima after the nuclear accident: The Fukushima health management survey. *International Journal of Disaster Risk Reduction*, *52*. https://doi.org/10.1016/j.ijdrr.2020.101962, 101962.

Hori, A., Takebayashi, Y., Tsubokura, M., & Kim, Y. (2020). PTSD and bipolar II disorder in Fukushima disaster relief workers after the 2011 nuclear accident. *BMJ Case Reports*, *13*(9). https://doi.org/10.1136/bcr-2020-236725, e236725.

Horney, J. A., Karaye, I. M., Abuabara, A., Gearhart, S., Grabich, S., & Perez-Patron, M. (2020). The impact of natural disasters on suicide in the United States, 2003–2015. *Crisis*, 1–7. https://doi.org/10.1027/0227-5910/a000723.

Konda, H. (2014). Application of bayesian method to estimating relative risks of suicide in Japan - making shrinkage estimators of standard Moratality rate using empirical bayesian models and hierarchical bayesian models. *Social Trust Studies*, *2*, 1–34.

Kuroda, Y., Orui, M., & Hori, A. (2021). Trends in suicide mortality in 10 years around the great East Japan earthquake: Analysis of evacuation and non-evacuation areas in Fukushima prefecture. *International Journal of Environmental Research and Public Health*, *18*(11), 6005. https://doi.org/10.3390/ijerph18116005.

Maeda, M., & Oe, M. (2017). Mental health consequences and social issues after the Fukushima disaster. *Asia-Pacific Journal of Public Health*, *29*(2_suppl), 36S–46S. https://doi.org/10.1177/1010539516689695.

Maeda, M., Ueda, Y., Nagai, M., Fujii, S., & Oe, M. (2016). Diagnostic interview study of the prevalence of depression among public employees engaged in long-term relief work in Fukushima. *Psychiatry and Clinical Neurosciences*, *70*(9), 413–420. https://doi.org/10.1111/pcn.12414.

Matsubayashi, T., Sawada, Y., & Ueda, M. (2013). Natural disasters and suicide: Evidence from Japan. *Social Science and Medicine*, *82*, 126–133. https://doi.org/10.1016/j.socscimed.2012.12.021.

Ministry of Health, Labour and Welfare. (2021a). *Number of suicides related to the Great East Japan earthquake*. https://www.mhlw.go.jp/stf/seisakunitsuite/bunya/hukushi_kaigo/seikatsuhogo/jisatsu/ jisatsu_new.html.

Ministry of Health, Labour and Welfare. (2021b). *Suicide statistics: Basic data on suicide in the community*. https://www.mhlw.go.jp/stf/seisakunitsuite/bunya/0000140901.html.

Murakami, M., Takebayashi, Y., Ono, K., Kubota, A., & Tsubokura, M. (2020). The decision to return home and wellbeing after the Fukushima disaster. *International Journal of Disaster Risk Reduction, 47*. https://doi.org/10.1016/j.ijdrr.2020.101538, 101538.

Murakami, M., Tsubokura, M., Ono, K., Nomura, S., & Oikawa, T. (2017). Additional risk of diabetes exceeds the increased risk of cancer caused by radiation exposure after the Fukushima disaster. *PLoS One, 12*(9). https://doi.org/10.1371/journal.pone.0185259, e0185259.

National Institute of Population and Social Security Research. (2008). Data by sex and age (5-year age group). In *Future population projections by region in Japan (2008 estimate)*. http://www.ipss.go.jp/pp-shicyoson/j/shicyoson08/5-sai/shosai.html.

National Police Agency. (2021). *Damage situation and police measures after the 2011 off the Pacific coast of Tohoku earthquake*. https://www.npa.go.jp/news/other/earthquake2011/pdf/higaijokyo.pdf.

Nomura, S., Blangiardo, M., Tsubokura, M., Ozaki, A., Morita, T., & Hodgson, S. (2016). Postnuclear disaster evacuation and chronic health in adults in Fukushima, Japan: A long-term retrospective analysis. *BMJ Open, 6*(2). https://doi.org/10.1136/bmjopen-2015-010080, e010080.

Oe, M., Maeda, M., Nagai, M., Yasumura, S., Yabe, H., Suzuki, Y., et al. (2016). Predictors of severe psychological distress trajectory after nuclear disaster: Evidence from the Fukushima health management survey. *BMJ Open, 6*(10), e013400.

Ohira, T., Nakano, Y., Okazaki, K., Hayashi, F., Yumiya, Y., Sakai, A., et al. (2018). Trends in lifestyle-related diseases before and after the great East Japan earthquake: The Fukushima health management survey. *Journal of the National Institute of Public Health, 67*(1), 34–41. https://doi.org/10.20683/jniph.67.1_34.

Orui, M., Nakajima, S., Takebayashi, Y., Ito, A., Momoi, M., Maeda, M., et al. (2018). Mental health recovery of evacuees and residents from the Fukushima Daiichi nuclear power plant accident after seven years—Contribution of social network and a desirable lifestyle. *International Journal of Environmental Research and Public Health, 15*(11), 2381. https://doi.org/10.3390/ijerph15112381.

Orui, M., Suzuki, Y., Maeda, M., & Yasumura, S. (2018). Suicide rates in evacuation areas after the Fukushima Daiichi nuclear disaster. *Crisis, 39*(5), 353–363. https://doi.org/10.1027/0227-5910/a000509.

Orui, M., Tanaka, E., Maeda, M., Yagi, J., Kondo, K., Nomura, K., et al. (2020). Mental health and suicide prevention during disasters. *[Nihon Koshu Eisei Zasshi] Japanese Journal of Public Health, 67*(2), 101–110. https://doi.org/10.11236/jph.67.2_101.

Safarpour, H., Sohrabizadeh, S., Malekyan, L., Safi-Keykaleh, M., Pirani, D., Daliri, S., et al. (2020). Suicide death rate after disasters: A meta-analysis study. *Archives of Suicide Research*, 1–14. https://doi.org/10.1080/13811118.2020.1793045.

Takebayashi, Y., Hoshino, H., Kunii, Y., Niwa, S. I., & Maeda, M. (2020). Characteristics of disaster-related suicide in Fukushima prefecture after the nuclear accident. *Crisis, 41*(6), 475–482. https://doi.org/10.1027/0227-5910/a000679.

Tanaka, T., & Okamoto, S. (2021). Increase in suicide following an initial decline during the COVID-19 pandemic in Japan. *Nature Human Behaviour, 5*(2), 229–238. https://doi.org/10.1038/s41562-020-01042-z.

Van Orden, K. A., Witte, T. K., Cukrowicz, K. C., Braithwaite, S. R., Selby, E. A., & Joiner, T. E. (2010). The interpersonal theory of suicide. *Psychological Review, 117*(2), 575–600. https://doi.org/10.1037/a0018697.

Chapter 10

Alcohol-related issues among affected people

Akiko Yagi
Radiation Medical Science Center, Fukushima Medical University, Fukushima, Japan

Introduction

On March 11, 2011, the Great East Japan Earthquake occurred, causing a tsunami and a subsequent accident at the Fukushima Daiichi Nuclear Power Plant (NPP). Survivors of the disaster survived difficult living situations during evacuation, such as fear of radiation exposure; loss of family, relatives, and property; and other life-threatening experiences.

The significant impact of the disasters on the mental health of people has been examined in many studies. For example, many victims showed increased symptoms of stress, sleep disturbance caused by anxiety, depression, posttraumatic stress disorder (PTSD), and suicidal tendencies, in both natural and man-made disasters (Grievink et al., 2007; Shinfuku, 2002; Suzuki et al., 2011; Tracy, Norris, & Galea, 2011; Wu et al., 2006). Disasters may result in a wide range of mental and physical health consequences for those who experience them (Norris, Friedman, Watson, et al., 2002). Severe psychological distress was likely to be found in those who experienced the earthquake, tsunami, and NPP accident in Fukushima (Kunii et al., 2016).

There have been abundant studies on associations between disasters and the use and abuse of alcohol. Many studies described that exposure to traumatic events caused increased alcohol consumption (Beaudoin, 2011; Cerdá, Tracy, & Galea, 2011; Keyes, Hatzenbuehler, & Hasin, 2011; National Hospital Organization Kurihama Medical and Addiction Center, 2011). Attending to alcohol-related issues could be an effective way to approach mental health issues of those who have experienced disaster. However, a limited number of people who have problems with alcohol actually receive treatment because the treatment of alcohol dependence in Japan focuses on abstinence from drinking while medical treatment is provided in psychiatric hospitals (Higuchi & Saito, 2013). Men who are able to drink large amounts of alcohol are often considered more masculine in Japanese society, and it is often difficult

Health Effects of the Fukushima Nuclear Disaster. https://doi.org/10.1016/B978-0-12-824098-4.00008-8

for them to seek help to stop binge drinking before their alcohol problems become severe (Shimizu, 2003).

In this article, I would like to introduce the following topics; *new findings on drinking behavior among evacuees in Fukushima*, and *community-based alcohol project on early intervention.*

Findings from the Fukushima Health Management Survey

The Fukushima Prefecture started the Fukushima Health Management Survey, a longitudinal survey to monitor the mental health status of the evacuees of the Fukushima NPP accident (Yabe et al., 2014; Yasumura et al., 2012). We used cross-sectional data from the mental health and lifestyle survey conducted in 2012, about 1 year after the accident. The target population of the survey was the entire population in the evacuation zone designated by the Fukushima Prefecture due to the NPP accident (210,189 people). The survey items varied according to the age category; for those who were 16 years or older at the time of the disaster, the self-administered PTSD Checklist Stressor-Specific Version (hereinafter PCL-S) of the questionnaire was used (Yabe et al., 2014). Among those who were aged 15 years or older as of March 11, 2011 ($n=180{,}604$), data on 73,569 participants were analyzed based on their valid responses (response rate: 40.7%) (Fukushima Prefecture, 2016). Because the legal drinking age is 20 years in Japan, we decided to exclude those who were aged less than 20 years at the time of the disaster and those who had a history of chronic hepatitis, so that the final sample size was $n=21{,}454$.

The behavioral change of turning from nondrinkers to drinkers ($n=2148$) was significantly related to being male, being younger, having less educational attainment, smoking, loss of family or relatives, change in work, having severe sleep problems measured by the Japanese version of the Athens Insomnia Scale, and the severity of traumatic symptoms measured by the PCL-S score. Having sleep problems and having more severe traumatic symptoms are significantly related to nondrinkers turning into drinkers (Yagi et al., 2020). As was shown in other studies (Lin et al., 2005; Tokyo Fire Department, 2017), male and younger respondents were more likely to become new drinkers. The results show that young males might have developed inadequate coping skills after the disaster, and the respondents who did not drink predisaster were likely to become new drinkers postdisaster if they had trouble sleeping and experienced more severe traumatic symptoms. Those who did not have a drinking habit before the disaster may have begun consuming alcohol to cope with their psychological reactions, such as difficulty sleeping or traumatic symptoms. It was also revealed that those who had high traumatic responses and/or sleep difficulties were considered to be vulnerable to alcohol misuse, endorsing findings obtained from past studies (Beaudoin, 2011; Cerdá et al., 2011; Keyes et al., 2011). Considering that posttraumatic stress responses are more likely to cause sleep difficulties, including nightmares related to traumatic events, in order to prevent affected people from starting to drink or reverting to their drinking problems after experiencing a disaster, we need to provide adequate care and psychoeducation to those who have experienced traumatic events.

As is well known, habitual drinking generally leads to physical illnesses such as diabetes, hypertension, and alcoholic hepatitis as well as other mental health issues such as depression, dementia, violence, and suicidal tendencies (Pompili et al., 2010). In Fukushima Prefecture, the standardized suicide mortality ratio decreased in the first 2 years after the disaster but increased in the third year after the disaster (Ohto, Maeda, Yabe, Yasumura, & Bromet, 2015). Also, the number of disaster-related suicide cases was remarkably higher than that in other areas affected mainly by the tsunami, such as Miyagi and Iwate Prefecture (Cabinet Office, Government of Japan, 2016). Evidently, urgent countermeasures to prevent suicide are needed in Fukushima (Maeda, Oe, Bromet, Yasumura, & Ohto, 2016). Given the significant relationship between suicide and alcohol abuse (Pompili et al., 2010), discouraging people from heavy drinking habits can be highly beneficial, especially for the prevention of suicide.

In addition, this study showed that smoking was significantly associated with drinking. In previous studies, being a male, being younger, having less educational attainment, changing jobs, and having traumatic symptoms were significantly associated with starting to smoke (Nakano et al., 2018). Both studies show that those who begin indulging in addictive behavior, that is, drinking and smoking, share very similar characteristics.

As described above, becoming a new drinker was correlated with age and gender, and younger males may be more prone to developing new, yet ineffective, coping strategies such as heavy drinking to face their difficulties. Introducing preventive approaches through medical professionals for this potentially high-risk group can be challenging because they do not develop physical illnesses that require medical attention or go for routine health check-ups as compared to the older population (Ministry of Health, Labour and Welfare, 2014). A study revealed that young people are especially vulnerable to alcohol marketing (Babor, Robaina, Noel, & Ritson, 2016). Therefore, collaborative efforts with the media and the alcohol industry, involving various types of educators, could be complementary measures to approach this group effectively.

According to the results, the newly developed problem of drinking was related to accumulated sleep problems and more severe traumatic symptoms. Experiencing a disaster, except loss of family or relatives and changing jobs, were not significantly related to the problem. The relationships between sleep problems, traumatic symptoms, disaster experiences, and changes in drinking behavior have been quite well researched. However, there has been no previous study on new drinkers. With its large sample size, this survey enabled us to analyze the behavioral change that occurred in those who had recently become problem drinkers and to examine how this was associated with other factors.

Prospective longitudinal research is required to monitor behavioral change over a longer period to see causality of factors such as drinking and sleep problems.

Fukushima "Alcohol Project"

The Fukushima Center for Disaster Mental Health (FCDMH) was founded in 2012 under contract with the Fukushima Committee for Mental Health and Welfare to provide comprehensive mental health support to those who suffered and continue to suffer from the aftermath of the Great East Japan Earthquake and the accident at the Fukushima Daiichi NPP. FCDMH has six branches spread across the Fukushima Prefecture, and consists of 50 multidisciplinary staff members, including psychiatrists, public health nurses, nurses, clinical psychologists, psychiatric social workers, certified social workers, and occupational therapists. FCDMH has provided a wide range of services such as medical consultation, outreach programs, and psychosocial education to local professionals and the public.

One of the programs in which the FCDMH has put a significant amount of time and energy is the Alcohol Project. It was launched in April 2014 to enhance community resources to attend to alcohol problems, and was sponsored by the Fukushima Prefecture.

At that time, many evacuees experienced drastic changes in their lifestyles and did not know when they would regain their usual life because of the nature of the NPP accident. It caused serious stress reactions and increased depressive symptoms. Many professionals who provided outreach services reported an increase in the number and complexity of cases related to alcohol problems, saying, "When we make home visits, clients are already tipsy and seem that they don't care." "Neighbors complain that there is a man who is always drunk, stays in temporary housing, and never comes out." "Clients said that there is nothing to do but drink, and there is no reason to live anymore, so I could only listen to them."

After the Hanshin-Awaji Earthquake, alcohol was found as one factor that leads to solitary death in temporary housing (Ueno, Asano, Nushida, Adachi, & Tatsuno, 1999). There was a concern in Fukushima 3 years after the disaster that alcohol problems had become more serious than what had happened in Kobe.

Under these circumstances, FCDMH collaborated with Fukushima Medical University and formed the Alcohol Project committee (Fukushima Center for Disaster Mental Health, 2014, 2015, 2016, 2017, 2018, 2019, 2020). The project aimed to: (1) train the community leaders on alcohol problems by providing training to local professionals regarding alcohol dependence and group supervision through case conferences, (2) organize open workshops for the community to raise awareness of alcohol problems, and (3) run a model project in a specific area to create an actual service network to help with alcohol problems because there was no medical facility specializing in alcohol treatment in Fukushima.

Trainings for local professionals

The FCDMH Alcohol Project provided face-to-face training and remote training using a TV conference system.

Face-to-face trainings were held in 3-h sessions, 2 days a year. Initially, the objectives of the trainings were to understand basic information on alcohol problems, and to learn about assessment, early intervention, and community approach. It was aimed to enhance the capacity of local professionals to deal with psychosocial problems, especially related to alcohol abuse, after a disaster. As time passed and the project developed, the objectives shifted to focus on the association between alcohol and health and lifestyle diseases, in which people tend to be more interested, and to make alcohol education part of health guidance. Also, they incorporated case presentation to learn practical techniques of the alcohol harm reduction model.

The participants in these workshops included medical and preventive health professionals, social welfare professionals, and local government workers; more than 800 of them attended these workshops over the years.

The topics of the trainings were as follows:

- Expand the support for alcohol problems; an alcohol harm reduction program, and how the actual treatment is done.
- How to support those with alcohol problems, an alcohol harm reduction program.
- Early intervention for alcohol problems; how to teach cutting back alcohol consumption for problem drinkers.
- 10-min education for alcohol harm reduction as a part of health guidance.
- Advanced course on alcohol harm reduction guidance with a case presentation.

According to the questionnaires filled out by the participants, they experienced difficulties such as:

- Intervene before clients develop dependence when there are no significant symptoms.
- Motivate clients to enter treatment.
- Refer clients to medical facilities when they refuse to use them.
- Support for those who live alone, without family and relatives.
- Relational problems among family members. Alcohol problems tend to be latent.
- Limitations on follow-ups via phone calls.
- No local medical facilities specializing in alcohol treatment.
- Support those who suffer from alcohol problems and depression.
- Support those who suffer from alcohol problems and personality disorder.
- Support middle-aged or older male clients who live alone; very difficult to change their habitual drinking and sustain the change.

- Support older adults in their 80s with signs of cognitive impairment.
- Support a client who gets hospitalized repeatedly and still goes out to local stores and causes a nuisance to buy alcohol.

The feedback from participants according to the questionnaires is as follows:

- Very helpful to learn about the harm reduction model.
- Harm reduction could be the way to reduce depression and prevent suicide.
- Able to practice harm reduction guidance through role play.
- Would like to try to motivate clients more to reduce alcohol intake.
- Able to understand the concept of "standard drink" and will utilize it at my work.
- Would like to learn more about how to use this technique during home visits.
- Became aware of my own drinking habit.
- I have worked with clients with severe alcohol dependence and became aware of problem drinkers through this workshop.
- I would like to disseminate this information at work.
- I would like to learn more about this approach and would like to attend more trainings.
- I have never thought about cutting back on alcohol as a part of health guidance. Very informative.
- Many of my clients say, "I hate to be rude to decline alcohol" or "We drink regularly as part of social activities in our neighborhood" as excuses. I was able to change my perspective and would like to start by giving out information.
- An alcohol harm reduction model seems less stressful than abstinence for both the clients and the clinicians. It would be easier for the clients to accept this approach.
- I realized that it is more important to respect the clients' own pace rather than try to change their lifestyle.
- I practiced alcohol harm reduction guidance with my family member after I attended this training last year, and it was successful. Your messages like "alcohol causes your brain to shrink and it shortens your lifespan" were very effective in my family.

The objectives of remote training were to learn effective intervention by learning techniques such as early intervention, motivational interviews, alcohol harm reduction programs, and building the capacity of local professionals to deal with alcohol problems. For remote trainings using a TV conference system, the National Hospital Organization Hizen Psychiatric Center took the initiative to create a network with hospitals leading in treating alcoholics, such as the National Hospital Organization Kurihama Medical and Addiction Center, the National Hospital Organization Ryukyu Hospital, and the National Hospital Organization Hanamaki Hospital.

The participants in the trainings were professionals who worked with local health promotion in the Iwaki area, where many evacuees moved and resources were stretched, and yet they were very far from other cities such as Fukushima or Koriyama. More than 30 training sessions have been held and more than 150 professionals have attended.

The topics of the trainings were as follows:

- Report on alcohol harm reduction guidance in the community.
- Hizen Alcoholism Prevention Program by Yuzuriha (HAPPY) at the Fukuoka municipal office.
- Alcohol intervention as a part of disaster support in Miyako City, Iwate.
- Group intervention on alcohol harm reduction.
- Alcohol harm reduction guidance without using the workbook.
- Sensible and Natural Alcoholism Prevention Program for You (SNAPPY) using a computer.
- Alcohol problems in disaster-affected areas.
- Evidence-based study on brief intervention at work.
- Alcohol problems in remote islands.
- Factors of HAPPY:
- Alcohol harm reduction guidance for local government workers on a remote island.
- Technical support for alcohol problems after the Great East Japan Earthquake.
- Preventive action in Okinawa.
- Alcohol and melanosis.
- Early intervention at Hanamaki Hospital.
- Early intervention with the Alcohol Use Disorders Identification Test (AUDIT).
- Alcohol harm reduction support among the communities having drink-for-work culture.
- Latest treatment on alcohol dependence; abstinence or harm reduction.
- Historical review of AUDIT.
- What is happening in the community and at the workplace regarding alcohol problems.
- Developing early intervention tools on the web; SNAPPY-CAT (computer advice technique).
- Comorbidity of addiction.
- Daijina hitono Abunai Seikatsu shuukan Henyou (lifestyle modification of those for whom you care, DASH) program.
- Alcohol Brief Counseling for Drinkers (ABCD) program (Figs. 1 and 2).

Open workshops for the community

In the first 2 years of the Alcohol Project, FCDMH organized open workshops for the local community on disseminating information on alcohol problems, including the dangers of drunk driving and health concerns. They also

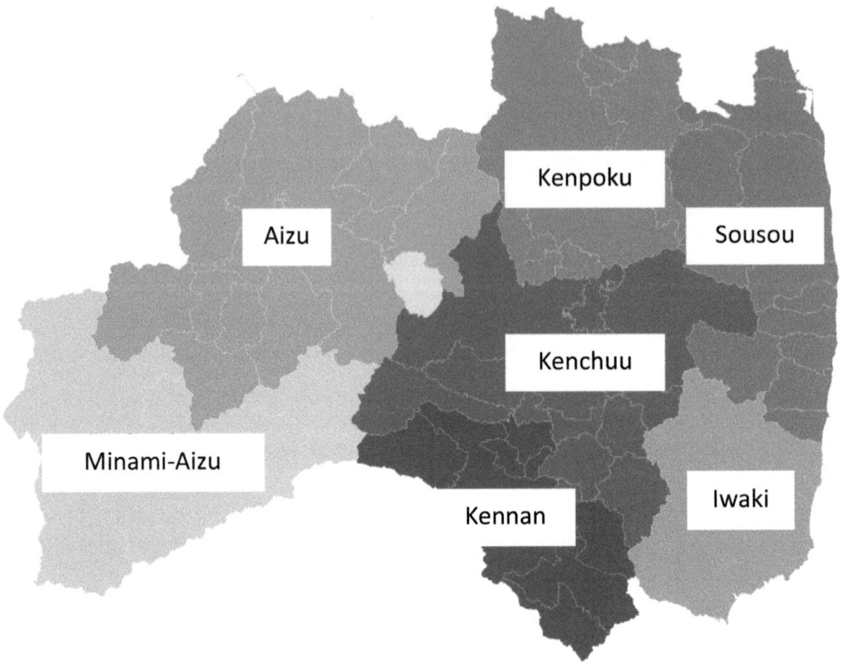

FIG. 1 Prefectual map of Fukushima.

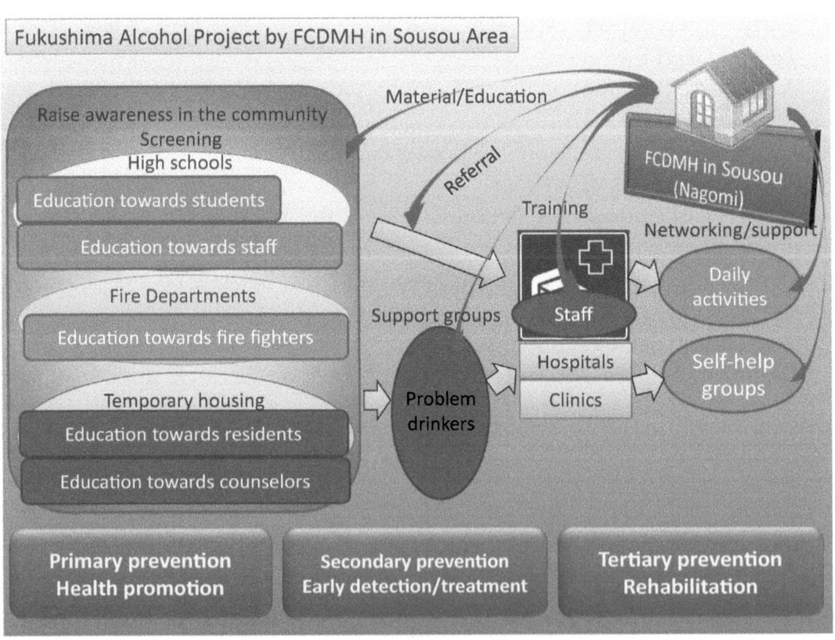

FIG. 2 Fukushima alcohol project by FCDMH in Sousou area.

introduced the idea of better ways of drinking, such as moderate drinking and tips for family members of those with alcohol problems.

From the third year, FCDMH broadened its approach and utilized different channels in the community. They participated in feedback sessions for health checkups sponsored by local governments, set up information booths at local events and health fairs, participated in the "war on drunk driving" campaign and handed out fliers, and developed educational leaflets on moderate drinking.

FCDMH staff also got dispatched as lecturers for family consultation sponsored by local governments. Topics included but were not limited to:

- What is Community Reinforcement and Family Training (CRAFT)?
- How to deal with violence.
- Understand the cycle of drinking and gambling.
- Communication training.
- Stop enabling.
- Enrich your own life.

After all these years, FCDMH was able to reach out to more than 7000 people.

Model projects in the local community

Fukushima Prefecture is the third largest of the 47 prefectures in Japan. It is 166 km (103 miles) in width from east to west, 133 km (83 miles) in length from north to south, and has an area of 13,781 km^2 (5321 sq. mi.). There were more than 2 million people living in Fukushima at the time of the disaster (Fukushima Prefecture, 2012), and it is ranked 40th of the 47 prefectures in terms of population density. As is customary, Fukushima Prefecture is divided into seven areas: Sousou (coastal north), Iwaki (coastal south), Kenpoku (north), Kenchuu (central), Kennan (south), Aizu (inland), and Minami-Aizu (inland south) (Fig. 3).

In Sousou, the Citizens Group to provide a new Psychiatric Care and Welfare System for SouSou (Nagomi) was formed in November 2011 to offer support to patients who were directly affected by the disaster and needed continuous treatment and psychological care. Nagomi is under contract with the Fukushima Committee for Mental Health and Welfare to provide services of the Alcohol Project with FCDMH.

The frame of the model project is as follows:

- As primary prevention, raise awareness among local residents and screen. Nagomi prepares materials and dispatches lecturers.
- At high schools, provide alcohol education directed toward students, teachers, and staff.
- At fire departments, provide alcohol education directed toward fire marshals.

FIG. 3 Outreach activities.

- At temporary housing, provide alcohol education directed toward the evacuees and the counselors.
- As secondary prevention, sponsor group activities for those who drink in large quantities, screened by the activities above. Nagomi facilitates the groups.
- As tertiary prevention, create a system and offer support. Nagomi sends lecturers to local hospitals such as Hibarigaoka Hospital to educate staff regarding alcohol problems. Nagomi also provides spaces to socialize during the day and sponsors meetings for the clients (Figs. 4–6).

Outreach for high school students

Nagomi chose high school students for their preventive intervention before they start habitual drinking, and started to collaborate with school nurses. Six high schools participated this project and agreed to have Nagomi contact their students and staff.

In Okawa's survey (Okawa, 2018), she distributed a questionnaire to 2241 people and received 1915 responses. Out of these, 973 (50.8%) were male and 942 (49.2%) were female. In terms of school grade, 674 (35.2%) were 10th graders, 609 (31.8%) were 11th graders, and 632 (33.0%) were 12th graders. Those who had experienced drinking were 312 (16.3%), and the breakdown

FIG. 4 Outreach activities.

FIG. 5 Outreach activities.

was as follows; 16.4% of male students and 16.1% of female students, 12.8% of 10th graders, 18.2% of 11th graders, and 18.2% of 12th graders. In terms of experience of disaster, 16.3% of those who did not evacuate, 15.5% of those who evacuated temporarily but returned home, and 21.7% of those who continued being evacuated at the time of the study had experienced drinking. In relation to the experience of loss of family and relatives due to disaster, 15.2% did not experience it while 18.2% of those who had lost loved ones had experienced drinking. Based on the Quantity Frequency Scale, 32 of 312 students

FIG. 6 Outreach activities.

(10.3%) who had experienced drinking exhibited a high risk of having health concerns. More than 75% of those who had experienced drinking obtained alcoholic beverages with family members' permission. The most popular reason for drinking alcohol for the first time was out of curiosity (47.1%), followed by being encouraged by family members (35.9%). Some of the students assumed that their family members had alcohol problems (9.2%) and were concerned about it (7.4%). In contrast, only 2.2% were concerned about their friends' alcohol problems. Only 22% knew about resources they could consult, and 72% did not know where to go to for help.

Okawa developed intervention programs for high school students, including three health and alcohol seminars at two high schools, using DVDs and graphics, and provided feedback on the questionnaire survey. One high school invited those in recovery as guest speakers. As a result, visual materials and evidence-based information were well received by students. Personal stories by those in recovery seemed to have a significant impact on the students. Okawa noted that it is critical to communicate not only with school nurses but with teachers and other school personnel prior to seminars as well as share awareness of alcohol-related problems. According to Okawa, to respond to the uniqueness of each school and community, it is important to prepare several types of materials and training contents and adapt these to specific needs.

Questionnaire and group intervention for public employees and helping professionals

As Maeda described (Maeda, 2016), one of the significant impacts of disasters we observed in Fukushima was that not only residents but also relief workers, public employees, and helping professionals were severely affected mentally. Many of the public employees experienced the disaster themselves, and yet they had to respond to the residents and were unable to attend to their own needs. Maeda's study revealed the serious mental status of the public employees in Fukushima, and he suggested providing adequate psychiatric intervention.

Ueda studied factors related to changes in drinking behavior among the evacuees who were helping professionals, using a questionnaire including disaster experiences, Japanese Patient Health Questionnaire Scale-9 (J-PHQ-9), and the Alcohol Use Disorders Identification Test (AUDIT) (Ueda, 2017). Of the 240 people who participated in the study, 139 (60.2%) were male and 92 (39.8%) were female; 9 (3.9%) were in their 20s, 18 (7.8%) were in their 30s, 43 (18.5%) were in their 40s, 54 (23.3%) were in their 50s, and 108 (46.6%) were over 60. In terms of disaster experience, 224 (93.3%) experienced the earthquake, 42 (17.5%) experienced the tsunami, and 182 (75.8%) experienced the NPP accident. Those who lost their family and relatives were 39 (16.3%), and those who were ordered to evacuate were 104 (46.8%).

Of those who drank habitually at the time of the study, 25 (17.1%) reported that they consumed more alcohol after the disaster. Their reasons for doing so were as follows: increase in occasions for social drinking (9, 36.0%); stress related to the living situation after the disaster (9, 36.0%); remedy for sleep problems (8, 32.0%); stress related to work (6, 24.0%); and lives alone with no one to keep them in check for drinking after the disaster (4, 16.0%). In contrast, 53 (36.3%) of those who drank less or quit drinking reported that they had fewer occasions for social drinking (30, 56.6%); were unable to drink as much due to aging (19, 35.8%); cut back due to health reasons (17, 32.1%); cut back due to pay cuts after changing jobs (15, 28.3%); were warned to cut back by family and relatives (5, 9.4%); and cut back due to serious alcohol-related problems after the disaster (1, 1.9%).

According to the AUDIT results, those who did not drink were 16 out of 176 (9.1%); those who were at low risk of alcohol dependence were 126 (71.6%); those who were at high risk of alcohol dependence were 29 (16.5%); and those with suspected alcohol dependence were 4 (3.8%). In terms of correlation with mental health, those with suspected alcohol dependence scored the worst on the J-PHQ-9, and it was significantly correlated.

With this result, Ueda provided group alcohol intervention programs based on HAPPY. HAPPY was created to help professionals to provide brief interventions for those who are at the high risk of developing alcohol dependence. The program consists of three sessions with lectures on adequate drinking and introduction of drinking journals to motivate participants to learn coping skills and set goals to cut alcohol consumption. In this study, the second session was done 2 weeks after the first session, and the third session was done 2 months after the second session. Ueda included topics such as "how to deal with alcohol during evacuation and relocation" and "how to manage residents who are heavy drinkers" to meet the needs of the participants. She also asked participants to answer a questionnaire for evaluation of the program as well as their alcohol consumption and health by AUDIT and J-PHQ-9.

There were 53 participants in HAPPY. Of those, 15 (31.3%) were male and 33 (68.3%) were female. One was in his/her 20s (2.0%), five (9.8%) were in their 30s, six (11.8%) were in their 40s, 12 (23.5%) were in their 50s, and 27 (52.9%) were over 60. In terms of disaster experience, 47 (88.7%) experienced the earthquake, 13 (24.5%) experienced the tsunami, and 39 (73.6%) experienced the NPP accident. Twelve (22.6%) lost their family and relatives and 26 (50.0%) were ordered to evacuate. According to the AUDIT results, those who did not drink were three of 36 (8.3%), those who were at low risk of alcohol dependence were 28 (77.8%), those who were at high risk of alcohol dependence were 5 (13.9%), and those with suspected alcohol dependence were 0 (0.0%).

According to questionnaires after the program, participants acquired knowledge on the health risks of drinking regardless of the AUDIT score. The average score of those who were at high risk of alcohol dependence became higher in "I want to cut back or quit drinking" after the program. The average amount of alcohol consumption among those who were at high risk of the alcohol dependence decreased to 2.4 drinks after the program compared to 4.1 drinks before the program. The average score of J-PHQ-9 among those who were at high risk of alcohol dependence improved after the program as well. The results of AUDIT and J-PHQ-9 were not significantly correlated.

Conclusion

Fukushima is well known for its rich natural environment, which provides good water, good rice, and especially good Japanese sake. According to the National Tax Agency, Fukushima was ranked fourth among 47 prefectures in consumption of sake per person in 2019 (Shokuhin Sangyo Newspaper, 2019). Alcohol is deeply embedded in people's daily lives in Fukushima. Under the circumstances, you could easily imagine how challenging it would be to introduce the idea of cutting back alcohol consumption.

However, the experience of getting through the major disaster actually gave an opportunity to change people's perception and behavior toward drinking in Fukushima. An analysis of large-scale research revealed that those who started drinking after the disaster presented more severe sleep problems and traumatic symptoms, and younger males were more prone to problem drinking as an ineffective coping strategy for their suffering. This result suggested that it is critical to intervene with new drinkers before they develop alcohol dependence after traumatic events and supported the idea of implementing early intervention provided in the community.

The Fukushima Alcohol Project done by FCDMH emphasized a community-based approach and collaboration with existing resources with integrated multilayer services to provide comprehensive support. Their activities included working with high schools, fire departments, local psychiatric hospitals, and public services. They also put time and energy in reaching out to residents who were at risk of problem drinking. The outreach project with high school students showed the importance of communication, not only with school nurses but also with teachers and other school personnel to provide information and services that fit with the specific needs of the community.

Public employees and helping professionals tend not to get attention as the target of support, but they are actually the ones who are in great need, especially after a major disaster. This is for two reasons: to be able to help residents more effectively and to get help for themselves. Presenting the idea of a harm reduction model by cutting back alcohol consumption relieved not only the residents but helped professionals because it appeared more attainable and realistic than complete abstinence.

Choosing a way to provide information in the manner that fits people's needs is the key for effective communication, and alcohol and problem drinking are topics that require a high level of sensitivity and awareness. Hopefully, the examples I presented will offer some insight for future research and practice.

References

Babor, T. F., Robaina, K., Noel, J. K., & Ritson, E. B. (2016). Vulnerability to alcohol-related problems: A policy brief with implications for the regulation of alcohol marketing. *Addiction*. https://doi.org/10.1111/add.13626 (Epub ahead of print).

Beaudoin, C. E. (2011). Hurricane Katrina: addictive behavior trends and predictors. *Public Health Reports*, *126*(3), 400–409.

Cabinet Office, Government of Japan. (2016, August 31). *Higashi nihon daishinsai ni kanrensuru jisatsushasuu, Heisei 27 nen* [Number of suicides related to Great East Japan Earthquake, 2015]. Available from: http://www.mhlw.go.jp/file/06-Seisakujouhou-12200000-Shakaiengokyokushougaihokenfukushibu/h27kakutei-shinsai_3.pdf.

Cerdá, M., Tracy, M., & Galea, S. (2011). A prospective population based study of changes in alcohol use and binge drinking after a mass traumatic event. *Drug and Alcohol Dependence*, *115*(1–2), 1–8 (Epub 2010 Oct 24) https://doi.org/10.1016/j.drugalcdep.2010.09.011.

Fukushima Center for Disaster Mental Health. (2014). *Fukushima Kokoro no Care Center Katsudo Kiroku 2014 Annual report of Fukushima center for disaster mental health*. https://kokoro-fukushima.org/activityreport/.

Fukushima Center for Disaster Mental Health. (2015). *Fukushima Kokoro no Care Center Katsudo Kiroku 2015 Annual report of Fukushima center for disaster mental health*. https://kokoro-fukushima.org/activityreport/.

Fukushima Center for Disaster Mental Health. (2016). *Fukushima Kokoro no Care Center Katsudo Kiroku 2016 Annual report of Fukushima center for disaster mental health*. https://kokoro-fukushima.org/activityreport/.

Fukushima Center for Disaster Mental Health. (2017). *Fukushima Kokoro no Care Center Katsudo Kiroku 2017 Annual report of Fukushima center for disaster mental health*. https://kokoro-fukushima.org/activityreport/.

Fukushima Center for Disaster Mental Health. (2018). *Fukushima Kokoro no Care Center Katsudo Kiroku 2018 Annual report of Fukushima center for disaster mental health*. https://kokoro-fukushima.org/activityreport/.

Fukushima Center for Disaster Mental Health. (2019). *Fukushima Kokoro no Care Center Katsudo Kiroku 2019 Annual report of Fukushima center for disaster mental health.* https://kokoro-fukushima.org/activityreport/.

Fukushima Center for Disaster Mental Health. (2020). *Fukushima Kokoro no Care Center Katsudo Kiroku 2020 Annual report of Fukushima center for disaster mental health.* https://kokoro-fukushima.org/activityreport/.

Fukushima Prefecture. (2012). *Census of Fukushima prefecture.*

Fukushima Prefecture. (2016). *Handout #4 for the 10th Prefectural Oversight Committee meeting for the Fukushima Health Management Survey: Report on Mental Health Lifestyle Survey.* (in Japanese). Available from: http://www.pref.fukushima.lg.jp/uploaded/attachment/6457.pdf (Cited 12 June 2016).

Grievink, L., van der Velden, P. G., Stellato, R. K., Dusseldorp, A., Gersons, B. P., Kleber, R. J., et al. (2007). A longitudinal comparative study of the physical and mental health problems of affected residents of the firework disaster Enschede, The Netherlands. *Public Health, 121*(5), 367–374 (Epub 2007 Feb 23).

Higuchi, S., & Saito, T. (2013). Inshuryo Teigen: Arukoru Izonshou no Chiryo Gouru [Reduction in alcohol consumption: Therapeutic goal in alcohol dependence treatment]. *Japanese Journal of Alcohol Studies & Drug Dependence, 48*, 17–31.

Keyes, K. M., Hatzenbuehler, M. L., & Hasin, D. S. (2011). Stressful life experiences, alcohol consumption, and alcohol use disorders: The epidemiologic evidence for four main types of stressors. *Psychopharmacology, 218*(1), 1–17. https://doi.org/10.1007/s00213-011-2236-1 (Epub 2011 Mar 5).

Kunii, Y., Suzuki, Y., Shiga, T., Yabe, H., Yasumura, S., Maeda, M., et al. (2016). Severe psychological distress of evacuees in evacuation zone caused by the Fukushima Daiichi Nuclear Power Plant accident: The Fukushima Health Management Survey. *PLoS ONE, 11*(7), e0158821 (eCollection 2016) https://doi.org/10.1371/journal.pone.0158821.

Lin, Y., Kikuchi, S., Tamakoshi, A., Wakai, K., Kawamura, T., Iso, H., et al. (2005). Alcohol consumption and mortality among middle-aged and elderly Japanese men and women. *Annals of Epidemiology, 15*(8), 590–597 (Epub 2005 Jan 23).

Maeda, M. (2016). Diagnostic interview study of the prevalence of depression among public employees engaged in long-term relief work in Fukushima. *Psychiatry and Clinical Neurosciences, 70*(9), 413–420. https://doi.org/10.1111/pcn.12414 (Epub 2016 Aug 2).

Maeda, M., Oe, M., Bromet, E., Yasumura, S., & Ohto, S. (2016). Fukushima, mental health and suicide. *Journal of Epidemiology and Community Health.* https://doi.org/10.1136/jech-2015-207086.

Ministry of Health, Labour and Welfare. (2014). *Health, labour and welfare whitepaper.* (in Japanese). Available from: http://www.mhlw.go.jp/wp/hakusyo/kousei/14/dl/1-02-1.pdf (Cited 10 January 2017).

Nakano, H., Ohira, T., Maeda, M., Yabe, H., Ohtsuru, A., Suzuki, Y., et al. (2018). Fukushima Health Management Survey Group. Associations of disaster-related and psychosocial factors with changes in smoking status after a disaster: A cross-sectional survey after the Great East Japan Earthquake. *BMJ Open, 8*(6). https://doi.org/10.1136/bmjopen-2017-018943, e018943.

National Hospital Organization Kurihama Medical and Addiction Center. (2011). *Saigaigo no inshu ni kansuru evidence* [Evidence on drinking behavior after disaster]. (Internet). Available from: http://www.kurihama-med.jp/shinsai/1_saigaigo.pdf (Cited 12 June 2016).

Norris, F. H., Friedman, M. J., Watson, P. J., et al. (2002). 60,000 disaster victims speak: Part I. An empirical review of the empirical literature, 1981–2001. *Psychiatry, 65*, 207–239.

Ohto, H., Maeda, M., Yabe, H., Yasumura, S., & Bromet, E. E. (2015). Suicide rates in the aftermath of the 2011 earthquake in Japan. *The Lancet*, *385*(9979), 1727. https://doi.org/10.1016/S0140-6736(15)60890-X.

Okawa, T. (2018). *A survey on revealing alcohol related problems of high school students in affected areas*. https://kaken.nii.ac.jp/en/file/KAKENHI-PROJECT-15K11886/15K11886seika.pdf.

Pompili, M., Serafini, G., Innamorati, M., Dominici, G., Ferracuti, S., Kotzalidis, G. D., et al. (2010). Suicidal behavior and alcohol abuse. *International Journal of Environmental Research and Public Health*, *7*(4), 1392–1431.

Shimizu, S. (2003). *Arukoru Kanren Mondai no Shakai-Byourigakuteki Kenkyu – Bunka, Rinsho, Seisaku* [Socio-pathological study on alcohol-related problems – Culture, clinical practice, and policy]. Tokyo: Minerva Shobo.

Shinfuku, N. (2002). Disaster mental health: Lessons learned from the Hanshin Awaji earthquake. *World Psychiatry*, *1*, 158–159.

Shokuhin Sangyo Newspaper. (2019, January 15). https://www.ssnp.co.jp/news/liquor/2019/01/2019-0125-1545-14.html.

Suzuki, Y., Tsutsumi, A., Fukasawa, M., Honma, H., Someya, T., & Kim, Y. (2011). Prevalence of mental disorders and suicidal thoughts among community-dwelling elderly adults 3 years after the Niigata-Chuetsu earthquake. *Journal of Epidemiology*, *21*, 144–150. https://doi.org/10.2188/jea.JE20100093.

Tokyo Fire Department. (2017). *Acute alcohol intoxication is your own concern*. (in Japanese). Available from: http://www.tfd.metro.tokyo.jp/lfe/kyuu-adv/201312/chudoku/ (Cited 9 January 2017).

Tracy, M., Norris, F. H., & Galea, S. (2011). Differences in the determinants of posttraumatic stress disorder and depression after a mass traumatic event. *Depression and Anxiety*, *28*, 666–675.

Ueda, Y. (2017).*Effects of a group alcohol intervention for evacuees after the Great East Japan Earthquake*. https://kaken.nii.ac.jp/en/grant/KAKENHI-PROJECT-15K17297/.

Ueno, Y., Asano, M., Nushida, H., Adachi, J., & Tatsuno, Y. (1999). "Solitary Deaths" in temporary housing units in the areas affected by the Southern Hyogo Earthquake. *Prehospital and Disaster Medicine*, *14*(S1). https://doi.org/10.1017/S1049023X00034385.

Wu, H. C., Chou, F. H., Su, C. Y., Tsai, K. Y., Ou-Yang, W. C., & Su, T. T. (2006). Survey of quality of life and related risk factors for a Taiwanese village population 3 years post-earthquake. *The Australian and New Zealand Journal of Psychiatry*, *40*, 355–361. https://doi.org/10.1080/j.1440-1614.2006.01802.x.

Yabe, H., Suzuki, Y., Mashiko, H., Nakayama, Y., Hisata, M., Niwa, S., et al. (2014). Psychological distress after the Great East Japan Earthquake and Fukushima Daiichi Nuclear Power Plant accident: Results of a mental health and lifestyle survey through the Fukushima Health Management Survey in FY2011 and FY2012. *Fukushima Journal of Medical Science*, *60*(1), 57–67.

Yagi, A., Maeda, M., Suzuki, Y., Yabe, H., Yasumura, S., Niwa, S., et al. (2020). Changes in drinking behavior among evacuees after the Fukushima Daiichi Nuclear Power Plant accident: The Fukushima Health Management Survey. *Fukushima Journal of Medical Science*, *66*(3), 133–142. https://doi.org/10.5387/fms.2019-24 (Epub 2020 Sep 25).

Yasumura, S., Hosoya, M., Yamashita, S., Kamiya, K., Abe, M., Akashi, M., et al. (2012). Study protocol for the Fukushima Health Management Survey. *Journal of Epidemiology*, *22*(5), 375–383. https://doi.org/10.2188/jea.JE20120105.

Chapter 11

Psychological impact on affected children in Fukushima

Tokio Uchiyama[a,b,c], Keiko Kawashima[b], Satomi Suzuki[c], and Yuka Karatani[d]

[a]*Clinical Psychology, Taisho University, Tokyo, Japan,* [b]*Fukushima Collage, Fukushima, Japan,* [c]*Institute of Counselling, Taisho University, Tokyo, Japan,* [d]*Faculty of Urban Science, Meijo University, Nagoya, Aichi, Japan*

Introduction

Ten years have passed since the Great East Japan Earthquake on March 11, 2011. The earthquake and the subsequent accident at the Fukushima Daiichi Nuclear Power Station (hereafter, referred to as the Fukushima nuclear disaster) still have a significant influence on life in Fukushima.

Fukushima Prefecture has suffered not only from damage to areas hit by the tsunami caused by the earthquake but also from the nuclear disaster and reputational damage. Along with the direct influence of radiation, the disaster also impacted various aspects of life, including a stable home environment that is essential for children, outdoor play, food selection, and changes in the school environment. With decontamination progress in Fukushima Prefecture, some of the evacuation orders have been lifted and an increasing number of residents are returning to Fukushima to rebuild their lives. Those born when the earthquake occurred are now in elementary school, those who were high school students are now adults, and some of them have become parents. Children who were in junior high school students are now in university or are working.

Psychological impact of the meltdown

The meltdown at Fukushima Daiichi Nuclear Power Station caused a problem: coping with anxiety regarding low-level radiation exposure (Maeda, Oe, & Suzuki, 2018). The government and power companies had been publicizing that a nuclear power station was safe, and many experts told the media that radiation would not leak from the reactors even if the power station was hit by the tsunami. However, as the nuclear authorities were making a categorical statement on TV that there would be no explosion, there was a hydrogen explosion

Health Effects of the Fukushima Nuclear Disaster. https://doi.org/10.1016/B978-0-12-824098-4.00005-2

(MEXT: Ministry of Education, 2011). Citizens started to harbor distrust in experts and government, and subsequently, started to voice their dissatisfaction not only with politicians and bureaucrats of the central government but also with officials on the ground, including local authorities, public health nurses, teachers, and doctors.

The government issued evacuation orders to designated areas, and families were evacuated from Fukushima even if they did not live in those areas. Specifically, families with children were more likely to be evacuated. Many evacuated from areas near the nuclear power station that were not designated as an order area to those in or outside the prefecture to coincide with the start of the new academic year in April 2011 or the summer break in August. They were considered to have voluntarily evacuated and treated differently regarding compensation from those who were ordered to evacuate.

Some experts such as doctors and nurses also evacuated with their family to other prefectures, which exacerbated residents' anxiety. Many medical and welfare professionals who evacuated have conflicting emotions and still blame themselves.

Anxiety about low-dose exposure to radiation spread throughout Fukushima Prefecture and surrounding areas. Many medical and nuclear experts expressed various opinions; some maintained it was safe for everyone while others suggested that children in the prefecture should be evacuated to other prefectures.

Outline of the early stage of evacuation

In the early morning on the day after the earthquake, an evacuation order was issued to those living within a 10-km radius from Daiichi Nuclear Power Station (Okuma-machi, Futaba-machi, Namie-machi, and Tomioka-machi). By the evening, the order was expanded to those living within a 20-km radius and on March 25, the order was issued to residents within a 20–30-km radius to evacuate (Shimura, Yamaguchi, Terada, Robert Svendsen, & Kunugita, 2015).

Many of those who evacuated in the initial stage invariably said, "I thought we could go back the next day, but it turned out to be the beginning of long-term and repeated evacuations." Many families with children evacuated without any preparation. On April 11, five municipalities outside a 20-km radius (Iitate-mura, Namie-machi, Katsurao-mura, Kawamata-machi, and parts of Minamisoma-shi) were designated as planned evacuation areas. Subsequently, Date-shi, Minamisoma-shi, and parts of Kawauchi-mura were designated as Specific Spots Recommended for Evacuation (SSRE, areas outside the restricted or the planned evacuation area where the accumulated amount of radiation for a year since the accident was estimated to exceed 20 mSv according to the Nuclear Emergency Response Headquarters). In these SSRE, no blanket evacuation order was issued and there was no restriction on economic activities, but pregnant women and families with children who were at more risk from radiation were encouraged to evacuate. In other words, the government's response was ambiguous. The designation was gradually lifted by December 2014.

We cannot ignore the mandatory evacuation and recommendation to evacuate when examining the impact of the nuclear disaster on the children of Fukushima. The number of evacuees living outside Fukushima was 29,359 as of November 2020, down from a peak of more than 164,000, and the evacuees are found not only in neighboring prefectures such as Ibaraki or Miyagi but also in Hokkaido (872 persons) and Okinawa (134 persons), showing that they are scattered across Japan.

The number of child evacuees under 18 years old, in and outside Fukushima, peaked at 30,968 in October 2012; as of 2018, this had dropped to 17,487, according to the government. Among those, 8096 are from Futaba-gun, 3654 from Minamisoma-shi, and 784 from Iwaki-shi, which means that coastal municipalities account for more than 70%. Moreover, from places far from the nuclear power station such as Fukushima-shi (1195) and Koriyama-shi (1571), thousands of children have evacuated (Fig. 1).

Children's mental health

The nuclear disaster has brought about changes to many aspects of children's lives (Otsuki, 2016). These changes have become more diverse and complex over time.

Children in general

In Fukushima Prefecture, relocation, changing schools, parent job situation, and moving houses have continued across the prefecture, which is not limited to the restricted and planned evacuation areas. In some areas, children could not play outdoors or use swimming pools for a long while. Even now, some parents do not let their children use a swimming pool. Many evacuees have moved more

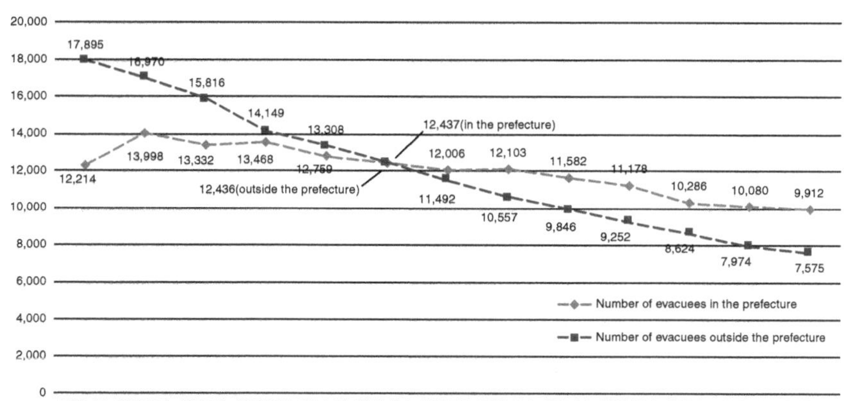

FIG. 1 The number of child evacuees due to the Great East Japan Earthquake (those recorded by municipalities) (from the Fukushima Prefecture website). *(From chrome-extension://efaidnbmnnnibpcajpcglclefindmkaj/viewer.html?pdfurl=http%3A%2F%2Fwww.pref.fukushima.lg.jp%2Fuploaded%2Fattachment%2F270489.pdf&clen=288201&chunk=true.)*

than once. Specifically, families with pregnant women and children were conflicted about whether they should evacuate, regardless of the evacuation order. The proportion of mothers having "negative feelings toward pregnancy" was significantly higher in Fukushima than other areas in Japan between 2011 and 2014. There was also a significant association between mothers' negative feelings toward pregnancy and depressive symptoms (Kuroda et al., 2021). Tsutsui et al. (Tsutsui, Ujiie, Takaya, & Tominaga, 2020) showed that even in 2015, psychological symptoms and radiation protection behaviors were higher for children and mothers within Fukushima Prefecture than for those in a control group living in regions outside the area. This suggests that the psychological effects in mothers and children living in low-dose radiation contaminated areas continued for at least 5 years after the accident. Thus, a situation under which children are exposed to strong stressors has continued in Fukushima Prefecture, and they have continuously experienced loss, which can lead to depression in the future. Specifically, children with developmental disorders have started to display disorder characteristics under stress, and their support needs have increased.

About evacuation and return (Do, 2020)

In adults, Murakami, Takebayashi, Ono, Kubota, and Tsubokura (2020) noted that those who had already returned had higher feelings of well-being than did those who did not know whether to return. As mentioned above, while there are children who are evacuees as of January 2021, more have returned to Fukushima. It is important to consider how these children evacuees perceive the "return" or "rebuilding life." For children who evacuated after the earthquake, *the life in evacuation is **a normal state***, and *the "return"* to a hometown or somewhere nearby and "*rebuilding life*" can be seen as a ***new change***. The length and destination of evacuation vary for children, and the environment in Fukushima has not returned to that of the prenuclear disaster period.

This is not a radiation issue. Many surveys revealed that for areas where the evacuation orders have been rescinded, radiation levels have fallen sufficiently and living there is permitted.

Hence, even if people wanted to regain their former life, it has been lost. For families whose houses were not damaged, living near a nuclear power station means choosing to live in a changed world. It is not the direct effect of radiation but the fear of radiation, propagated by media and some experts, that has made this problem more complex.

Interviews

To understand how children were coping, we interviewed 30 parents of children with developmental disorders and their supporters in the Hamadori region. We focused on children with developmental disorders to understand the demand for support but while conducting interviews, we learned about the state of parents of children without developmental disorders or local supporters. We interviewed parents and supporters who were from Futuba-gun and had encountered the disaster near the nuclear power station.

We interviewed them about their life since the disaster, the support they needed, and their future prospects. They actively and willingly shared their thoughts with us.

Interview data were divided into three categories:

1. In the immediate aftermath of the nuclear disaster (moving to evacuation centers such as the gym of a nearby school).
2. During evacuation (living in temporary housing and so on).
3. After the return and rebuilding life (returning to the hometown, building a life in a new place).

The classification has helped us comprehend that in a large-scale disaster situation, which could reoccur, it is important for support recipients and providers to have an idea about their current stage and future prospects to maintain their mental health. It is important to organize information chronologically.

Characteristics of Futaba-gun

The main industry of Futaba-gun, where Fukushima Daiichi Nuclear Power Station is located, is agriculture and fishery, as it faces the Pacific Ocean and has a temperate climate. However, it is also a major source of electricity in Japan with six nuclear reactors at the Fukushima Daiichi Nuclear Power Station, four at the Fukushima Daini Nuclear Power Station, and five turbines at the Hirono Thermoelectric Power Plant. Due to the accident at the power station, the majority of the areas were designated as evacuation areas and many residents were evacuated. This consisted of eight municipalities (six "machi" and two "mura") with a total population of 72,818 and a total of 25,466 households (2010 Census). Its total area is 865.12 km^2 stretching 30 km from east to west and 40 km from north to south. Many areas in Futaba-gun were hit by the tsunami that was higher than 10 m. Most of the residents were engaged in agriculture or fishery, or with work related to the nuclear power stations.

Evacuation process

Municipalities near the nuclear power station were small, and relatively intensive support was provided to children, mothers, and the elderly by public health nurses who visited them at home. We interviewed the public health nurses engaged with mother-and-child health support in the town halls in Futaba-gun. To protect the anonymity of the nurses, we ensured that the individuals or municipalities could not be identified while discussing the interviews with them.

There was confusion regarding the command system in the immediate aftermath of the earthquake. The municipalities near the nuclear power station were supposed to obtain information from the prefectural emergency broadcast system. However, as the prefectural office was damaged by the earthquake, the system that had been recently updated did not function well. Town hall officials on the ground had to order the evacuation of residents without sufficient information from the central or prefectural governments about details of the state of the nuclear power station or the reason for evacuation. One of the officials told us they were instructed "to evacuate to the west" by someone and acted on it based on their instinct. In Katsurao-mura, which is the farthest from the nuclear power station among municipalities of Futaba-gun, the mayor decided not to evacuate until they received orders from the central or prefectural governments. At 9:05 p.m. on March 14, someone who had access to the fire service's radio system informed the town hall that "personnel have evacuated from the offsite center (local emergency response headquarters)," and at 9:15 p.m., the mayor decided to evacuate. They left the town hall at 10:15 p.m. and broadcast the decision to the residents with sirens. Under these circumstances, local authority staff were separated from their families, guided the residents to evacuation centers, and worked there to support the residents. Although the staff was affected by the disaster, they often became the target of the residents' complaints and anger. In the beginning, Katsurao-mura accepted evacuees from Okuma-machi and Tomioka-machi, but insufficient information led to confusion. The confusion was even more severe in areas closer to the nuclear power station.

It is important to build a system that maintains the physical and mental health of local authority staff who run the evacuation centers. One of the interviewees said that when a doctor who came to provide support at the evacuation center told them to take a rest, they felt for the first time that they needed to rest. In an emergency, *it is important to maintain the supporters' health.*

Communication with residents

On the day after the earthquake, a public health nurse felt caring for pregnant women should be prioritized and went to the town hall to call them using the register. However, as it only had home phone numbers, she could not speak to them. She then learned that mobile phone numbers should be recorded for an emergency, but it was too late.

Meanwhile, in a small village in the mountainous area where residents were aging fast, Internet protocol telephony (IP telephony) had been introduced to all homes. While telephone connections were unstable after the earthquake, IP telephony lines were stable even after the earthquake, and the village officials could confirm the safety of all residents and quickly guide them to the evacuation centers.

Many public health nurses who worked at the town hall became staff of the first-aid station set up at the evacuation center. They did not treat infants or children often but were busy dealing with infections such as influenza and gastroenteritis. As the evacuation center was more than 10 km away from the town hall, they found it difficult to work in an area that was unknown to them. Under these circumstances, *help from the local public health nurses was effective*. Many commented that information on medical or welfare services in the locality as well as support was helpful. Moreover, nearby hospitals sent out *the same doctors regularly* to work in the evacuation center and they also accepted some in-patients. *It is also important that the local authority from the evacuees' areas effectively liaise with that of the areas where they moved.*

With time, support for children with developmental disorders was explored. At each evacuation center, parents and children were supported by the staff's ingenuity. Not all children with developmental disorders had been diagnosed, and no one reported to the evacuation center.

At an evacuation center, a play area for all children was set up and monitored by staff. Through watching them play, staff identified children with developmental disorders and provided support. At another evacuation center, parents were asked to register their mobile phone numbers so that information about services for children could be disseminated rapidly. An evacuation center assigned a staff member to support children who appeared to have a developmental disorder. Many evacuation centers were set up in and outside the prefecture, and the range of social resources available for each evacuation center differed. Staff at each enter continued to support children by generating their ideas.

Interviews with parents of children with developmental disorders

In the immediate aftermath of the earthquake

Many mothers continued to work while some went back home after the earthquake and others evacuated to a nearby center. However, as the mobile phone network was down, they could not communicate with family members. In many areas where roads were destroyed or closed due to the earthquake or tsunami, people had to spend a long time evacuating by car. As family members could not be updated about each other's locations, they had to think and act for themselves. As a result, a few family members went to collect children from the kindergarten or schools in the car only to return home after not finding them. *It is important to have an agreement in the family before a disaster regarding the person responsible for picking up children during an emergency and the mode of communication.* After the earthquake, many institutions started to provide training to ensure communication between parents and teachers/nursery staff and to ensure the smooth handover of children in an emergency.

Complex evacuation

In the immediate aftermath of the nuclear disaster, the evacuation area was designated based on the understanding that radiation would spread concentrically. However, radioactive material was scattered in the form of radioactive clouds, and it did not spread concentrically because of the climate and topography (Diagram 1). Due to the dearth of information at the time of the nuclear disaster,

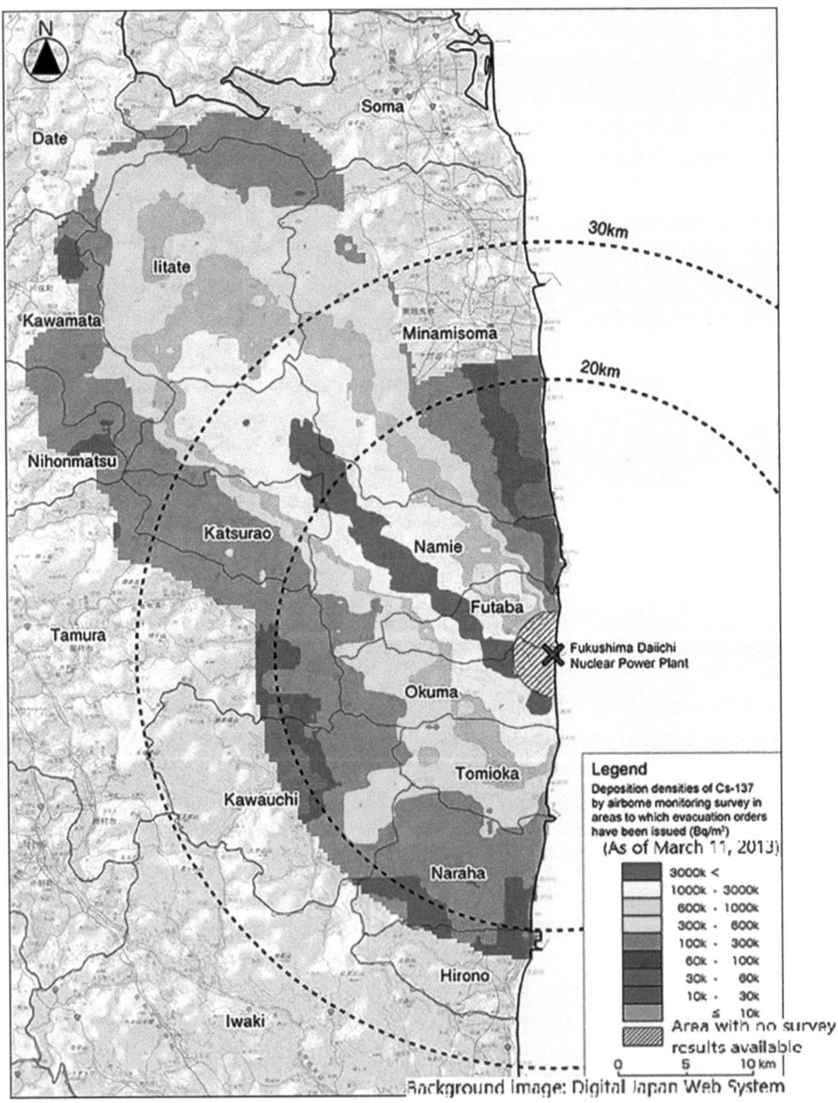

DIAGRAM 1 The location of Futaba-gun and the amount of radiation (converted as of March 11, 2013).

residents tried to move as far away from the nuclear power station as possible. As a result, some ended up evacuating to the area with a higher level of radiation than the place where they began. Some families who evacuated with children still carry self-blame and guilt.

A major feature of the evacuation due to the nuclear disaster is that confusion in the evacuation command system and information led to a repeated change of the destination. In the beginning, residents were ordered to evacuate as far away as possible from the nuclear power station. As a result, on the first day, residents went to a nearby evacuation center and then to an evacuation center more than 20 km away from the nuclear power station. As the order was repeatedly changed, they had to move without any preparation. Transport for evacuation varied; some used buses organized by the town hall while others used their vehicles. Not all families evacuated together. Many families were separated at the time of evacuation: in some families, while fathers remained at work, mothers and children evacuated and in others, grandparents evacuated separately from the younger generation. It was difficult for the entire family to live in a confined space at the evacuation center. Consequently, sometimes, grandparents were left at the center while the younger generation moved to a farther evacuation center.

There were pregnant women but as this was unclear due to the early stages of pregnancy, some reported that it was difficult to receive consideration or support at the evacuation center. Specifically, meals at the evacuation center were not nutritious and some needed to be mindful as to not become anemic. *This shows that in preparing for a future disaster, it is not enough to provide food. Nutrition should also be considered.*

Many families with children with developmental disorders did not use the evacuation center but lived in their car because they worried that their children would be noisy and disruptive. They judged that living in their car would be more comfortable. As many evacuation centers were based in a big gym or auditorium, it was tempting for children to run around, scream, or jump from a higher place. As children with autism spectrum disorder (ASD) or attention-deficit hyperactivity disorder (ADHD), whose disability is not explicit, were not provided a secluded space, many parents left the evacuation center worrying about the reactions of other evacuees.

Because there will always be some evacuees who have to live in their car, it is important to ensure there is no inequality in services provided for evacuees. The support system should be adjusted to provide a large parking space dedicated to those who live in their car, an outdoor supply station, and information to those living in their car.

However, there was a report regarding children with severe autism with learning difficulties that when the diagnosis and the need for a separate room were communicated at the evacuation center, individual support was provided. Moreover, some parents looked for evacuation centers that were sympathetic through information exchange (social media) among mothers of children with developmental disorders. *The building of a communication network and networking in normal times can be very useful in a time of crisis as well.*

Some parents of children with developmental disorders brought their children's favorite games and toys to the evacuation center to calm them. They also said that it was important to ensure that their siblings also have their favorite items. This is because controlling the level of stress of the children with developmental disorders and their siblings contributes to a comfortable environment for the family.

If the frequency of children fighting decreases, adult scolding also decreases, which reduces children's crying. They do not have to worry about the reactions of other evacuees. All family members might be able to spend time peacefully. *Support for siblings is equally important as that for children with developmental disorders.*

As we interviewed the support providers and recipients, we found that support provided at the evacuation center improved from the initial confused state due to trial-and-error by the staff on the ground. The support providers whom we interviewed were professionals dealing with mother-and-child health and developmental disorder support. It would be useful to develop a framework that enables the staff to make the most of their expertise at the evacuation center, to quickly share information about effective support, and to send professionals with experience in dealing with disaster to the affected areas in an emergency or disaster. Some of the interviewees told us they would work better next time.

Prolonged evacuation

What is life as an evacuee?

When the evacuation becomes prolonged, some evacuees experience major life events such as marriage, pregnancy, giving birth, and child rearing at the place of evacuation.

While there is some variation according to the local authority, in Japan when pregnancy is confirmed, women often apply for the mother and child health handbook at the local authority. Upon receiving a handbook, they have check-ups at the hospital and join antenatal classes for mothers or fathers and wait for the child's birth. In other words, they are engaged with the local professionals and public health nurses from the early stages of the pregnancy. After the birth, each municipality carries out a comprehensive home visit as part of the infant project ("Hello, Baby"). In this system, staff with a specialist qualification such as a public health nurse or midwife would visit all families with an infant up to 4 months old. After this, parents and infants are invited to the infant health check-ups held at the local health center at the age of 4 months, 10 months, 1.5 years, and 3 years. Moreover, many local authorities provide a dental check-up for 2 year olds. This means parents and infants would see the local public health nurse once every few months. If they have any concerns about their children, they can consult with the public health nurse at the health center. If the child is certified as needing support, a dedicated public health nurse is appointed who continuously provides support. In the field of mother-and-child health in Japan, these measures have been taken.

However, when the nuclear disaster triggered evacuation, the local authority moved to the place of evacuation while the residents were scattered, which meant an infant health check-up could not be carried out. Consequently, the government put in place the Act on Special Measures for the Evacuees due to the Nuclear Disaster (hereafter, the Special Measures Act) so that children could receive an infant health check-up and immunization at the place of evacuation. However, when the infant health check-up identified a developmental disorder, it was often referred to the local authority of origin and not followed up by the local authority of the place of evacuation. This means problems were likely to emerge about the person responsible for providing support for the family, the public health nurse of the place of origin, or that of the place of evacuation. As for the service provided by the local authority where residents have evacuated, there might be limited service for evacuees. It is necessary to meet the mother to understand the child's situation to refer the child for appropriate treatment, but as there is a lack of clarity regarding the person responsible and as parents do not know whom to consult with, they might end up being isolated.

Parents who need a high level of support

Fukushima Prefecture set up outreach posts in the areas where many evacuees live to support parents and children. The demand for support has not decreased since the immediate aftermath of the earthquake and now with more complex problems, there is a need to support parents and children.

Various kinds of support are provided. Moreover, mothers are increasingly diagnosed with a developmental disorder. This is because while mothers with a mild developmental disorder had managed before the earthquake, the level of stress has increased because they no longer have help from grandparents or their husbands or have to adapt to a new environment. It is necessary to improve the situation for parents and children by assessing parents' development and mental health and by providing welfare services (including help for household chores and childcare) based on the assessment. The demand for support regarding developmental disorders has become more complex and diverse since the earthquake.

Reputational damage and discrimination: Support by a homogeneous group

Not all evacuees have been provided support; in many cases, mothers are worried about delay and inconsistency in their children's development. Many evacuees do not want to be known as that in the place to which they have evacuated.

The reason for this varies. It could be inequality in compensation for the nuclear disaster, or prejudice and discrimination due to the radiation. As a result, mothers tend to hide that they are evacuees and avoid making friends or communicating with neighbors in the new area. They are even worried about the home visit by the public health nurse of the place of evacuation. Some were worried that their mail might carry their old address. The impact of such long-term stress on families is concerning.

We met Mother A who was worried about her child's development delay but she could not speak to anyone until the public health nurse from her hometown whom she knew visited. Although about once a month a public health nurse sent from a welfare organization visited her and asked how she was doing and if she needed help, she could not confide in them. The support provider might have been able to ensure the family was functioning but could not provide appropriate support. It is difficult for any worried mother to consult someone she does not know.

As they are exposed to stress such as severe worry and anxiety for a long time while being evacuated, they tend to complain about physical symptoms such as "body aches" or say "I easily get tired" and "I cannot do household chores." It is important to gradually build trust with mothers by accepting and empathizing with their complaints. As the evacuees tend to have a higher level of anxiety and tension in general due to evacuation, prejudice, discrimination, and conflict about compensation along with the direct influence of the disaster, it is necessary to respond to them carefully. While it may take time, communicating with mothers is often the quickest way to provide support for children with developmental disorders.

Newly revealed problems

In a case, a withdrawn adult male had not been known to the authorities until the evacuation. He had lived with his elderly parents but neither their neighbors nor the local town hall knew he was withdrawn to the house. This was revealed when all residents evacuated due to the nuclear disaster and when the local authority checked the safety of the residents. Some withdrawn people have developmental disorders or mental health issues, and it is important to provide appropriate support.

Response of local authorities

As soon as a disaster hits, the local authority has to ensure the citizens gather at a common place and meet them there. In many areas, the evacuees made a phone call to the municipality call centers to register their new address and phone number. It was difficult to collect this kind of information unless the evacuees got in touch. The local authorities that managed to collect information about their residents set up an outpost and temporary schools in areas where many residents evacuated, and arranged a school bus to ensure that evacuated residents could live with some convenience and comfort. Many of the officials working at the town hall were busy with work that was different from usual even though they were qualified public health nurses. In the immediate aftermath of the earthquake, securing minimum food, water, and places to sleep were top priorities. However, as the evacuation was prolonged, securing housing to accommodate residents, setting up schools and the town hall function, and securing appropriate personnel became more important. The interviewees said, "public health nurses on a long-term secondment." In other words, external support was very useful. There were some problems when many supporters came from outside as the local officials were exhausted after dealing with them. "Personnel on a long-term secondment" is useful in an affected area. External supporters coming in on a fortnightly, monthly, or yearly basis and voluntarily organizing "work handover" when the external supporters leave was welcomed. In an emergency, people who can think and make decisions for themselves are required. Moreover, they should not be insistent about their way of doing things, but to learn the local ways and to work collaboratively. In other words, because external supporters are required to have a high level of skills including communication skills, they tend to develop conflicts and stress as they provide support on a long-term basis. To maintain support activities for places with high demand, it is important to build a support system for external supporters.

Characteristics of nuclear disaster evacuees as of 2020

One of the characteristics of the evacuation was that they were forced to change destinations repeatedly. Many evacuees first evacuated to the nearby evacuation center in the immediate aftermath of the earthquake, then moved to an evacuation center that was more than 20km away from the nuclear disaster point. After approximately 1month, they moved to inns and hotels to prepare for a prolonged evacuation, followed by life in temporary housing or accommodation rented by the local authority.

A questionnaire about evacuation administered to parents of children with developmental disorders found that the most frequent response was that they had to change the evacuation destination four times or more. The affected area used to have many extended families with grandparents living together, but evacuation has accelerated the shift to nuclear families. This means no help from grandparents was available for childcare.

The decision on the ultimate destination of evacuation depends on what each evacuee perceives as important for their situation. The amount of radiation, convenience, links with the hometown, and children's educational environment were considered, and we have identified the following four patterns (Table 1):

- Pattern1(U-turn): Hometown ⇥ Hometown (returning home after the evacuation order was lifted) (Fig. 2).
- Pattern 2 (J-turn): Hometown ⇥ A place near the hometown with a high degree of convenience (Fig. 3).

TABLE 1 Evacuation patterns.

	Area characteristics	Community interaction	Convenience (shopping, hospitals, schools etc.)	Nuclear families	School characteristics (special support)	Access to welfare services	High-risk children
U-turn	Home	Good	Poor	Predominant	Small class size (available)	Poor	Many
J-turn	Near home, similar culture	Strained	Adequate	Predominant	Large class size (few)	Poor	Many
I-turn	Distant from home, differing in topography and local government policies	Normal	Adequate	Predominant	Chosen by carers	Adequate	Unknown
II-turn (move far away)	New area in another prefecture, but few concerns about radiation	Difficult	Adequate	Predominant	Large class size (few)	Adequate	Unknown

U-turn

Characteristics Many children with ASD find it hard to cope in the evacuation area and parents and children require support. Although there is no nearby ASD treatment service, pre-school and elementary school-class sizes are small enabling extra personalized care.
One advantage: Not overcrowded area so with few problems with neighbors.

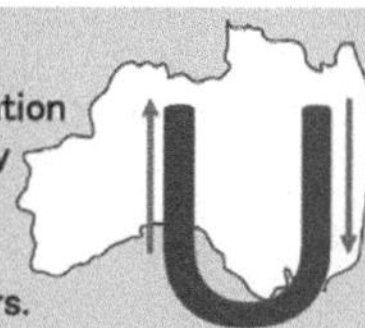

Example Two siblings diagnosed with ASD and attending special schools, in year 3 of elementary school and year 1 of middle school respectively. Their father remained alone for work, but they and their mother were evacuated to locations within and outside the prefecture. In temporary accommodation when they were younger, there were complaints about noise and difficulty getting them to and from two separate schools. Their mother found coping difficult but she was reluctant to ask for support. A support professional arranged escorting and social services. Family life regained stability only when they returned to their old home and the children started at special schools.

FIG. 2 Evacuation pattern: U-turn.

J-turn

Characteristics This type involves the most problems. The resettlement area has the attractions of a coastal location, similar to their hometown, and urban conveniences of good transport, shopping etc. However, the local residents often resent the compensation received by evacuees, causing friction. Both parents and children may experience mental health problems. Health and social services may be unable to cope with the marked rise in population due to the resettlement (see Graph 1).

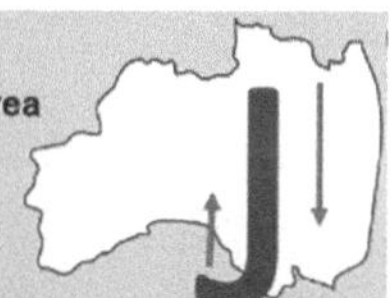

Example Boy in year 3 of special class in elementary school at the evacuation destination. New houses were built for resettlement. Originally a large family but split up. When revealed as evacuees they were shunned by their neighbors. His mother felt unable to tell other mothers that they were evacuees and was unable to speak Openly to the child's public health nurse. Only when a public health nurse from their home area telephoned was she able to discuss his problems. She later found friends and confidants when her son began to use an evacuees-only educational facility.

FIG. 3 Evacuation pattern: J-turn.

- Pattern 3 (I-turn): Hometown ⍈ Moving to where the local authority relocated (Fig. 4).
- Patter 4 (II-turn): Hometown ⍈ A place that is far away from the nuclear disaster (outside the prefecture) (Fig. 5).

Characteristics and typical cases are shown in Table 1 and Figs. 2–5.

We would like to draw the reader's attention to Pattern 4 (J-turn). Many evacuees from Futaba-gun chose J-turn. It was because the new place was close to their hometown with similar weather and climate, and with a higher degree of convenience such as ease of transport and shopping. However, there was a serious hidden problem: conflict among residents due to inequality in compensation for the nuclear disaster (Uchiyama, Kawashima, & Karatani, 2018).

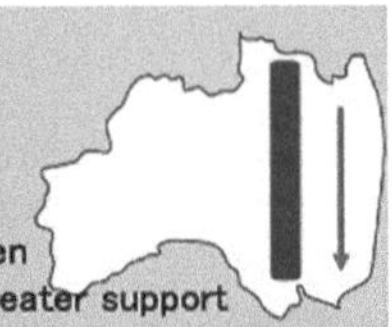

I-turn

Characteristics An entire population was moved from the EOZ with local government functions transferred at the same time. Since the community has been kept together, people find familiar faces in the temporary accommodation but there is little contact with the local population. Children who can adjust well attend local elementary schools but those requiring greater support often choose to attend pre-school and elementary schools run by the EOZ town government.

Example Girl, currently year 2 of elementary school, diagnosed with ASD at specialized consultation after evacuation and attends an ASD treatment service in the evacuation area. She has also attended pre-school and elementary school run by the EOZ town government where she received support. Since there is a lack of special school provision, she has had to attend a regular elementary school but since this has a good pupil-teacher ratio there are no problems. She uses local ASD treatment services and medical facilities which share information with health and other professionals from her previous home area.

FIG. 4 Evacuation pattern: I-turn.

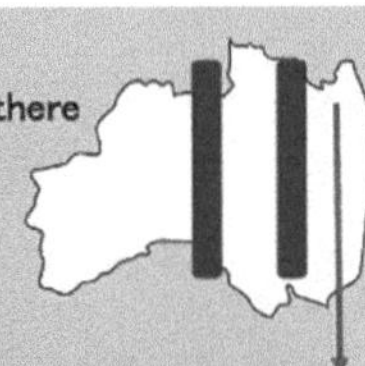

II-turn

Characteristics many evacuees have moved to another prefecture where there are few concerns about radiation. They do not reveal they are evacuees unless absolutely necessary. Later however, the lack of people who sympathize and those who can offer appropriate support has caused problems with bullying and school refusal and several families have returned to Fukushima.

Example Three siblings in years 1, 3 and 5 of elementary school. Three years after their initial evacuation within Fukushima Prefecture, they moved to another prefecture and were resettled in newly-built accommodation. The first and third children were diagnosed with developmental disorders before the move to the other prefecture and the second child was noted as being of concern before the move but was only offered standard support by the kindergarten teacher. After the move the second child refused to attend school but no diagnosis was made by a local clinic. A consultation with specialists from the Fukushima EOZ led to a diagnosis of ASD, resulting in the provision of support by her elementary school. Information from the EOZ was extremely important in understanding the course of her development from infancy.

FIG. 5 Evacuation pattern: II-turn.

Radiation is invisible, and it is difficult to measure the degree of damage appropriately and quantitatively. However, the authorities have to issue an evacuation order to the designated area and compensate the evacuees. It is impossible to draw a line on which everyone agrees, and the difference in the distance from the Daiichi Nuclear Power Station has resulted in a huge variation in the size of compensation. Such inequality in compensation has made relationships among residents difficult. This is one of the reasons why evacuees had to hide their identity. A mother raising children said that she lived with anxiety and stress that she would be spoken ill of if she got a new bag. Moreover, to not reveal that they were evacuees, they avoided being friendly with other mothers and children at school. Therefore, all family members had to live as if they were guilty of something. Some evacuees have built a new house in the place of evacuation to start a new life but are psychologically suffering as their relationship with the neighbors became tense when it was revealed that they were evacuees. For mental health care, it is important to build a system through which accurate information about the nuclear power station, radiation, and the life of evacuees is shared by all to prevent prejudice and discrimination.

It appears that stress experienced by the evacuees due to the nuclear disaster subsequently develops the following three-layer structure:

- Primary stress (experience of the disaster).
- Secondary stress (change in the living environment due to evacuation).
- Tertiary stress (anxiety, worries, and difficulties during evacuation).

The nuclear disaster's impact on the mental health of parents and children has become more complex, and it will be a long-term effect. Ten years have passed but it has not been resolved.

Toward the return/rebuilding of life: Isolation

As Diagram 2 shows, a characteristic of evacuation is a prolonged period in which residential registration does not match their current address. This is particularly common among those who chose J-turn. Many people have started to rebuild their lives in a new place but are still conflicted about residential registration and permanent residence. As described above, as the residential registration of those who are still evacuated does not often match their current address, they tend to miss out on services provided by local authorities (mother-and-child health, welfare, education, and so on).

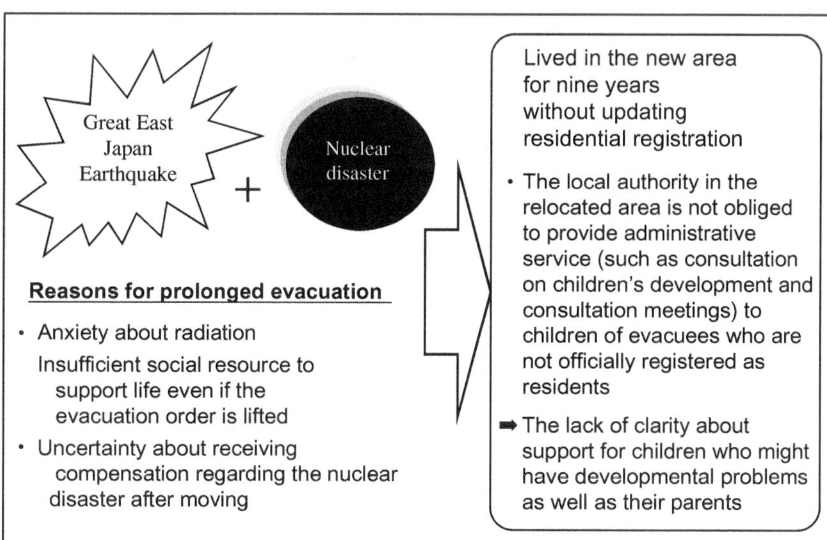

DIAGRAM 2 Problems in providing support for people with developmental disorders among evacuees.

Marriage, pregnancy, giving birth, and child rearing in evacuation

In normal times, mothers can go to a nearby health center whenever they want to consult with a familiar public health nurse. Getting to know the public health nurse makes it easier to ask for advice. Having a professional nearby who helps with raising the child reduces anxiety for parents and the risk of maltreatment of children. Mothers and children who are still evacuated find it difficult to receive this support.

If residential registration remains in the hometown, it is difficult to receive services from the local authority where they have evacuated. Some local authorities provide a wide range of services to evacuees, but it is difficult for areas that have accepted a large number of evacuees. For example, places like Iwaki-shi, which are likely chosen in J-turn, have seen a rapid increase in population due to evacuation since around 2014 and various social resources have started to run out. Furthermore, when evacuation is prolonged, the family structure changes. There is a report that the proportion of births by evacuees over a year accounts for about 10% of the annual number of births in the area to which evacuees have moved. The central and prefectural governments have provided support such as sending personnel but the provision of professionals and support for medical institutions are not sufficient.

There are cases wherein a person with an address in the evacuation order area marries someone whose address is outside these areas. For example, if a man who has evacuated from the evacuation order area marries a woman who lives in a municipality that is not in the evacuation area, and if she joins his family register, she becomes a resident of the evacuation order area. Childcare service has to be provided by the local authority from the area the woman has evacuated but as the municipality is far away, she receives minimum service such as an infant health check-up if she has evacuated under the Special Measures Act. Subsequently, if she finds there is a problem regarding the child's development, she will have to consult with the public health nurse in her husband's hometown. The mother has to communicate with a public health nurse she does not know and who is in a faraway area, probably by phone, regarding very sensitive topics such as the child's development and worries about childcare. The problem of distance hinders the building of trust between the mother and the public health nurse and support for the family would also be delayed. As more time is needed to solve the problem, it may develop into a serious problem.

There is a lack of regular communication between mothers and the kindergartens, schools as well as care institutions where children attend in the area of evacuation. So the support providers such as local authorities have to spend more time and energy than normal because they have to liaise with various institutions by sharing information and holding a case conference.

The residents have evacuated to a wide range of areas from the neighboring municipality to Aizu, Tokyo, and Okinawa. It is impossible to send outreach personnel to or visit all the areas. The biggest challenge here is to decide who will primarily be in charge of support provision.

Family's mental health

We have been conducting psychological assessments such as development and child psychiatric assessments as part of our support for evacuated children. There are many cases in which family members such as mothers, fathers, and grandparents are found to have mental health issues, characteristics of developmental disorders, and learning difficulties. It appears that if parents or children are poor at adapting to the place of evacuation, they tend to go back home even if their home is close to the nuclear power station.

For example, we encountered a case in which the mother's child rearing was close to maltreatment and she had problems in carrying out household chores, managing money and work, and maintaining a relationship with neighbors. We carried out an assessment focusing on the child's problems, but we also suspected that the mother might have a developmental disorder. Therefore, we assessed the mother, which showed that she had mild intellectual disability and ASD characteristics. By providing welfare service to the mother (help with household chores), the health of the mother and child improved. This is a case in which medical diagnosis has identified the necessary support, improved the relationship between the parents and supporter, and facilitated smooth provision. This is not a rare case.

Before the disaster struck, parents who needed support may not have needed diagnosis or support because there was general support provided by the community. It may also be the case that because the expectation for them has increased due to marriage or giving birth, their condition might have become worse. The need for support is likely increased due to changes in the environment, the birth of the child, or the burden of child rearing.

It is important to detect and provide the type of support early. Nearby supporters should be continuously involved to regularly gather detailed information about the situation in which the family lives and to share it among supporters.

Survey on the infant health check-up (Uchiyama, Kawashima, & Suzuki, 2020)

We have analyzed data from the infant health check-up in the affected areas. We have compared the answers of the mother and public health nurses regarding infants aged 1.5 and 3 years from 2012 (including data from 2011 when the disaster hit) to 2018.

For 3-year-olds, the public health nurse asked their name and checked if they could give their full name. The proportion of children who could not give their full name was on the increase compared to the predisaster period (Fig. 6). The proportion of children whom the public health nurse identified as "restless" based on observation on the day and that of mothers who chose "being restless" as a problem of their children are also increasing. In the last 9 years, such problems have not been resolved. However, regarding the physical development of

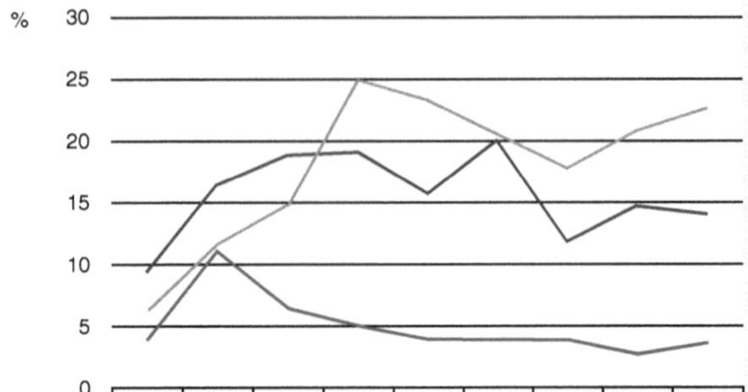

	2010	2011	2012	2013	2014	2015	2016	2017	2018
Can not say Full name	6.3	11.6	14.8	25	23.2	20.4	17.8	20.8	22.6
Poor eye contact	4	11.1	6.5	5.1	4	4	3.9	2.8	3.6
Hyperactivity/impulsiveness	9.4	16.4	18.7	19.1	15.7	20.1	11.8	14.6	14

Fiscal year	number
2010	505
2011	207
2012	230
2013	215
2014	249
2015	298
2016	304
2017	356
2018	358

FIG. 6 Assessment by public health nurse.

children such as height, we have not found any particular impact of the earthquake. Regarding weight, in the immediate aftermath of the earthquake, we noted a slight increase in the number of overweight children but this has improved since then.

To assess the mother's mental health, they were asked if they enjoyed each day and told to answer "yes" or "no," and the proportion of mothers who said "no" was slightly higher than before the earthquake; it remains high.

We continue to administer this survey to provide material to examine the demand for support for families. It is required to capture individual demand for support, to examine the family's situation in its totality, and to contribute to mother-and-child health and projects to provide developmental disorder support.

Conclusions

The authors assessed children and provided support within the framework of child psychiatry. In providing support to areas that have been hit by the earthquake, the assessment of family members, including the mother, and support are naturally required. We have found that changes in the environment due to the nuclear disaster have exerted considerable influence on the mental health of the family. To provide support for children, mothers, fathers, grandparents, and siblings, and support for the supporters of children including public health nurses, teachers, and networking with nurseries, kindergartens, schools, care facilities, and local authorities are needed. By capturing an accurate picture of the situation on the ground and the reality, we identified actions that need to be taken. We believe that capturing accurate information leads to the provision of better support.

We have clarified what needs to be done to prepare for future disasters by learning how evacuated children with developmental disorders and their parents as well as their supporters have lived since the earthquake. We would like to express our deep gratitude to the evacuees and their supporters who have participated in this study.

Support provided in normal times can offer many useful insights that can be applied during an emergency. We sincerely hope that our study will be useful in providing support for families as well as for children and people with developmental disorders.

References

Do, X. B. (2020). Return migration after the Fukushima Daiichi nuclear disaster: The impact of institutional and individual factors. *Disasters*, *44*(3), 569–595. https://doi.org/10.1111/disa.12381.

Kuroda, Y., Goto, A., Koyama, Y., Hosoya, M., Fujimori, K., Yasumura, S., et al. (2021). Antenatal and postnatal association of maternal bonding and mental health in Fukushima after the Great East Japan Earthquake of 2011: The Japan Environment and Children's Study (JECS). *Journal of Affective Disorders*, *278*, 244–251. https://doi.org/10.1016/j.jad.2020.09.021.

Maeda, M., Oe, M., & Suzuki, Y. (2018). Psychosocial effects of the Fukushima disaster and current tasks: Differences between natural and nuclear disasters. *Journal of the National Institute of Public Health*, *67*(1), 50–58. https://doi.org/10.20683/jniph.67.1_50.

MEXT: Ministry of Education, Culture, Sports, Science and Technology. (2011). *Results of airborne monitoring by the Ministry of Education, Culture, Sports, Science and Technology and the U.S.* Department of Energy. Retrieved from https://www.mext.go.jp/en/incident/title01/detail01/sdetail01/sdetail01/1373084.htm.

Murakami, M., Takebayashi, Y., Ono, K., Kubota, A., & Tsubokura, M. (2020). The decision to return home and wellbeing after the Fukushima disaster. *International Journal of Disaster Risk Reduction*, *47*, 101538. https://doi.org/10.1016/j.ijdrr.2020.101538.

Otsuki, M. (2016). Raising children in Fukushima. *Annals of the ICRP*, *45*(2 Suppl), 65–70. https://doi.org/10.1177/0146645316666495.

Shimura, T., Yamaguchi, I., Terada, H., Robert Svendsen, E., & Kunugita, N. (2015). Public health activities for mitigation of radiation exposures and risk communication challenges after the Fukushima nuclear accident. *Journal of Radiation Research*, *56*(3), 422–429. https://doi.org/10.1093/jrr/rrv013.

Tsutsui, Y., Ujiie, T., Takaya, R., & Tominaga, M. (2020). Five-year post-disaster mental changes: Mothers and children living in low-dose contaminated Fukushima regions. *PLoS One*, *15*(12). https://doi.org/10.1371/journal.pone.0243367, e0243367.

Uchiyama, T., Kawashima, K., & Karatani, Y. (2018). The return from evacuation and the support needs of ASD children and their families following the Fukushima nuclear accident. In *23rd World Congress of the International Association For Child and Adolescent Psychiatry and Allied Professions (IACAPAP2018)*.

Uchiyama, T., Kawashima, K., & Suzuki, S. (2020). Psychological effects on children in Fukushima and their care. In *Paper presented at the International Symposium of the Radiation Medical Science Center for the Fukushima Health Management Survey, Fukishima*. http://kenko-kanri.jp/en/news/2nd_intl_symposium_report_published.html.

Chapter 12

Disaster-related health issues among older adults

Mayumi Harigane[a,b] and Seiji Yasumura[a,b]
[a]*Radiation Medical Science Center for the Fukushima Health Management Survey, Fukushima Medical University, Fukushima, Japan,* [b]*Department of Public Health, Fukushima Medical University School of Medicine, Fukushima, Japan*

Introduction

The Fukushima nuclear power plant disaster has caused health problems different from other natural disasters, and even now, 10 years after the disaster, disaster-related deaths still occur. It is said that the physical and mental health of older adults is greatly affected by disasters (Cherniack, 2008; Jia et al., 2010), and that they need consideration during and after disasters (Basic Act on Disaster Management, 1961; Wisner & Adams, 2002). In addition, because older adults in Fukushima Prefecture have accounted for more than half of the disaster-related deaths following the Great East Japan earthquake, it can be said that the health issues of older adults after a nuclear disaster are an important public health concern. In this chapter, we will discuss how the nuclear disaster in Fukushima differs from other natural disasters, how the health problems of older adults in nuclear disasters differ from those of the young, and what support has been provided.

Older adults are more likely to suffer from poor health during a disaster than younger people and are considered to be a population that requires special consideration. The World Health Organization (WHO) defines vulnerable populations as those who should be given more attention than others during a disaster: "Vulnerability" is the degree to which a population, individual, or organization, namely children, older adults, and the disabled, is unable to anticipate, cope with, resist, or recover from the impacts of disasters (Wisner & Adams, 2002). As shown in Fig. 1, lack of access (e.g., to health services, credit, information, etc.) and lack of resources (e.g., income, assets, reserves, social support, etc.) affect exposure to hazards and threats and lead to reduced capacity to cope and to recover, which in turn increases vulnerability during disasters (Wisner & Adams, 2002). In Japan, the Basic Act on Disaster Management lists

Health Effects of the Fukushima Nuclear Disaster. https://doi.org/10.1016/B978-0-12-824098-4.00018-0

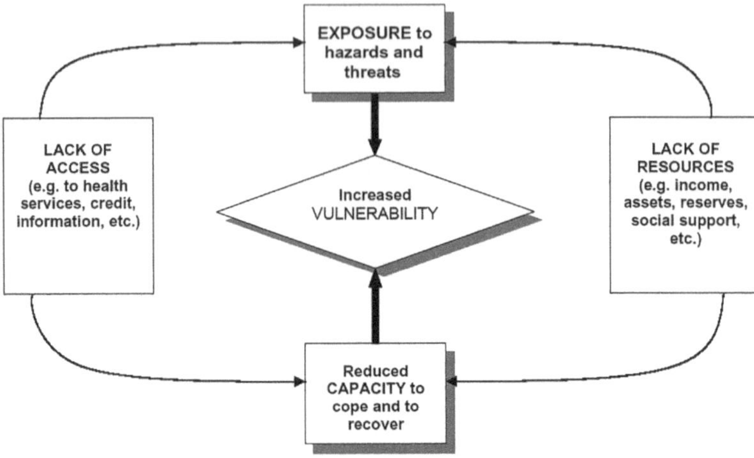

FIG. 1 Disaster vulnerability as a function of exposure to hazards and threats as well as a reduced capacity to cope and recover (Wisner & Adams, 2002).

"older adults, disabled persons, infants, and others requiring special care under a single umbrella term that includes all 'persons requiring special care'" (Basic Act on Disaster Management, 1961). As part of an effort to prevent and mitigate disasters, municipalities are establishing systems to identify these vulnerable people and support them in times of disaster.

Older adults are considered vulnerable because they are more likely to experience loss than younger people. According to previous research, growing old means experiencing four kinds of loss: loss of physical and mental health, loss of financial independence, loss of family and social ties, and loss of purpose in life (Hasegawa, 1975). While younger people have the strength and time to compensate for loss, older adults have little functional reserve and must endure loss. These complex losses that occur in old age can cause great stress and pressure for older adults, and can easily lead to depression and persecutory reactions (Hasegawa, 1975). Even in the absence of a disaster, older adults are likely to experience these losses over time, but in the case of a nuclear disaster, they are likely to experience multiple losses in a shorter period of time. After the Fukushima nuclear disaster, the government ordered the evacuation of residents to avoid life-threatening dangers to their lives due to exposure to radioactive materials released by the damaged nuclear reactors.

Evacuating from one's home due to a nuclear disaster increases the possibility of experiencing the loss of one's home, work, social relationships, and purpose in life as well as changes in lifestyle, decline in physical and cognitive functions, and loss of physical and mental health. In other words, aged evacuees may lose their physical and cognitive functions, and their chronic diseases worsen due to evacuation. They also lose their economic independence due to the loss of their homes and jobs. They lose their connections with their families and friends in the neighborhood due to evacuation. They may also lose their purpose in life because they may be unable to take care of their grandchildren or to engage in farming, for example.

In other natural disasters, people experience a temporary loss of social ties with family and neighbors, but in the case of nuclear disasters, the loss of social ties is prolonged. This overlapping experience of loss in older adults can affect their grief reactions and lead to depression. Therefore, it is important to consider older adults as vulnerable after a nuclear disaster, assess their health status and social resources as early as possible, and provide support for them.

Actual conditions and support for older adults in the Fukushima nuclear disaster

Disaster-related deaths and older adults in the Fukushima nuclear disaster

Most disaster-related deaths in the Great East Japan earthquake consisted of older adults in Fukushima prefecture. As of September 30, 2020, 3767 deaths of all ages attributed to this disaster had been identified in Tokyo and nine prefectures (Fig. 2). This included 3335 people (88.5%) aged 66 and above. Limiting the number of people to Fukushima prefecture only, the count includes 2313 deaths (61.4% of the total). The cumulative number of people over 66 that died from disaster-related conditions in Fukushima prefecture totals 2079 (55.2% of the total) (Reconstruction Agency, 2020). According to a previous study, the residents forced to evacuate due to the nuclear accident included hospitalized patients (Tanigawa, Hosoi, Hirohashi, Iwasaki, & Kamiya, 2012). Approximately 840 people were transported to different hospitals or nursing care facilities, of which 12 patients died during transportation and 50 patients

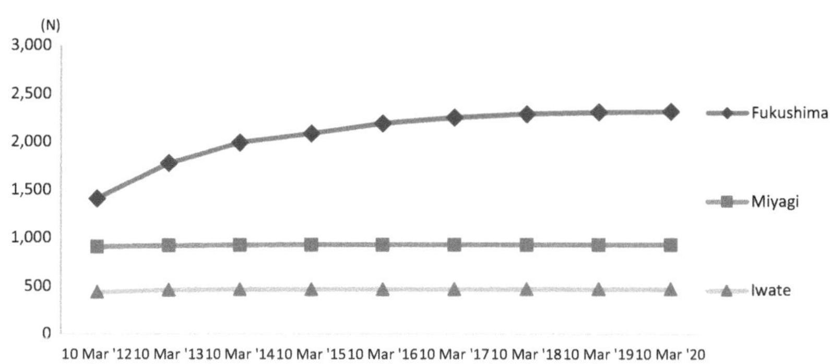

FIG. 2 Cumulative number of disaster-related deaths from the Great East Japan earthquake. This figure displays a comparison between Miyagi, Iwate, and Fukushima prefectures from March 2012 to March 2020. Only the Fukushima prefecture shows a continued rise in deaths. This figure was drawn by the authors based on the values in the table (Reconstruction Agency, 2020).

died at the destination immediately after transportation (Tanigawa et al., 2012). Furthermore, the results showed that the mortality rate of older adults in institutions after the nuclear accident was 2.4 times higher than the previous year (Yasumura, Goto, Yamazaki, & Reich, 2013). In fact, the first 3 months after the disaster were especially high (3.1 times higher than the previous year), and continued to be 1.4–1.8 times higher in the following 9 months than in the previous year (Yasumura, 2014).

The Reconstruction Agency conducted a survey on the causes of the deaths of 35 people who died in Fukushima Prefecture more than 1 year after the Great East Japan Earthquake (Reconstruction Agency, 2013). The mean number of times the evacuees moved was seven. In some cases, evacuees fell ill after returning home temporarily (their physical strength deteriorated due to long hours at the reception desk, etc.). There was also one documented suicide. In total, about 50% of the deaths were due to "physical and mental fatigue from living in evacuation centers," approximately 20% were due to "physical and mental fatigue during transportation to evacuation centers," and approximately 10% were due to a "delay in initial treatment due to hospital shutdown." In addition, most patients gradually became weak due to fatigue, stress, lack of exercise, and medical conditions caused by the evacuation immediately after the disaster and throughout evacuation life.

What is special about nuclear disasters for older adults?

A nuclear power plant accident is a disaster that affects older adults in different ways than a natural disaster such as an earthquake, tsunami, or flood. How a nuclear disaster differs from other natural disasters will be outlined first, and then explained in detail.

The first difference is spatial. A nuclear power plant accident requires evacuation to distant places due to the spread of radioactive materials. Compared to natural disasters, evacuations due to nuclear disasters are characterized by the fact that many residents are forced to evacuate to places farther from their home towns than they would for natural disasters. The spread of radioactive materials forced 164,865 residents to evacuate during the peak period of May 2012. Of these, 102,827 residents (62.4%) evacuated to places within Fukushima prefecture and 62,038 (37.6%) evacuated to places outside Fukushima prefecture (Fukushima Prefecture, 2020).

The second difference is cognitive. Invisible radioactive materials cause people to feel vague anxiety. Residents were not easily reassured, even though the government and scientists repeatedly showed results that suggested that the radioactive materials were unlikely to have any negative health effects. As a result, there were large differences in how people perceived the effects of radioactive materials on their health. In particular, there were differences in how people perceived the health effects of radioactive materials depending on their age, which is unique compared to natural disasters. In a study on the perception of how nuclear disasters impact health, the results indicated that respondents in the age group 15–49 years were more concerned about delayed effects while older age groups (50–64 and ≥65 years) were more concerned about genetic effects (Suzuki et al., 2015).

The third difference is time. The period of time from the start of evacuation until evacuation ends is longer for nuclear disasters than for other disasters. As the evacuation period continues, the health of older adults may deteriorate, they may become ill, or their illnesses may worsen. In the aftermath of the Great East Japan Earthquake, there was a worsening of the level of care required for those who needed support and care (Yasumura, 2013). In addition, prolonged evacuation may make it difficult for people to return to their original areas for reasons that will be discussed later. Thus, it can be said that the prolonged evacuation process is a unique feature of nuclear disasters when compared to natural disasters.

① Spatial dislocation: Evacuating to distant places

Compared to other age groups, older adults are more likely to have chronic diseases, and many require nursing care. In 2017, the medical treatment rates (per 100,000 population) for older adults in Japan were 2734 for inpatient and 10,369 for outpatient, compared to treatment rates of 1036 for inpatient and 5675 for outpatient across all other age groups (Ministry of Health, Labour and Welfare, 2019). In terms of long-term care insurance, as of the end of April 2019, there were 35,278,142 Category 1 insured persons over 65. Of these, 6,593,779 were certified as requiring long-term care or support, and 5,594,206 were recipients of long-term care. This means that 15.9% of older adults were using nursing care services (Ministry of Health, Labour and Welfare, 2020).

When older adults evacuate to distant places due to disasters, it becomes difficult for them to continue receiving medical and nursing care services. Older adults are no longer able to rely on their physicians and care managers as they did before the disaster. In the Great East Japan Earthquake, medical institutions were also damaged by the tsunami and medical information was lost, making it extremely difficult to care for older adults, especially those with chronic diseases (Tanaka, 2012). Based on the lessons learned from this disaster, the local medical system has been information-technology enabled, and the sharing of medical information has been progressing in Fukushima Prefecture. Because of this, the risk of losing medical information due to disasters is expected to decrease in the future. It should be noted, however, that for older adults, evacuating to a distant place means that medical and nursing care services may be interrupted; thus, there is a risk that they may become ill until new places for hospitalization, hospital visits, residential care, and nursing care services are determined.

Most older adults have been living in their areas of residence for decades. Therefore, the local culture and climate become part of their lives. For example, whether it is warm or cold, whether it is near the sea or the mountains, what kinds of food they often eat, and what dialect they speak all have to do with their area of residence. If the evacuation area is far away, older adults may experience greater cultural and climate differences than if the evacuation destination was closer. An unfamiliar climate can make them sick easily, and being surrounded by unfamiliar dialects can reduce communication. Older adults can easily go shopping if they are in a familiar area, but if they are in an unfamiliar area, they may go shopping less frequently.

In addition, most older adults have social relationships that they have nurtured over long periods of time in the community in which they reside. Kahn and Antonucci proposed the Convoy Model, which is analogous to a convoy in which people are protected by the various relationships that surround them as they go through phases of their lives (Fig. 3). In this model, concentric circles show how individuals live surrounded by three layers of people with different levels of closeness (Kahn & Antonucci, 1980). The innermost circle shows stable relationships with people who are not dependent on roles, and the outermost circle shows unstable relationships with people who are dependent on roles. As individuals age, relationship numbers tend to decrease, and people in the innermost circle tend to become the center of relationships among older adults. Nevertheless, nuclear disasters threaten the relationships within peoples' innermost circles. Evacuees may be separated from their family members or from their neighbors who have helped each other like family for years.

A case study: An older adult separated from good friends

A woman in her 80s thought that she would live in the place where she was born and raised until the end of her life. Due to the nuclear disaster, however, she was forced to evacuate and remain away for a long period of time, with no prospect for her return to her hometown. Before the disaster, she enjoyed her life with her neighbors; if someone stepped out of the house for a moment, a circle of chatter would spontaneously form. Since the evacuation, she has been living in a rented apartment provided by the government and does not even know who lives next door. She has become completely isolated. She laments that if it had not been for the nuclear accident, she would have been able to live in a familiar place surrounded by mutually compatible friends until the end of her life.

This case concerns the loss of social ties and its resultant isolation due to evacuation from the nuclear disaster. When evacuating from the nuclear disaster, some residents near the evacuation site evacuated together while others evacuated individually. Because social networks become smaller with age after adolescence (Carstensen, 1991; Kahn & Antonucci, 1980; Wrzus, Hänel, Wagner, & Neyer, 2013), if given a choice when evacuating, it is better for older adults to evacuate with their neighbors to mitigate the loss of their social relationships. For evacuees who are forced to

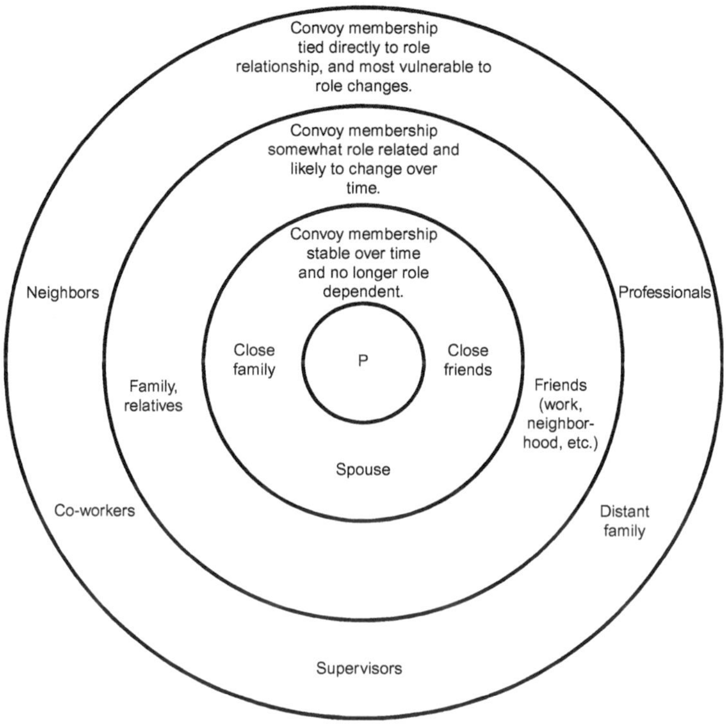

FIG. 3 Hypothetical example of a convoy (Kahn & Antonucci, 1980 p. 273).
This figure shows the Convoy Model proposed by Kahn and Antonucci (1980). In this model, concentric circles show how individuals live surrounded by three layers of people with different levels of closeness. The innermost circle shows stable relationships with people who are not dependent on roles, and the outermost circle shows unstable relationships with people who are dependent on roles. As individuals age, relationship numbers tend to decrease, and people in the innermost circle tend to become the center of relationships among older adults.

evacuate individually and who are likely to be socially isolated from their supporters, it is still possible to mitigate their isolation. One way to do this is by transferring information from the person in charge of the municipality before evacuation to the person in charge of community health after evacuation. However, as will be discussed later, some evacuees do not want people in their postevacuation communities to know that they are evacuees. For this reason, it is desirable for supporters from the preevacuation area to use effective means of communication, even in remote areas. This may entail using phone calls or letters to understand the needs of the individual and to connect them to social resources that can be used in the postevacuation area in a way that will adequately meet their requirements.

It is not easy for older adults to establish new social relationships. In Japan, however, participation in care prevention programs provided by local governments and activities at Silver Human Resources Centers have been established in each municipality. These programs provide "temporary, short-term, or other light work" and "ikigai" to older adults, such as retirees, and provide opportunities to build new social relationships. There seems to be greater appeal for care prevention programs aimed at preventing dementia and the need for nursing care than there is for other hobby groups. For older men who tend to not participate in social activities, their social participation is encouraged in the form of employment at the Silver Human Resources Center. Older adults who have been evacuated due to a disaster can work at the Silver Human Resources Center closest to their place of residence after the evacuation, even if they have not moved their resident registration.

What has become apparent due to the nuclear disaster in Fukushima is that many older adults seem to have lost their ikigai, or purpose in life. The loss of ikigai is said to lead to depression. The word ikigai refers not only to an important object that is valuable and meaningful for one's continued existence, but also to the feeling that one's life is meaningful because of the existence of that object. There are two types of ikigai in old age: namely, self-actualizing ikigai and interpersonal ikigai. Self-actualizing ikigai refers to people's engagement in activities that allow them to realize the meaning of their lives as individuals, apart from their previous occupational activities where they earned a living and carried out family maintenance functions such as child rearing. Interpersonal ikigai refers to being with family, friends, and like-minded people where strong feelings of ikigai come about. As a result of the nuclear disaster, older adults lost their ikigai of working, their self-actualizing ikigai such as hobbies and volunteer work, and the interpersonal ikigai they had gained from living with their children, grandchildren, and neighbors. For example, older adults who used to work as farmers and grew produce for a long time have been unable to grow vegetables and fruits after the evacuation. Many older adults who used their bodies and minds every day while farming now spend their postevacuation lives watching television.

② Cognition: Vague anxiety about radioactive materials

Many evacuees worry about the effects of the radioactive materials that were released after the nuclear power plant accident. They worry about the long-term effects of these radioactive materials on their bodies rather than the direct physical effects. According to data in the Mental Health and Lifestyle Survey, a detailed component of the Fukushima Health Management Survey conducted in fiscal year (FY) 2011, 48.1% and 60.2% of the residents that responded felt that the likelihood of delayed effects and genetic effects were "likely" or "very likely," respectively. In FY 2018, the percentages were 33.5% and 36.0%, respectively (Figs. 4 and 5). Comparing the percentages at the two time points, it appears that the percentage of respondents who answered this way was decreasing. However, while it had been decreasing until FY 2014, there has been little change since FY

2015 (Radiation Medical Science Center for the Fukushima Health Management Survey, 2020). Thus, we can see that while some people's perceptions of radiation have changed after the nuclear accident, others have not. It is possible to say, then, that the perception of the effects of radiation on health ranges from "very unlikely" to "very likely." Moreover, such individual differences in perception may occur between family members; for example, some residents may want to eat mushrooms picked from nearby mountains while others may only want to eat produce that has been tested for radioactivity and marketed in this way. Such a disparity is often seen across generations. Older adults may want to eat mushrooms they have picked themselves while their children's family members only want to eat those that have been purchased.

A generational disparity is also apparent in the response to air radiation. For example, in the case of radiation dose and cancer risk, if a person is exposed to a low dose of radiation over a long period of time, the longer period equates to a higher total dose exposure. By this calculation, for example, the total exposed dose from age 80–120 and the total dose exposed from birth to 120 are quite different; more radiation exposure is assumed from birth than from age 80, and the risk is therefore perceived to be higher for younger people. The relationship between radiation dose and cancer incidence remains unclear, however. Despite the fact that it is unknown whether the risk increases linearly for doses below 150 mSv (Ministry of the Environment, 2020), a few family separations occurred in areas with doses much lower than 150 mSv; older people stayed and younger people left. This is an example in

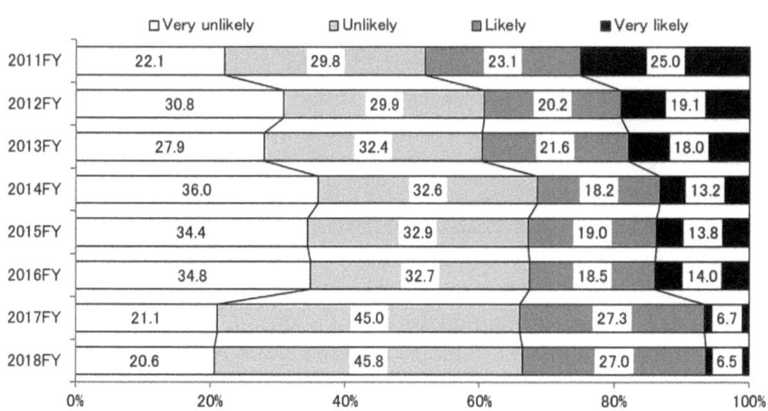

FIG. 4 Trends in the percentage of the perception of radiation risk among people aged 16 and older (delayed effects).
This figure shows the trends from FY 2011 to FY 2018 in the Mental Health and Lifestyle Survey, a detailed component of the Fukushima Health Management Survey. In FY 2011, 48.1% of residents reported delayed effects as "likely" or "very likely"; by FY 2018, the percentage was 33.5%. The percentage of "highly likely" or "very likely" decreased only from FY 2011 to FY 2014, and has remained virtually unchanged since FY 2015 (Radiation Medical Science Center for the Fukushima Health Management Survey, 2020).

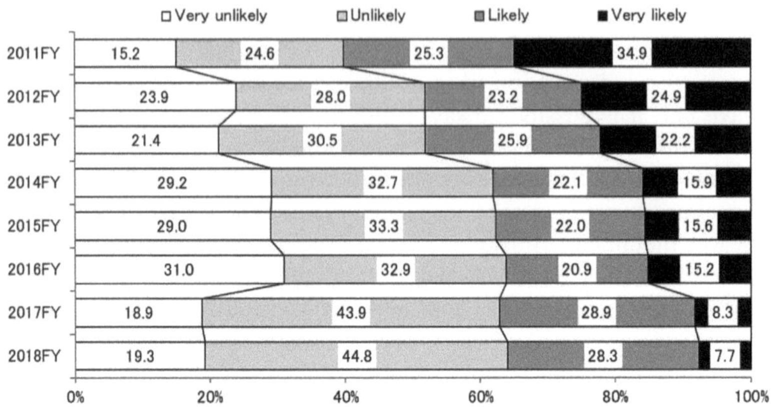

FIG. 5 Trends in the percentage of the perception of radiation risk among people aged 16 and older (genetic effects).
This figure shows the trends from FY 2011 to FY 2018 in the "Mental Health and Lifestyle Survey," a detailed component of the Fukushima Health Management Survey. In FY 2011, 60.2% of residents reported that a genetic effect was "likely" or "very likely," but by FY 2018, the percentage was 36.0%. The percentage of "highly likely" or "very likely" decreased only from FY 2011 to FY 2014, and has remained virtually unchanged since FY 2015 (Radiation Medical Science Center for the Fukushima Health Management Survey, 2020).

which different age groups choose different behaviors due to the same risk perception. A similar generational gap also occurred when people chose whether to return to areas where evacuation orders had been lifted. The actual number of residents who returned was relatively high among older adults. For example, in Naraha Town, all residents evacuated and the evacuation regulation for the entire area was eventually lifted. As of December 31, 2020, the Basic Resident Ledger estimated the residence rate to be 59.6% with residents over 65 years old accounting for 37.4% of the total (Naraha Town, 2021). Compared to the predisaster rate of 25.9% of residents as of October 1, 2010 (Fukushima Prefecture, 2017), this represents an 11.5% increase in the proportion of older residents. However, when considering the national population of residents 65 years and over as of December 1, 2020, it constituted 28.8% of the population after the earthquake (Statistics Bureau, 2020) and 23.0% before the earthquake as of October 1, 2010 (Statistics Bureau, 2011), representing a 5.8% increase. When these two proportions are compared (11.5% versus 5.8%), it appears that Naraha town is experiencing a relatively large influx of older adults returning to the area.

Information gap between evacuees and others due to the nuclear power plant accident.

Nuclear accidents spread not only radioactive materials but also many rumors. As a result, there were many evacuees who were very cautious when talking to nonevacuees. This is because being known as an evacuee from a nuclear power plant accident may cause issues that are difficult for the evacuees to deal with, such as misunderstandings about radiation exposure and compensation.

A case study: An older adult who lost ikigai due to the nuclear accident

A woman in her 80s had been farming for a living until the earthquake. She had to leave her farm because of the nuclear accident and has had nothing to do in her new place after the evacuation, so she is now homebound. Her children repeatedly suggested she go out for a walk, but she did not want to do so. In truth, she does not want her neighbors to ask her where she moved from or to know that she is an evacuee, if they want to speak to her while she is taking a walk.

③ Temporality: The length of evacuation

The longer a postdisaster evacuation takes, the greater the hurdle for evacuees to return to their predisaster area. In the nuclear disaster, some people had to evacuate because of the government's evacuation order while others voluntarily evacuated to safer areas that were not under the evacuation order. For those who evacuated due to the evacuation order, the lifting of the order is not seen as a simple joy of being able to return to their predisaster area. There are many worries that may come to an older resident's mind when considering returning. Some reasons include: I want to return to my familiar home, but I wonder if the financial support I have been receiving will continue; I wonder if the other people in my neighborhood will return; I wonder if the hospitals, nursing homes, schools, and grocery stores will return; I wonder if I will be able to repair my home that has been destroyed by wild animals; I wonder if I will have to leave the friends I have made at the evacuation site; I wonder what will happen if I leave the workplace I have become accustomed to; I wonder if I will be able to adapt to life in my new home after leaving the life I have become accustomed to; and I wonder if I will have the physical strength to endure the moving process again. Their worries are warranted. For example, in municipalities where evacuation orders have been lifted and people are able to return to their predisaster area, and despite the fact that welfare needs are increasing due to the increase in older adults population, a lack of younger residents returning has made it difficult for nursing staff to provide adequate nursing care services (Ogawa, 2020).

A case study: An older adult who wants to go back to where they lived before evacuation, but cannot

A man in his 70s was allowed to go back to his hometown after the evacuation order had been lifted. However, he had already demolished his house because it had been damaged by wild animals, was dilapidated, and the deadline for compensation of demolition costs was approaching. He has no home to go back to even if he wanted to. He regrets that if he had not been evacuated for so long, he may not have had to dismantle his house.

On the other hand, there is also a dilemma for those who have evacuated from a difficult-to-return zone where evacuation orders have not yet been lifted. This is the dilemma of whether to permanently settle down in the postevacuation area or to continue living in temporary housing until they are able to return home someday. The longer the evacuation, the more likely they are to settle down in the area after the evacuation. It is important to note that the longer the evacuation period, the more likely it is that older adults will experience an accelerated decline in their independence and activities of daily living. People who were living independently before the disaster were more likely to become frail due to inactivity and to require support and care due to reduced social relationships and fewer outings due to evacuation (Okawa, 2012). As a result, the longer the evacuation period, even if older adults want to return to the area where they used to live before the disaster, they will not be able to do so alone as they will have to live with their children who will take care of them, or they will be placed in a nursing home near their children. In addition, it is important to note that the lack of prospects for the future makes it difficult for these residents to have hope for the future. Some older adults have taken their own lives. Their mental health deteriorates because of this lack of hope that they may 1 day return to their hometowns.

In the event of a nuclear disaster, when evacuation of older adults is inevitable, it is necessary to assess the size of their social networks. It is also necessary to assess whether older adults have places to go out to after evacuation to prevent them from becoming isolated or homebound, called "Tojikomori" in Japanese. In the Mental Health and Lifestyle Survey, clinical psychologists, public health nurses, and nurses from the Mental Health Support Team provided outreach via telephone interventions to people who were at high risk for mental health and lifestyle issues based on their responses to the questionnaire. Of the high-risk individuals targeted by the telephone interventions in FY 2018, the majority (more than 60%) were over 60 years old (Radiation Medical Science Center for the Fukushima Health Management Survey, 2020). In addition to providing feedback on the results of the questionnaire-based health survey, outreach via telephone intervention plays a role in assessing the current situation (assessment) and assisting people to appropriately use local medical care and consultation services, or to connect them when necessary (Horikoshi, Oikawa, & Maeda, 2020; Kashiwazaki et al., 2016). The method of telephone intervention, which can intervene with older adults who have evacuated to distant places, will play an important role as a method of support after a nuclear disaster.

Summary

This chapter focused on how the nuclear disaster in Fukushima differed from natural disasters, how the health problems of older adults in nuclear disasters differ from those of younger people, and how support has been provided. In the future, after a nuclear disaster, it will be important to improve the assessment of older adults and ensure that measures are in place to support these individuals even in distant areas.

The cases introduced in this chapter are fictitious cases based on a composite of several cases supported by the Mental Health Support Team of the Fukushima Health Management Survey.

References

Basic Act on Disaster Management. (1961). http://www.japaneselawtranslation.go.jp/law/detail/?id=3322&vm=&re.

Carstensen, L. L. (1991). Socioemotional selectivity theory: Social activity in life-span context. *Annual Review of Gerontology and Geriatrics, 11*(1), 195–217.

Cherniack, E. (2008). The impact of natural disasters on the elderly. *American Journal of Disaster Medicine, 3*(3), 133–139.

Fukushima Prefecture. (2017). *Annual report of the survey of current population in Fukushima Prefecture, 2010 edition: Table 1 population by city, age, and sex*. https://www.pref.fukushima.lg.jp/uploaded/attachment/22316.xls.

Fukushima Prefecture. (2020). *Steps for reconstruction and revitalization in Fukushima Prefecture*. Fukushima Prefecture. https://www.pref.fukushima.lg.jp/uploaded/attachment/394169.pdf.

Hasegawa, K. (1975). General remarks. In K. Hasegawa, & T. Kashu (Eds.), *An approach to the psychology of aging (Rojin-Shinri eno approach)* (pp. 10–24). Igaku Shoin (written in Japanese).

Horikoshi, N., Oikawa, Y., & Maeda, M. (2020). Higashinihondaishinsaigo no kadengata (autoriichigata) denwashien no kokoromi [Outreach-type telephone support after the great East Japan earthquake]. In M. Maeda, M. Momoi, & Y. Takebayashi (Eds.), *Enkakushinrishien sukirugaido [Guide for telepsychological support skills]* (pp. 61–69). Seishin Shobo (in Japanese).

Jia, Z., Tian, W., Liu, W., Cao, Y., Yan, J., & Shun, Z. (2010). Are the elderly more vulnerable to psychological impact of natural disaster? A population-based survey of adult survivors of the 2008 Sichuan earthquake. *BMC Public Health, 10*, 172.

Kahn, R. L., & Antonucci, T. C. (1980). Convoys over the life course: Attachment, roles, and social support. In P. B. Baltes, & O. Grim (Eds.), *Vol. 3. Life span development and behavior* (pp. 253–286). Academic Press.

Kashiwazaki, Y., Maeda, M., Yagi, A., Fujii, S., Takahashi, N., Yabe, H., et al. (2016). Effectiveness of telephone-based intervention for people living in Fukushima disaster area/Fukushima health management survey. *Seishinigaku, 58*(5), 433–442.

Ministry of Health, Labour and Welfare. (2019). *Summary of patient survey, 2017: Outline of survey results 2 rates of estimated patients*.

Ministry of Health, Labour and Welfare. (2020). *Annual health, labour and welfare report 2020: (10) health and welfare services for the elderly*. Ministry of Health Labour and Welfare. https://www.mhlw.go.jp/wp/hakusyo/kousei/19-2/dl/10.pdf.

Ministry of the Environment. (2020). Health effects of radiation. In *Booklet to provide basic information regarding health effects of radiation* (1st ed., pp. 77–151). https://www.env.go.jp/en/chemi/rhm/basic-info/1st/03.html.

Naraha Town. (2021). *Number of people living in Naraha Town*. https://www.town.naraha.lg.jp/admin/cat337/006099.html.

Ogawa, S. (2020, January 18). *Shrinking local governments: Data on the return of residents (10 years after the Fukushima nuclear accident)*. Tokyo Shimbun. https://www.tokyo-np.co.jp/article/79669.

Okawa, Y. (2012). 'Disuse syndrome' as a target for care management in disaster situations: What can be learned from care prevention in normal situations? *Care Manager*, *14*(3), 78–82.

Radiation Medical Science Center for the Fukushima Health Management Survey. (2020). *The 38st prefectural oversight committee meeting for Fukushima Health Management Survey; Mental Health and Lifestyle Survey for FY 2018 Result report*. https://www.pref.fukushima.lg.jp/uploaded/attachment/386028.pdf.

Reconstruction Agency. (2013). *Report on the prevention of earthquake related deaths in Fukushima Prefecture*. Reconstruction Agency. https://www.reconstruction.go.jp/topics/20130329kanrenshi.pdf.

Reconstruction Agency. (2020). *Data on the disaster-related death*. Reconstruction Agency. https://www.reconstruction.go.jp/topics/main-cat2/sub-cat2-6/20201225_kanrenshi.pdf.

Statistics Bureau. (2011). *Population projection, base population by 2010 national survey: Reference table 2*. https://docs.google.com/viewer?url=https%3A%2F%2Fwww.stat.go.jp%2Fdata%2Fjinsui%2Fkijun%2Fzuhyou%2F02.xls.

Suzuki, Y., Yabe, H., Yasumura, S., Ohira, T., Niwa, S., Ohtsuru, A., et al. (2015). Psychological distress and the perception of radiation risks: The Fukushima health management survey. *Bulletin of the World Health Organization*, *93*, 598–605. https://doi.org/10.2471/BLT.14.146498.

Tanaka, H. (2012). Healthcare IT system in the midst of and after great East Japan earthquake disaster : Grand design for reconstruction of Tohoku-region healthcare IT system. *Journal of Information Processing and Management*, *54*(12), 825–835. https://doi.org/10.1241/johokanri.54.825.

Tanigawa, K., Hosoi, Y., Hirohashi, N., Iwasaki, Y., & Kamiya, K. (2012). Loss of life after evacuation: Lessons learned from the Fukushima accident. *The Lancet*, *379*(9819), 889–891. https://doi.org/10.1016/S0140-6736(12)60384-5.

Wisner, B., & Adams, J. (2002). *Environmental health in emergencies and disasters: A practical guide*. Geneva: World Health Organization. https://apps.who.int/iris/handle/10665/42561; (1961).

Wrzus, C., Hänel, M., Wagner, J., & Neyer, F. J. (2013). Social network changes and life events across the life span: A meta-analysis. *Psychological Bulletin*, *139*(1), 53–80. https://doi.org/10.1037/a0028601.

Yasumura, S. (2013). Elderly person support in disaster. In S. Yasumura, & I. Kai (Eds.), *Manual of elderly person health welfare* (pp. 129–137). Nanzando.

Yasumura, S. (2014). Evacuation effect on excess mortality among institutionalized elderly after the Fukushima Daiichi nuclear power plant accident. *Fukushima Journal of Medical Science*, *60*(2), 1–4.

Yasumura, S., Goto, A., Yamazaki, S., & Reich, M. R. (2013). Excess mortality among relocated institutionalized elderly after the Fukushima nuclear disaster. *Public Health*, *127*(2), 186–188. https://doi.org/10.1016/j.puhe.2012.10.019.

Chapter 13

Psychosocial support for anxiety related to thyroid cancer: A team approach during the secondary confirmatory examination in thyroid ultrasound examination

Noriko Setou[a,b], Satoru Suzuki[a], and Hiroki Shimura[a,c]
[a]*Radiation Medical Science Center for the Fukushima Health Management Survey, Fukushima Medical University, Fukushima, Japan,* [b]*Department of Disaster Psychiatry, School of Medicine, Fukushima Medical University, Fukushima, Japan,* [c]*Department of Laboratory Medicine, School of Medicine, Fukushima Medical University, Fukushima, Japan*

Introduction

The Great East Japan Earthquake on March 11, 2011, caused an accident at the Fukushima Daiichi Nuclear Power Plant (NPP) that dispersed radioactive substances into the environment. The nuclear hazard severely affected the daily lives and health of Fukushima's residents, industries, and agriculture in various ways. Approximately 165,000 residents of Fukushima Prefecture were forced to evacuate right after the earthquake. Today, more than 40,000 people are estimated to still be exiled from their homes, in and out of the prefecture (Fukushima Prefecture, n.d.).

Another major problem was the health effect of radiation on children. This accident was rated as a Level 7 (serious accident) as per the International Atomic Energy Agency's (IAEA) established scale (INES). Level 7 is an evaluation equivalent to the level of the Chernobyl Nuclear Power Plant Accident in 1986 in which childhood thyroid cancer frequently occurred due to exposure (Ministry of the Environment, n.d.). Consequently, there was great concern about the increase in childhood thyroid cancer caused by radiation exposure similar to the Chernobyl disaster. As a result, starting in October 2011, Fukushima Prefecture began conducting the Thyroid Ultrasound Examination (TUE)

Health Effects of the Fukushima Nuclear Disaster. https://doi.org/10.1016/B978-0-12-824098-4.00007-6

program for children (Radiation Medical Science Center for the Fukushima Health Management Survey, n.d.-a).

The TUE program is a part of the Fukushima Health Management Survey conducted by the Radiation Medical Science Center of Fukushima Medical University, commissioned by the Fukushima Prefecture (Radiation Medical Science Center for the Fukushima Health Management Survey, n.d.-a). Approximately 385,000 children aged 0–18 who were living in Fukushima Prefecture at the time of the earthquake are examined once every 2 years. The survey was divided into two parts: the Preliminary Baseline Survey and the Full-Scale Survey. By 2020, the third Full-Scale Survey (the fourth-round survey) was completed, and the fourth Full-Scale Survey (the fifth-round survey) had gotten under way.

The examinations consist of primary and secondary confirmatory examinations (Radiation Medical Science Center for the Fukushima Health Management Survey, n.d.-a). The primary examination consists of an ultrasound examination of the thyroid. If nodules measuring >5.0 mm and/or cysts measuring >20.0 mm in diameter are found, the secondary confirmatory examination is recommended (Fig. 1). Often, examinees and their families show high levels of concern and fear of thyroid cancer during the secondary confirmatory examination. Therefore Fukushima Medical University, which has conducted the highest number of secondary confirmatory examinations in Japan, established the Thyroid Support Team in November 2013 to provide psychosocial support to those examinees and their families (Setou et al., 2021).

In this chapter, we introduce details of psychosocial support by the Thyroid Support Team during the secondary confirmatory examination at Fukushima Medical University and review what support has been required in accordance

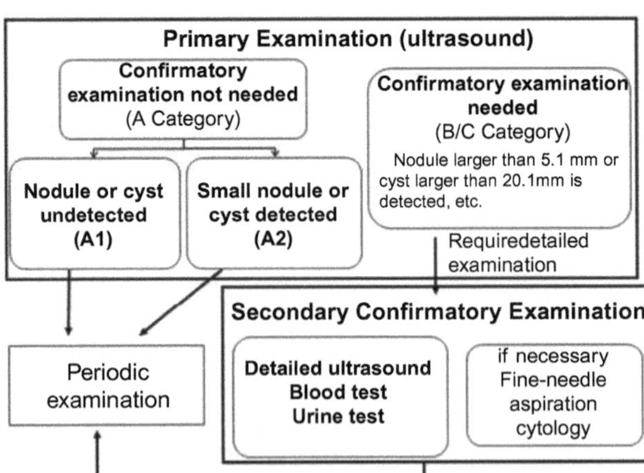

FIG. 1 Primary and secondary confirmatory thyroid examination in TUE.

with the accumulating evidence of thyroid examinations and changing social context as well as which tasks are currently performed.

Roles of thyroid support team

Secondary confirmatory examinations are conducted twice a week at Fukushima Medical University. An examinee usually visits at least twice during the secondary examination. On the first visit, the examination procedure is explained prior to the actual examination. After the medical and sonographic examinations, the results of the ultrasound examination are explained by the doctors on the same day. If sonographic findings do not meet the guideline for fine-needle aspiration cytology (FNAC), the secondary confirmatory examination is concluded, and information is given on taking future thyroid examinations. As a result of the Preliminary Baseline Survey and past Full-Scale Surveys, FNAC was required only in a small number of examinees (Suzuki, 2016). However, FNAC was sometimes recommended for nodules if the sonographic findings met the guideline for FNAC (Suzuki, 2017; Suzuki et al., 2016). Surgical treatment is sometimes required, depending on the results of FNAC diagnosis. On the other hand, some examinees are advised to take an ultrasound examination before the next survey, such as when benign nodules tend to increase in size. Some examinees are also found to have abnormal thyroid function, such as Hashimoto's thyroiditis. These individuals will be followed up in clinical departments of endocrinology or endocrine surgery (Fig. 2).

The Thyroid Support Team at Fukushima Medical University has four staff members: three nurses and one psychologist. The team members provide seamless support from the onset of the secondary confirmatory examination through

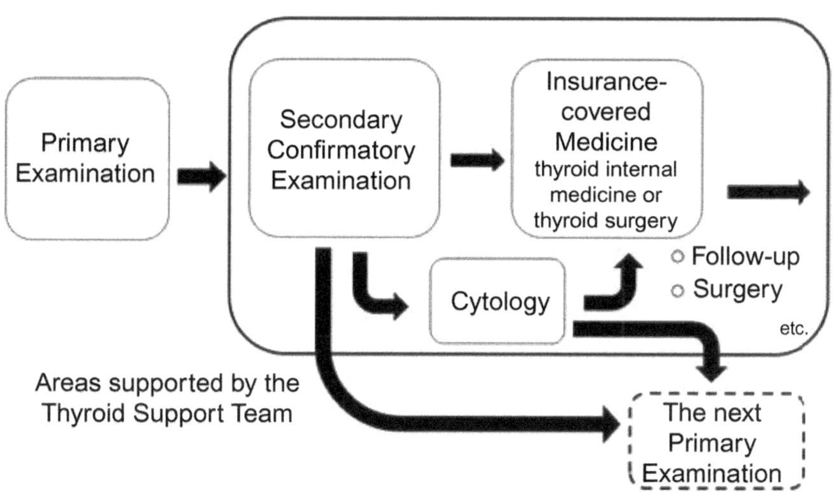

FIG. 2 Areas of thyroid support team at Fukushima Medical University.

endocrinology or endocrine surgery if treatment or medical follow-up is provided at Fukushima Medical University Hospital (Setou et al., 2021).[a]

There are four main roles of the Thyroid Support Team. The first is alleviation of anxiety. Examinees and their families often have strong concerns when called in for the secondary examination, especially because the percentage of examinees referred to the secondary examination is as low as about 0.8% of those who receive a primary examination (Fukushima Prefecture, n.d.). The second is to explain the procedures and contents of each subsequent examination. Those who are afraid of having blood drawn for exams or FNAC are given sufficient orientation about the purpose and procedures of the exams. The third is decision-making support. The support team members provide appropriate information to help examinees and their families make a decision on whether they undergo FNAC and future rounds of TUE. Finally, we provide support to examinees with suspected thyroid cancer or other thyroid diseases by explaining the information they need in an understandable way so that they will be able to accept the situation. Also, in regard to surgery or other treatment, the team member helps the examinees exactly understand the disease and current condition, and gives them a pre- and postoperational outlook so that they can self-manage their health.

Details of thyroid support

A team of doctors, nurses, laboratory technicians, and clerical staff works with the Thyroid Support Team to provide care for the examinees during the secondary confirmatory examination at Fukushima Medical University. Some examinees are accompanied to the exams by family members and some visit alone, especially if they are college students or older. Examinees usually visit the hospital at least twice during the secondary confirmatory examination. Fig. 3 shows the flow of examinations and thyroid support for the first and second visits as well as the second visit with FNAC as needed and third visit with the explanation of the FNAC results.

On the first visit (Fig. 3A), the examination procedure is explained before the doctor's examination (ultrasound exam) followed by blood and urine exams. The results of the ultrasound examination are delivered by the doctor on the same day.

Because many examinees and their families feel nervous when they come to the hospital, the Thyroid Support Team member tries to offer words of kindness,

---

a. Even if the secondary confirmatory examination is conducted at Fukushima Medical University, medical treatment may be undertaken at other nearby hospitals.

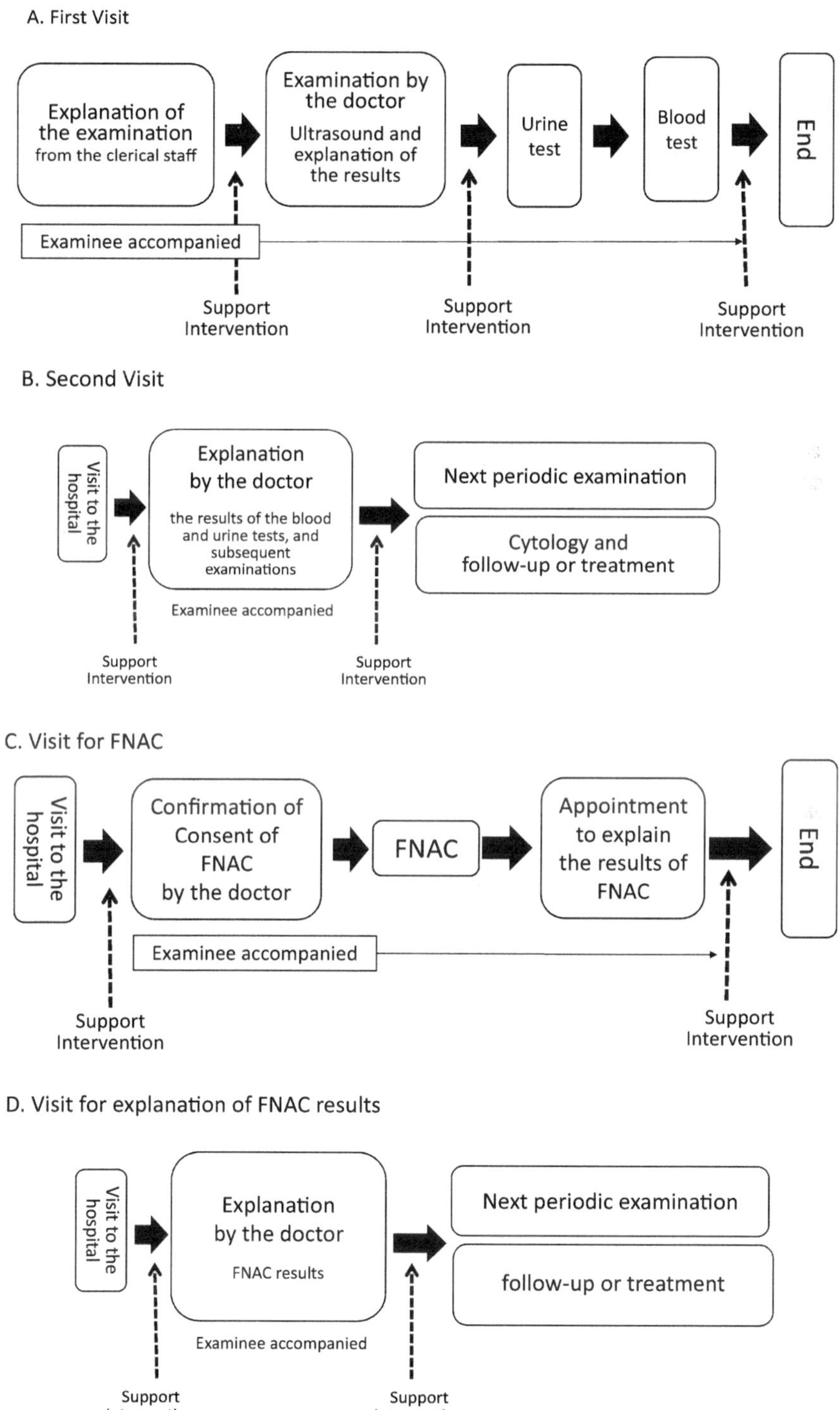

FIG. 3 Examination flow and thyroid support at the secondary confirmatory examination.

listen to the patients carefully, and communicate with them to ease tension and anxiety. Examinees are encouraged to ask questions of doctors and staffs about any issue before the examination starts. Then, a support member always accompanies the examinee, as the member listens to the doctor's explanations of the ultrasound examination results, confirms the patient's understanding, and then supplements that with necessary explanations. After all tests are completed, we check to see if they have any questions, and provide further information as needed.

At the second visit (Fig. 3B), a support member conducts an orientation for the day's schedule to the examinee and fields any new questions. They will then listen to an explanation by a doctor. The team member accompanies the examinee to the end. The doctor gives him/her the results of the blood and urine tests, and explains the next round of TUE if FNAC was not required, including the ultrasound test. On the other hand, depending on the results of the ultrasound, doctors will explain the specifics of FNAC and possible outlook. FNAC is conducted as one of the tests in the secondary confirmatory examination, and a laboratory technician provides additional details about it. After all explanations, the support member checks to confirm that the examinee and his/her family members understand the explanations, listens to their thoughts, and provides them with the information they need.

FNAC is a voluntary and consensual test. Some examinees and their families wonder when and whether they should take the test if the doctor recommends it. Also, some examinees and parents may have divergent opinions. Therefore, if the second visit with FNAC is scheduled (Fig. 3C), FNAC will be booked on a tentative basis. On the day of FNAC, prior to the doctor's consultation, a support member would confirm that the family discussed the FNAC, and ask for a final decision and their thoughts. Based on the information, the doctor reconfirms their decision, and lets them fill out the consent form when they agree. Then the FNAC is performed. The fine needle is usually inserted twice to increase diagnostic accuracy, and the pain from the puncture varies individually. After the FNAC, the examinee rests for a while and eventually a support member checks on them and explains some notes for the day.

On the visit for the FNAC results explanation (Fig. 3D), a support member always meets with the examinees and their families in advance and asks about their thoughts and feelings about listening to the results. When the doctor explains the results, they are relieved if it is benign, but if not, their shock is often great. In the former case, the next round of TUE may be recommended for the examinee. In the latter case, the doctor will tell them the diagnosis, pathology, expected symptoms, and the prospects for the future. The support member, along with the examinee and his/her family, listens to the explanations from the doctor. After the consultation, we reexplain slowly for the examinee and his/her family, and respond carefully to their thoughts and questions. We also let them know that support will continue to be provided by the support staff during treatment, and try to make them feel as relieved as possible.

Frequently asked questions during the secondary confirmatory examination

Ten years have passed since the start of TUE, and the results of the past TUE program have been reported (Fukushima Prefecture, n.d.; Ohtsuru et al., 2019; Shimura, 2019; Shimura et al., 2014). In the Preliminary Baseline Survey, of the 2091 examinees who had the secondary confirmatory examination, 26.2% required cytology, of which 5.5% were malignant or suspected malignant. In the first Full-Scale Survey, of the 1826 examinees who had the secondary confirmatory examination, 11.3% required cytology, of which 3.9% were malignant or suspected malignant. In the second Full-Scale Survey, of the 1038 examinees who had the secondary confirmatory examination, 6.9% required cytology, of which 2.8% were malignant or suspected malignant (Fukushima Prefecture, n.d.). Thus, the results of the secondary confirmatory examinations showed that many examinees have nonmalignant results. Murakami reported that TUE was significantly associated with not only a reduction in radiation anxiety but also an increase in emotional stress (Murakami et al., 2018). One of the reassurance factors is to undergo a secondary examination. On the other hand, even if the results of the tests are acceptable, examinees and their families are often concerned about their future progress. Therefore, it is important to carefully identify and respond to concerns and questions after the results explanation. In addition, we take the time to support the examinee and his/her family so they have a better outlook for the future, especially when FNAC results suggest thyroid cancer.

The Thyroid Support Team at Fukushima Medical University is currently analyzing the support during the secondary confirmatory examination based on our records after being approved by the ethics committee (Setou et al., 2021). One of them is an analysis of frequently asked questions (FAQ). There are many questions from examinees and their families who have a high level of anxiety, especially at the first visit of the secondary confirmatory examination. Careful response to these questions often reduces the anxiety of the examinee and his/her family. Our analysis classified four categories of questions: Findings and causes of nodules, nodule changes, thyroid cancer, and TUE. In this section, we picked out some of the FAQ from examinees and their families, and described how the support team member responds to them. However, our explanations are based on previous studies and some may change in the future, depending on the accumulation of research and social contexts.

Q1: "Are my symptoms related to this nodule?"

This is one of the questions about "findings and causes of nodules." The examinee and his/her family tend to think that his/her current symptoms are caused by a nodule in the thyroid gland. The symptoms include swelling of the neck, fatigue, choking phlegm, dizziness, susceptibility to catching colds, etc. Some people are sensitive to discomfort around the neck or throat, especially after they are called for the secondary confirmatory examination.

As almost cysts and nodules found in TUE are small, less than 20 mm, the doctor often explains that the symptoms are not related to thyroid nodules based on the ultrasound diagnosis. On the other hand, if the examinee's symptoms overlap with symptoms of thyroid disease, such as Hashimoto's disease or Graves' disease, the results of blood tests as well as the ultrasound exams will be taken into account. After these explanations from the doctor, the support member checks with the examinee and his/her family for their understanding and concerns. If they ask about the symptoms that are likely to appear if the nodule has grown, we provide additional explanations, such as feeling the nodules touching the surface of his/her neck or feeling a little difficulty in swallowing.

Q2: "Why did the nodule form?"

This is another one of the questions about "findings and causes of nodules." In answering this question, we begin by explaining what cysts and nodules are and look like on ultrasound images, so that the examinee and his/her family can have a better understanding of them. Many people are relieved when they are informed that cysts and nodules are sometimes prone to form even in the young generation, and that their growth often slows down as the examinees age, even though they can grow larger in younger people.

Even today, some people can still associate the nodules found on the ultrasound examination and the thyroid cancers that have been found in previous rounds of TUE with radiation effects from the accident at the Fukushima Daiichi Nuclear Power Plant. Though they often don't express these concerns verbally, it's explained by a support member that nodules and cysts are very common in healthy people who undergo mass screening of the thyroid gland (Japan Breast-Thyroid Ultrasound Medical Association, Thyroid Terminology Diagnosis Standard Committee, 2016), that the dose of radiation exposure from the Fukushima nuclear accident was much lower than that from Chernobyl (Kim et al., 2016, 2020), and that experts suggested that the effects of radiation exposure are unlikely to be present in the thyroid cancers found until now (Prefectural Committee Meeting for Fukushima Health Management Survey, 2016). These explanations are very important, particularly for examinees who are suspected of having thyroid cancer on ultrasound images, because the cause of thyroid cancer is a matter of concern. In recent years, explanations have been carefully given to examinees who were too young to understand TUE at the time of the disaster to try to decrease their concerns.

Q3: "Can benign nodules turn into thyroid cancer?

This is one of the questions about "nodule changes." Some examinees and their family worry about future changes in the nodule, even if the results of the ultrasound examination are not problematic at the secondary confirmatory examination.

If the ultrasound images show benign findings, the doctor explains that the current nodule is not something to be particularly worried about. But some examinees or their families have concerns about changes in the nodules after the consultation. Therefore, the support member accompanies them during the doctor's explanation of the ultrasound results. The member then reconfirms their understanding based on the explanation, and this may be added to the supplementary information even if the nodule is benign. In addition, the support member explains that even benign nodules may rarely be treated in the future, but that future rounds of TUE may help with follow-up of the nodule. If necessary, we provide feedback about their further concerns and/or questions to the doctor and ask for reexplanation to the examinee and his/her family.

Q4: "How many people are diagnosed as malignant cases?"

This is one of the questions about "thyroid cancer" and is asked regardless of the results of the ultrasonography. In addition to this question, we also receive questions such as "Are there many people with thyroid cancer in Fukushima Prefecture?" In response to these questions, they are shown two graphs: one has the results of previous TUE (Fukushima Prefecture, n.d.) and the other has the results of a three-prefecture survey (Ministry of the Environment, 2017). It's explained that the relationship between thyroid cancers and radiation exposure was unlikely based on the results of the preceding TUE, and that there was no statistical difference in the detection rates of thyroid cancer or suspected thyroid cancer between Fukushima Prefecture and other prefectures (Hayashida et al., 2013; Taniguchi et al., 2013).

In the early days of TUE, Fukushima residents had a great deal of concern about the effects of radiation exposure of the nuclear disaster and so we have been asked a great many of these questions, but recently these questions have been decreasing. Examinees and their families are gradually receiving accurate information such as the results of TUE through the Thyroid Newsletter delivered to their home regularly and on the website of the Fukushima Health Management Survey: Thyroid Ultrasound Examination. On the other hand, because the long-term effects of radiation exposure are still unclear and the residents of the prefecture still have some concerns, the support members try to provide information and listen carefully to their anxiety.

Q5: "What kind of treatment is given for thyroid cancer?"

This is also one of the questions about "thyroid cancer," and some examinees and their families ask this question, regardless of the results of the ultrasound examination.

In particular, if the result of FNAC is thyroid cancer or suspected thyroid cancer or indeterminable at that time, we always provide an explanation of the prospects for treatment and follow-up. In the case of thyroid cancer or suspected thyroid cancer, the examinee and his/her family may often be upset by the diagnosis and have strong anxiety for the future. Therefore, the support member tries to slowly explain it again, confirming their understanding after the doctor's explanation. We are also careful to interactively communicate and to have sufficient time so that the examinee and his/her family can ask what they want to know.

Even if one has a diagnosis of malignancy or suspected malignancy, the details of subsequent treatment will be determined after the visit to the thyroid surgeon. Therefore, we are giving him/her some general additional information about treatment at the time of the secondary confirmatory examination; for example, we give them information that thyroid cancer diagnosed at this time is usually small and fairly early stage, thyroid cancer is generally treated with surgery, the length of hospital stay for surgery is short (within a week in the case of a single lobectomy of the thyroid at our hospital), and after surgery, many patients are able to return to their school and work like before.

Many examinees and their families feel a little relieved when they learn that thyroid cancer has a better prognosis than other cancers and that they can participate in sports and other activities without problems after surgery. In addition, seamless support from the same support member continues after surgical referral, and we try to make patients feel as relieved as possible.

Due to the examinee's age and circumstances, the treatment may coincide with exams, schooling, or employment. In such cases, we listen to the thoughts and wishes of the patient and his/her family and consult with the doctor.

Q6: "Will this examination continue?" "Is it better for me to undergo this examination?"

We are now in the fifth rounds of TUE, and there have recently been more such questions about the "Program of TUE." Because most examinees of school age undergo primary examination at school, some examinees and their families believe that thyroid screening must be required until high school, or that the TUE program is not available after high school. On the other hand, there are those who strongly advocate for the continuation of TUE.

So, the support member informs them as neutrally as possible that the future rounds of TUE will be available after the examinees graduate from high school, and that it is a voluntary examination. We explain where, when, and how the next round of TUE can be taken after graduation from high school. In particular, the voluntary nature of thyroid monitoring has been more strongly recommended (International Agency for Research on Cancer, 2018) and the support member has a more careful explanation of the advantages and disadvantages[b] of undergoing TUE in Fukushima (Shimura et al., 2020).

Even if the nodule has a benign finding, the examinee is more likely to be called back for the secondary confirmatory examination in the next round when the nodule again exceeds the size of the criteria. Some people who are repeatedly called in for the secondary examination feel relieved by the confirmation that there is nothing wrong with the results of the tests while others feel burdened by continuing the examination itself. The degree to which examinees are encouraged to visit the next round of TUE is flexible because it depends on the findings of the current conditions and changes in the size of the nodules to date. A support member is always in attendance when the doctor explains the next round of examination. After that, the support member again explains the voluntary nature and the advantages and disadvantages of TUE, and helps the examinee and his/her family fully consider future TUE. The next examination would be 2 years later (or 5 years later for those who are over 25), so both the TUE situation and the individual's circumstances may have changed. For this reason, we encourage them to read the announcement letter for the next round of examination at that time and to think for themselves about whether to undergo TUE as an opportunity to manage their own health.

Recent considerations and tasks of psychosocial support during TUE

TUE, which began as a result of the traumatic events of the nuclear accident, is now in its 10th year. At the beginning of TUE, young children resisted the examination and cried because they didn't understand the purpose of the ultrasound examination. But most of them are now teenagers or older and there is no longer confusion in the examination scene. In addition, there was great relief for Fukushima residents when the experts reported that results from the previous TUE indicated that the effects of radiation exposure due to the nuclear accident were unlikely. A support system is also being developed for the secondary confirmatory examination, where anxiety and concern are often induced. One of

---

b. The advantages of having a thyroid examination are that (1) if the examination results show no problems, it can lead to peace of mind for those who are concerned about the health effects of radiation, (2) if changes that require treatment are found, it may lead to early diagnosis and treatment, and (3) the possibility of clarifying radiation risk in Fukushima by epidemiological analyses. On the other hand, disadvantages include (1) the possibility of diagnosing and treating a harmless thyroid cancer that may go unnoticed for the rest of one's life, (2) the possibility of some complications associated with treatment, and (3) the discovery of nodules and cysts, which may lead to anxiety even if they are benign.

these efforts is the Thyroid Support Team in collaboration with specialists in clinical thyroidology at Fukushima Medical University.

On the other hand, the psychosocial support during TUE has had the following unique difficulties that differed from support in other medical clinical scenes. In response to these difficulties, support for the examinees and their families has continued through a trial-and-error process, and new considerations have been needed recently.

Consideration for the organization of information and communication in the context of TUE

The nuclear disaster is a CBRNE[c] disaster that causes strong fear and anxiety among people, and the TUE program has characteristics related to the anxiety about radiation in itself because it was initiated due to a nuclear disaster. In particular, there was a historical background of the high incidence of childhood thyroid cancer after the Chernobyl nuclear power plant accident, which had a strong impact on the public. There was also medical background of the lack of evidence on the epidemiological features and methodology of support for patients with childhood thyroid cancer. For this reason, there was a strong public request for TUE, and these examinations had the major purpose of accumulating evidence and determining the presence and extent of radiation effects. Regarding the short-term effects of radiation, as mentioned above, it is currently considered unlikely that children will be affected by radiation. But there is a need for continuous examination and analysis of the results with regard to long-term radiation effects. However, on a different level from its original purpose, overdiagnosis has been pointed out as a negative effect of mass thyroid examination (Midorikawa et al., 2017; Schnadig, 2018), therefore, increasing demands for a more voluntary examination (International Agency for Research on Cancer, 2018). This is likely to cause a decrease in the number of examinees. Nevertheless, the TUE program itself can help to alleviate anxiety about the health effects of radiation exposure among Fukushima residents. Taking these circumstances into account, there would be a lot of work to be done and tasks in the future TUE to meet the demands of the Fukushima residents who want the long-term effects of radiation on children to be examined.

It is not easy for the general public to understand these complicated TUE histories, as they require medical knowledge such as thyroid monitoring and thyroid cancer characteristics. However, examinees and their families who undergo secondary confirmatory examinations are particularly concerned about thyroid cancer and may have additional future examinations, so they need support to facilitate their understanding of the nature and background of the TUE,

---

c. CBRNE stands for chemical, biological, radiological, nuclear, and explosive, and the disasters caused by them are called CBRNE disasters. Terrorism, nuclear accidents, and other infectious disease pandemics are treated in the same way as CBRNE disasters.

so they understand how to manage their current thyroid condition and accurately understand the medical findings (Setou et al., 2021; Shimura et al., 2020). In addition, because thyroid cancer is age-dependent, it is expected to increase spontaneously with the age of the examinee in the future (Shimura, 2016). Therefore, the need for support is even greater.

In addition, even after we convey the information, it may be important to confirm how the examinees and their families perceive and understand the risk of radiation effects, because generally perceptions of risk of radiation exposure vary greatly from person to person. At present, the amount of information, their experience with the nuclear accident, and their perceptions of the risks of radiation effects are very diverse among the examinees and their families. The support members need to verify and organize their information and then provide the necessary information. In particular, nearly 40% of the people who lived in the area close to the nuclear power plant accident and who actually experienced fear and trauma are still concerned about both the effects of radiation in later years and the effects on future generations, according to the results of the Fukushima Health Management Survey (Radiation Medical Science Center for the Fukushima Health Management Survey, n.d.-b). Therefore, it is necessary to continue to communicate adequately with examinees and their families during the secondary confirmatory examination to support them in organizing information and reducing their anxiety. Further work needs to be done to determine what kind of support would be more effective in the short time available for support during the secondary examination.

Changes in the age of the examinees and consideration for family support

In the 10 years that have passed since TUE started, the children who were 0–18 years old at the beginning of the examination are now 10–29 years old. In another 8 years, all the examinees will be over 18. Children change their level of interest and understanding of matters depending on their age. Each young individual has his or her own developmental tasks at that age. In all cases in the support scenes, therefore, it is important to take into account the characteristics of the examinee's age and to anticipate their later developmental stages.

For the past few years, our support team has more focused on confirming their understanding and explaining not only to the family, but also to the examinees themselves, taking into consideration the increased numbers of the AYA generation. Many young people in Fukushima Prefecture move out of the prefecture after graduating from high school for higher education, employment, or marriage. Therefore, it is necessary to consider support systems to anticipate an increasing number of out-of-prefecture examinees in the future (Setou et al., 2021). For example, we may need to increase the number of out-of-prefecture medical institutions for TUE and strengthen support for examinees and their families there. As we have described, psychosocial support for TUE requires understanding the background of previous thyroid examinations and the anxiety of the examinees and their families. Therefore, it is necessary to consider TUE support outside the prefecture more specifically, including how to assign and train support staff in the future, taking into account the wishes of the examinees and convenience.

In addition, as the examinee gets older, there is a need to consider whom to respect and how to reconcile opinions when there is a disagreement between the examinee's thoughts and his/her family's opinions. During the secondary examination, the families are more concerned about both current examinees' conditions and future ones, and they ask us more questions than the examinees. High levels of family anxiety are similar to those reported after past nuclear disasters (Bromet, 2014; Bromet, Havenaar, & Guey, 2011). Several previous studies have also reported that the mental health level of family members (mothers) after a disaster affects their children (Bromet, Guey, et al., 2011; Schonfeld & Demaria, 2015; Linet et al., 2018). However, it is the examinee himself/herself who actually should decide whether to continue to undergo examinations and may have examinations or treatments in the future. In addition, some of the examinees may not share their results or feelings with their families because they do not want their parents to worry. While communication within the family is especially important when there are disagreements among family members or if someone in the family feels like they have to deal with it alone, the AYA generation is also at a stage when conflicts and considerations arise with their parents. For this reason, the role of the support member can be crucial in coordinating family opinions (Rolland, 2018). In this context, psychosocial support, especially during the secondary confirmatory examination, includes the important element of "family support." It will be necessary to further consider the form of family support in future TUE.

References

Bromet, E. J. (2014). Emotional consequences of nuclear power plant disasters. *Health Physics*, *106*(2), 206–210. https://doi.org/10.1097/HP.0000000000000012.

Bromet, E. J., Guey, L. T., Taormina, D. P., Carlson, G. A., Havenaar, J. M., Kotov, R., et al. (2011). Growing up in the shadow of Chornobyl: Adolescents' risk perceptions and mental health. *Social Psychiatry and Psychiatric Epidemiology*, *46*(5), 393–402. https://doi.org/10.1007/s00127-010-0203-5.

Bromet, E. J., Havenaar, J. M., & Guey, L. T. (2011). A 25-year retrospective review of the psychological consequences of the Chernobyl accident. *Clinical Oncology (Royal College of Radiologists (Great Britain))*, *23*(4), 297–305. https://doi.org/10.1016/j.clon.2011.01.501.

Fukushima Prefecture. (n.d.). Status of the evacuation zone and support for the affected people. Retrieved November 11, 2020, from https://www.pref.fukushima.lg.jp/site/portal/list271.html.

Hayashida, N., Imaizumi, M., Shimura, H., Okubo, N., Asari, Y., Nigawara, T., et al. (2013). Thyroid ultrasound findings in children from three Japanese prefectures: Aomori, Yamanashi and Nagasaki. *PLoS One*, *8*(12). https://doi.org/10.1371/journal.pone.0083220, e83220.

International Agency for Research on Cancer. (2018). *IARC expert group on thyroid health monitoring after nuclear accidents: Vol. IARC Technical Publication No. 46.*

Japan Breast-Thyroid Ultrasound Medical Association, Thyroid Terminology Diagnosis Standard Committee. (2016). *Thyroid ultrasound diagnostic guidebook.* (Revised 3rd Edition).

Kim, E., Tani, K., Kunishima, N., Kurihara, O., Sakai, K., & Akashi, M. (2016). Estimation of early internal doses to Fukushima residents after the nuclear disaster based on the atmospheric dispersion simulation. *Radiation Protection Dosimetry, 171*(3), 398–404.

Kim, E., Yajima, K., Hashimoto, S., Tani, K., Igarashi, Y., Iimoto, T., et al. (2020). Reassessment of internal thyroid doses to 1,080 children examined in a screening survey after the 2011 Fukushima nuclear disaster. *Health Physics, 118*(1), 36–52. https://doi.org/10.1097/HP.0000000000001125.

Linet, M. S., Kazzi, Z., & Paulson, J. A. (2018). Pediatric considerations before, during, and after radiological or nuclear emergencies. *Pediatrics, 142*(6). https://doi.org/10.1542/peds.2018-3001.

Midorikawa, S., Tanigawa, K., Suzuki, S., & Ohtsuru, A. (2017). Psychosocial issues related to thyroid examination after a radiation disaster. *Asia-Pacific Journal of Public Health, 29*(2_suppl), 63S–73S. https://doi.org/10.1177/1010539516686164.

Ministry of the Environment. (2017). *TUE—Comparison of a preliminary baseline survey and surveys in other prefectures.*

Ministry of the Environment. (n.d.). Assessment of international nuclear and radiological event scale. Retrieved November 9, 2020, from https://www.env.go.jp/chemi/rhm/h28kisoshiryo/h28kiso-06-02-04.html.

Murakami, M., Takebayashi, Y., Takeda, Y., Sato, A., Igarashi, Y., Sano, K., et al. (2018). Effect of radiological countermeasures on subjective well-being and radiation anxiety after the 2011 disaster: The Fukushima Health Management Survey. *International Journal of Environmental Research and Public Health, 15*(1). https://doi.org/10.3390/ijerph15010124.

Ohtsuru, A., Midorikawa, S., Ohira, T., Suzuki, S., Takahashi, H., Murakami, M., et al. (2019). Incidence of thyroid cancer among children and young adults in Fukushima, Japan, screened with 2 rounds of ultrasonography within 5 years of the 2011 Fukushima Daiichi Nuclear Power Station Accident. *JAMA Otolaryngology. Head & Neck Surgery, 145*(1), 4–11. https://doi.org/10.1001/jamaoto.2018.3121.

Prefectural Committee Meeting for Fukushima Health Management Survey, 2016. Interim report on the Fukushima health management survey. In Report of the Fukushima health management survey (2019). http://kenko-kanri.jp/en/img/report_r1.pdf (Accessed 9 November 2020).

Radiation Medical Science Center for the Fukushima Health Management Survey. (n.d.-a). Report of the Fukushima Health Management Survey (FY2019). Retrieved November 9, 2020, from http://kenko-kanri.jp/img/report_r1.pdf.

Radiation Medical Science Center for the Fukushima Health Management Survey. (n.d.-b). Report on the Results of the FY 2018 Survey on Mental Health and Lifestyle Habits. Retrieved November 9, 2020, from https://www.pref.fukushima.lg.jp/uploaded/attachment/386028.pdf.

Rolland, J. S. (2018). *Helping couples and families negative illness and disability. An integrated approach: A integrated approach.* The Guilford Press.

Schnadig, V. J. (2018). Overdiagnosis of thyroid cancer: Is this not an ethical issue for pathologists as well as radiologists and clinicians? *Archives of Pathology & Laboratory Medicine, 142*(9), 1018–1020. https://doi.org/10.5858/arpa.2017-0510-ED.

Schonfeld, D. J., & Demaria, T. (2015). Providing psychosocial support to children and families in the aftermath of disasters and crises. *Pediatrics, 136*(4), e1120–e1130. https://doi.org/10.1542/peds.2015-2861.

Setou, N., Suzuki, S., Matsuzuka, T., Iwadate, M., Maeda, M., Namekata, Y., et al. (2021). Psychosocial support for the examinees and their family during the secondary confirmatory examination: Analyses of support records at first visit. *Fukushima Journal of Medical Science*, *67*(2), 1–11.

Shimura, H. (2016). Characteristics and prevalence of thyroid incidentaloma. *Journal of Breast and Thyroid Sonology*, *5*(3), 5–9.

Shimura, H. (2019). Current status of thyroid ultrasound examination in Fukushima. *Journal of Breast and Thyroid Sonology*, *8*(1), 3–7.

Shimura, H., Suzuki, S., Fukushima, T., Midorikawa, S., Suzuki, S., Hayashida, N., et al. (2014). Prevalence of thyroid nodular lesions in children and adolescents. *Fukushima Journal of Medical Science*, *60*(2), 196–202. https://doi.org/10.5387/fms.2014-28.

Shimura, H., Yokoya, S., & Kamiya, K. (2020). An accurate picture of Fukushima's thyroid ultrasound examination program. *Archives of Pathology & Laboratory Medicine*, *144*(7), 797.

Suzuki, S. (2016). Childhood and adolescent thyroid cancer in Fukushima after the Fukushima Daiichi nuclear power plant accident: 5 years on. *Clinical Oncology (Royal College of Radiologists (Great Britain))*, *28*(4), 263–271. https://doi.org/10.1016/j.clon.2015.12.027.

Suzuki, S. (2017). *The features of childhood and adolescent thyroid cancer after the Fukushima nuclear power plant accident. Thyroid cancer and nuclear accidents: Long-term after effects of Chernobyl and Fukushima. 15* (pp. 155–163). Academic Press.

Suzuki, S., Suzuki, S., Fukushima, T., Midorikawa, S., Shimura, H., Matsuzuka, T., et al. (2016). Comprehensive survey results of childhood thyroid ultrasound examinations in Fukushima in the first four years after the Fukushima Daiichi nuclear power plant accident. *Thyroid : Official Journal of the American Thyroid Association*, *26*(6), 843–851. https://doi.org/10.1089/thy.2015.0564.

Taniguchi, N., Hayashida, N., Shimura, H., Okubo, N., Asari, Y., Nigawara, T., et al. (2013). Ultrasonographic thyroid nodular findings in Japanese children. *Journal of Medical Ultrasonics (2001)*, *40*(3), 219–224. https://doi.org/10.1007/s10396-013-0456-1.

Chapter 14

Radiation risk perception after the Fukushima disaster

Michio Murakami[a,*], Makiko Orita[b], and Naoya Sekiya[c]
[a]*Fukushima Medical University, Fukushima, Japan,* [b]*Atomic Bomb Disease Institute, Nagasaki University, Nagasaki, Japan,* [c]*The University of Tokyo, Tokyo, Japan*

Introduction

For many years, radiation risk perception, or radiation anxiety, has been a central part of research on risk perception in multiple academic disciplines. One is the psychological field. Starr investigated the social acceptance of various technologies in terms of risk and benefit–nuclear safety in particular (Starr, 1969). Slovic determined that risk perception can be summarized into dread risk and unknown risk dimensions, and that stronger regulation is required when dread risk perception and unknown risk perception are higher (Slovic, 1987). Radiation-related technologies are perceived as high dread risk among the various hazards—a finding that has been observed in several countries (Bronfman & Cifuentes, 2003; Kleinhesselink & Rosa, 1991). Interestingly, research on dread risk perceptions conducted after the Fukushima disaster (March 2011) confirmed that people with higher dread risk perception demanded stronger regulations (Machida et al., 2020). These studies imply that the public perceives risk on a different scale from experts, which does not necessarily mean that they are incorrect. For example, consider radioactive materials in foods after a nuclear disaster and traffic accidents (Murakami, Kumagai, & Ohtsuru, 2018): even though the mortality rate for traffic accidents is higher than for radioactive materials in foods, the delayed effects of radioactive materials can cause anxiety about it after consumption of food. In this context, risk perception is uniquely human and not illogical.

Another psychological approach is to organize the factors associated with risk perception. For example, per Sjoberg's (2000) model, attitude, risk sensitivity, and specific fears are risk perception factors. Furthermore, salient value similarity—via trust—governs risk (and benefit) perceptions (Siegrist et al.,

* Current address: Osaka University, Osaka, Japan

Health Effects of the Fukushima Nuclear Disaster. https://doi.org/10.1016/B978-0-12-824098-4.00015-5

2000; Sjoberg, 2000). For example, in a Swiss study, trust via risk (and benefit) perceptions governed the acceptance of a nuclear power plant. The model structure itself had been consistent before and after the Fukushima disaster, although the magnitude of each factor's contribution was different (Visschers & Siegrist, 2013).

Another major academic discipline is psychiatry, in which radiation risk perception has been investigated from the perspective of promoting mental health. The Chernobyl accident's effect on mental health has been a major risk (World Health Organization, 2006) especially because radiation risk perception is associated with poor mental health (Bromet, 2012). In this regard, a strong association was found between radiation risk perception and psychological distress in the aftermath of the Fukushima disaster (Suzuki et al., 2015).

The objectives of this study are twofold. First, we summarize the main survey items used after the Fukushima disaster, and introduce the differences in radiation risk perceptions among areas and how they changed annually following the disaster. Second, we systematically summarize the factors and effects of radiation risk perception, post-Fukushima.

Takebayashi et al. conducted a systematic review of papers published up to May 16, 2017 (Takebayashi et al., 2017). In this paper, we surveyed papers published from there on and added new findings. Based on those findings, this paper discusses the essential implications of investigating and discussing radiation risk perception after a nuclear disaster.

Radiation risk perception: Survey/questionnaire items and changing perceptions after the Fukushima disaster

Radiation risk perception: Questionnaire items

The questionnaire items used by different authors in the aftermath of the Fukushima disaster (Table 1) included: dread and unknown risk perception by Slovic, risk perception for delayed and genetic effects by Lindell, and seven-item radiation anxiety (subscales include radiation anxiety related to health and radiation anxiety related to discrimination and prejudice) by Umeda et al. (Fukasawa et al., 2017, 2019; Kuroda et al., 2018; Lindell & Barnes, 1986; Murakami et al., 2016; Slovic, 1987; Suzuki et al., 2015; Umeda et al., 2014). There are single-item scales on general or specific radiation anxiety (Hino et al., 2016; Suzuki, Murakami, et al., 2018). Some studies included a questionnaire item on food intake avoidance rather than on radiation anxiety itself (Takebayashi et al., 2020; Tsubokura et al., 2020). In this regard, food avoidance differs from the questionnaire items for other risk perceptions; it is used as a proxy for radiation risk perception and investigates actual behavior.

Radiation risk perception: Geographic differences

Fig. 1A shows the differences in the proportion of people with high radiation risk perception regarding their living environment among the three regions, Hamadori (eastern coastal region, closest to the Fukushima Daiichi nuclear

TABLE 1 Questionnaire items used post-Fukushima disaster, including examples of application to the Fukushima disaster and original literature

Risk perception scale	Original reference	Examples of application after Fukushima disaster	Questionnaire
Dread risk	Slovic (1987)	Murakami et al. (2016)	• It is difficult to reduce the effects of radiation on health • Radiation may have a fatal effect on health • The effects of radiation on health are increasing following the Great East Japan Earthquake • Effects of radiation on future generations will occur • Radiation is intuitively dreaded • Cancer risk will increase as a result of radiation • Radiation kills many people at once
Unknown risk	Slovic (1987)	Murakami et al. (2016)	• The effects of radiation on health are unknown • Health risks from radiation are known to science. (reversed) • The effects of radiation on health are immediate. (reversed) • The people surrounding you have correct knowledge about radiation. (reversed)
Delayed effect	Lindell and Barnes (1986)	Suzuki et al. (2015)	• What do you think is the likelihood of damage to your health (e.g., cancer onset) in later life as a result of your current level of radiation exposure?
Genetic effect	Lindell and Barnes (1986)	Suzuki et al. (2015)	• What do you think is the likelihood that the health of your future (i.e., as-yet unborn) children and grandchildren will be affected as a result of your current level of radiation exposure?

Continued

TABLE 1 Questionnaire items used post-Fukushima disaster, including examples of application to the Fukushima disaster and original literature—cont'd

Risk perception scale	Original reference	Examples of application after Fukushima disaster	Questionnaire
Seven-item radiation anxiety: Radiation anxiety regarding health	Umeda et al. (2014)	Fukasawa et al. (2017, 2019); Kuroda et al. (2018)	• I am concerned about getting a serious illness in the future due to the effects of radiation • Every time I feel ill, I am afraid this is caused by radiation exposure • I am concerned that radiation effects can be inherited by the next generation such as children and grandchildren • I feel strong anxiety when I see news reports concerning the nuclear power plant accident
Seven-item radiation anxiety: Radiation anxiety regarding discrimination and prejudice	Umeda et al. (2014)	Fukasawa et al. (2017, 2019), Kuroda et al. (2018)	• I have had the experience of being discriminated against (or unfairly treated) because I lived in the area that is reported to have high levels of radiation • I try not to tell others that I am a resident of that area as far as possible • I have experienced conflicts and trouble with my family members over the radiation health effects

power plant [NPP]), Nakadori (central region), and Aizu (western mountainous region) (Fukushima Prefecture, 2020c). In terms of the differences in radiation risk perception by place of residence, the proportion of people with high radiation risk perception was highest in Hamadori, followed by Nakadori and Aizu in the Fukushima Prefecture in 2019. Radiation risk perception regarding the living environment is generally strongly and positively associated with proximity to the NPP and weakly or insignificantly with air dose rates in residence (Suzuki, Murakami, et al., 2018).

With regard to relocation and risk perception, residents of 13 municipalities, including the evacuation order area, were divided into two groups based on where they lived after the Fukushima disaster: inside and outside Fukushima Prefecture (2020b). Those who relocated from Fukushima Prefecture after the disaster had higher radiation risk perceptions than those who lived in Fukushima Prefecture (Fig. 1B).

Furthermore, residents of Fukushima Prefecture were classified into four categories: evacuated (now, in the past, or voluntarily), and not evacuated (Murakami et al., 2016). The dread risk perceptions of radioactive materials in foods differed largely between those who evacuated (or had been evacuated) and those who did not evacuate. Compared to the radiation risk perception of residents of the Tokyo and Osaka Prefectures, the risk perception of Fukushima residents who evacuated was much higher, and those who did not evacuate, lower (Fig. 1C). Interestingly, there was no difference in radiation risk perception between those who evacuated voluntarily and involuntarily. This implies that the direction only of high radiation risk perceptions of those moved far away from the NPP cannot explain the causal relationship between location and radiation risk perception, but that those who experienced evacuation have a higher radiation risk perception.

Internationally, Fig. 1D shows that the proportion of people who had high radiation risk perception was higher in Asian countries (excluding Japan), and lower in Western countries (including France, the United Kingdom, and the United States) (Sekiya, 2018).

In summary, the radiation risk perception differences as per place of residence indicate that in Fukushima Prefecture, people who live near the NPP have a higher perceived radiation risk to their living environment. Those who have experienced a voluntarily or involuntarily evacuation—especially those who evacuated from Fukushima Prefecture—have higher radiation risk perception. In contrast, residents of Fukushima Prefecture who did not evacuate had a lower perceived risk of radiation in foods than residents of other Japanese prefectures. Furthermore, people among Asian countries had a higher perceived risk of radiation in foods produced in Fukushima than people in Japan itself.

Radiation risk perception: Annual changes among residents in Fukushima Prefecture

Fig. 2A shows the yearly changes in radiation risk perceptions among residents in Fukushima Prefecture (Fukushima Prefecture, 2020a, 2020b, 2020c;

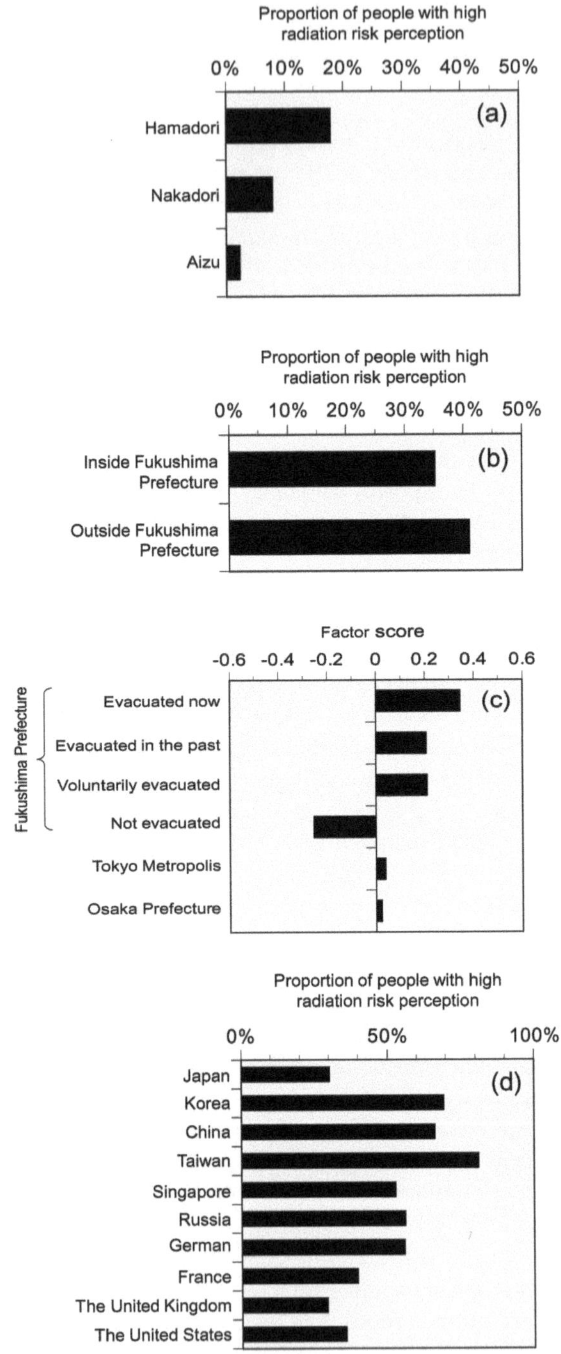

FIG. 1 Differences in high radiation risk perception among place of residence. (A) Radiation risk perception regarding life environment among three regions of Fukushima Prefecture in 2019 (target population: residents in Fukushima Prefecture) (Fukushima Prefecture, 2020c), (B) Radiation risk perception of genetic effect between inside and outside Fukushima Prefecture in 2019 (target population: residents in 13 municipalities including evacuation order areas) (Fukushima Prefecture, 2020b), (C) Dread risk perception of radioactive materials in foods in 2015 (target population: Fukushima, Tokyo, and Osaka Prefectures) (Murakami et al., 2016). (D) Radiation risk perception of crops produced in Fukushima in 2017 (target population: residents in the largest city in each country) (Sekiya, 2018).

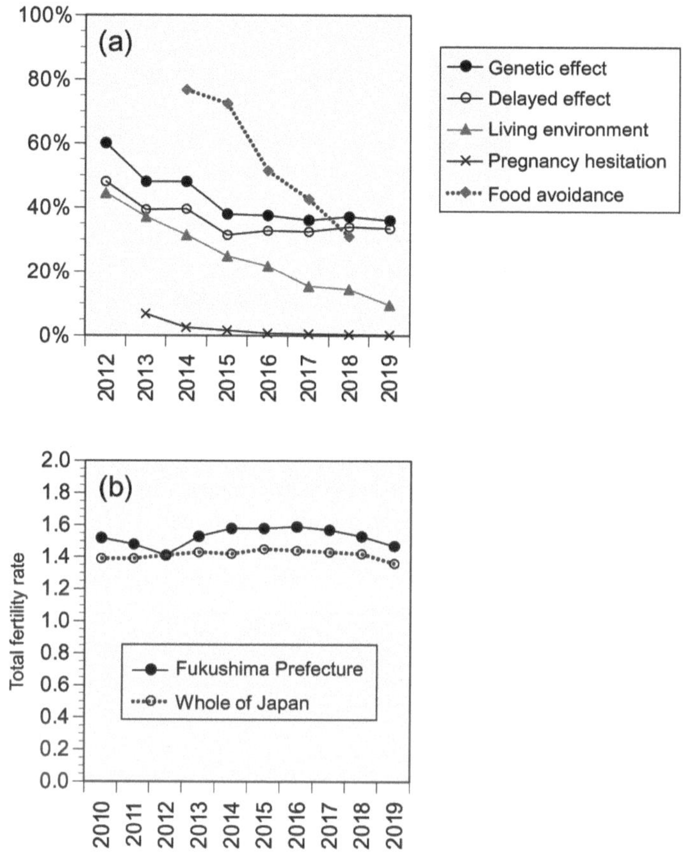

FIG. 2 Yearly changes of radiation risk perception and total fertility rates in Fukushima Prefecture. (A) Radiation risk perception, (B) total fertility rate in Fukushima Prefecture and whole of Japan (Ministry of Health Labour and Welfare, 2020). Genetic effects and delayed effects: Residents in 13 municipalities (Fukushima Prefecture, 2020b). Living environment: residents in Fukushima Prefecture (Fukushima Prefecture, 2020c). Pregnancy hesitation (no pregnancy intention next time due to concern about radiation): pregnant and parturient women in Fukushima Prefecture (Fukushima Prefecture, 2020a). Food avoidance (avoidance of rice collected in Fukushima): Parents of children aged 6 months to 6 years in Minamisoma City (Tsubokura et al., 2020).

Tsubokura et al., 2020). Despite the international scientific consensus that there has been no discernible increase in the genetic effects after the Fukushima disaster (United Nations Scientific Committee on the Effects of Atomic Radiation, 2014), in 2012 approximately 60% of the affected people perceived that the genetic effect was very likely (Fukushima Prefecture, 2020b). Despite the gradual decline of risk perceptions of genetic effects ever since, as of 2019, more than 30% of the population still held high risk perceptions. A similar trend was observed for risk perceptions of delayed effects.

On the other hand, Fukushima Prefecture residents' perceptions of radiation risk to their living environment have consistently declined (Fukushima Prefecture, 2020c). Furthermore, a questionnaire to parents of children aged 6 months to 6 years in Minamisoma City (a municipality largely affected by the Fukushima disaster) showed a large decrease in the avoidance of rice produced in Fukushima (Tsubokura et al., 2020). Moreover, in a survey of pregnant women across this prefecture, the proportion of pregnancy hesitation (i.e., the proportion of those who answered that they had no intention to fall pregnant again due to radiation concerns) was about 15% in 2013 and less than 1% in 2019 (Fukushima Prefecture, 2020a). Thus, it is important to note that the trends vary widely depending on the target population. It also depends on the radiation risk perception scales used. In general, results from Lindell's scale, which addresses perceptions of possible radiation health effects, are characterized by a small decline in radiation risk perception, both for delayed effects and for genetic effects.

While radiation risk perceptions of genetic effects remain constantly high, actual fertility rates show a different trend. In Fukushima Prefecture, it was higher than the national rate prior to the disaster, but it declined to a level similar to the national rate in 2012 (after the disaster). (Ministry of Health Labour and Welfare, 2020). The lower fertility rate may have been due to the radiation risk perception, especially because there was a negative correlation between future pregnancy intention and concerns about radioactive materials in breast milk in 2012 and 2013 (Goto et al., 2019). However, the total fertility rate has been increasing since 2013, and in particular, since 2014 it has been higher than prior to the Fukushima disaster. It is likely that parents who were reluctant to give birth immediately after the disaster did so only after about 3 years had passed. These findings suggest that while some people perceive that the genetic effects after the Fukushima disaster are likely, many people in Fukushima Prefecture do not hesitate to have children. This highlights the need to carefully interpret the meaning and significance of radiation risk perception surveys after a nuclear disaster.

Radiation-risk perception: Factors and effects post-Fukushima

Method

Radiation risk perception surveys highlight the diversity of the effects of radiation risk perceptions. The effects also include psychological distress (as found in the Chernobyl accident) (Bromet, 2012; Suzuki et al., 2015) and pregnancy intention (Goto et al., 2019). Takebayashi et al. systematically reviewed 105 papers (published by May 16, 2017) regarding radiation risk perception among Japanese people in the aftermath of the Fukushima disaster (Takebayashi et al., 2017) and summarized the factors and effects of radiation risk perception. Therefore, we screened articles published since then to review the factors and effects of radiation risk perception using a mostly similar methodology (Takebayashi et al., 2017). The study selection criteria were:

A. Survey or intervention participants were living in Japan.
B. Surveys included measurement of risk perception factors or anxiety regarding radiation after the 2011 NPP accident.
C. Factors governing risk perception and anxiety were investigated.
D. The effects of risk perception and anxiety were investigated.

Criteria further required that both A and B and C or D were met. Books as well as review and letter articles were excluded from the selection. We used the MEDLINE database for screening engines because the previous study accepted only articles from the MEDLINE database for study selection among the MEDLINE database, the PsycINFO database, and the Journal of Risk Research (Takebayashi et al., 2017). We searched for ([perception or anxiety or anxieties] AND [radiation or radionuclides or radionuclides or radioactive] AND [Fukushima]) as keywords on July 3, 2020, and found 46 new articles.

Two individuals with different specialties, MM (risk science) and MO (nursing), independently screened articles based on the title and the abstract, and 27 articles remained ($\kappa=0.91$). Then the full text of the articles was assessed to select those that investigated the factors or effects of radiation risk perception, and 25 articles were selected (Fukasawa et al., 2017, 2019, 2020; Goto et al., 2019; Hayashi et al., 2020; Hidaka et al., 2019; Hori et al., 2019; Ito et al., 2018; Kuroda et al., 2018; Kusama et al., 2018; Kusumi et al., 2017; Matsunaga et al., 2019; Miura et al., 2017; Murakami, Hirosaki, et al., 2018; Murakami, Takebayashi, et al., 2018; Nakayama et al., 2019; Orita et al., 2020; Orui et al., 2017; Orui, Fukasawa, et al., 2020; Orui, Nakayama, et al., 2020; Shigemura et al., 2018; Shirai et al., 2019; Suzuki et al., 2019; Suzuki, Murakami, et al., 2018; Suzuki, Takebayashi, et al., 2018). Furthermore, since Umeda et al. (2014) investigated radiation anxiety regarding discrimination and prejudice (Umeda et al., 2014), we added two references on bullying, discrimination, and social division using a hand (manual) search (Murakami et al., 2019; Sawano et al., 2018). In addition to the findings in the previous study (Takebayashi et al., 2017), significant associations were added for factors of radiation risk perception in two or more articles, and effects in one or more articles.

Results

The results of the new organization of factors and effects on radiation risk perception are shown in Fig. 3. It should be noted that the directions of causality are based on the hypothesis of the selected articles and may not be proven sufficiently.

New factors in radiation risk perception include trust in local government (Nakayama et al., 2019; Orui, Nakayama, et al., 2020), air dose rates

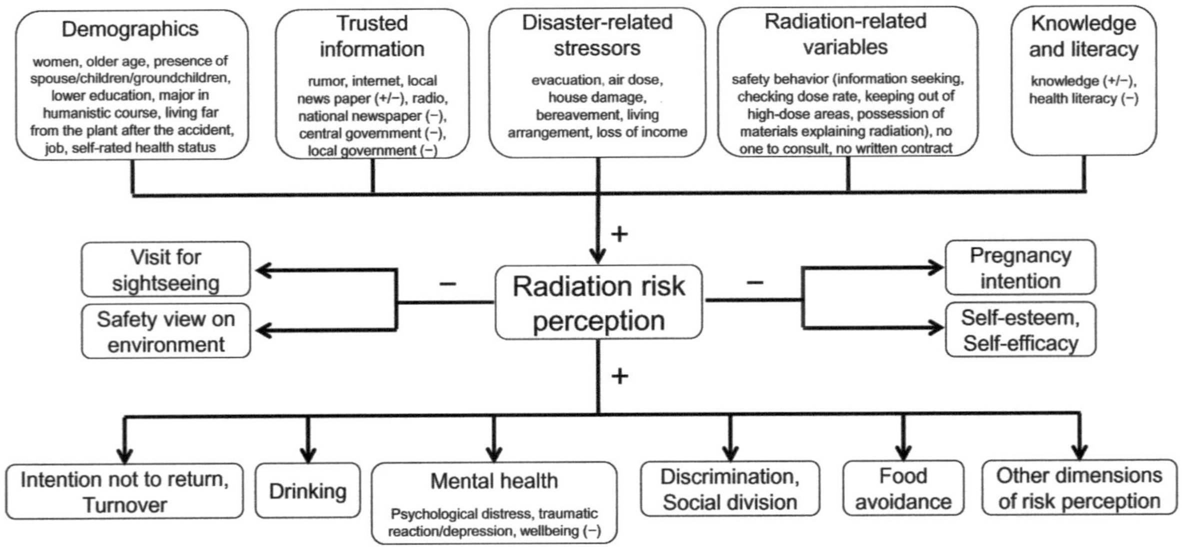

FIG. 3 Factors and effects of radiation risk perception. Findings based on new screening were added to the previous study (Takebayashi et al., 2017). Here, "+" indicates a positive association. For example, women have a higher radiation risk perception. On the other hand, "−" indicates a negative association: e.g., a high radiation risk perception affects no pregnancy intention.

(Fukasawa et al., 2017; Suzuki, Murakami, et al., 2018), knowledge (Hidaka et al., 2019; Nakayama et al., 2019; Shirai et al., 2019), health literacy (Kuroda et al., 2018; Nakayama et al., 2019), and self-rated health status (Kuroda et al., 2018; Nakayama et al., 2019).

Factors were classified into five categories: demographics, trusted information, disaster-related stressors, radiation-related variables, and knowledge and literacy. Of these, the majority (demographics, disaster-related stressors, trusted sources of information, and radiation-related factors such as information-seeking behavior) were unmodifiable or difficult to modify because they reflect values and norms that are closely related to individual or group identity. However, the knowledge (or literacy) category was modifiable, despite the association between knowledge and radiation risk perception being weak. Moreover, inconsistent findings were observed for knowledge: some studies showed positive and negative associations with radiation risk perception (Hidaka et al., 2019; Nakayama et al., 2019; Shirai et al., 2019). In particular, knowledge about physics is positively associated with radiation risk perception, and social and health-related knowledge is negatively associated with radiation risk perceptions (Shirai et al., 2019). Two other studies (Kuroda et al., 2018; Nakayama et al., 2019) also reported a negative association between health literacy and radiation risk perception.

Our screening confirms that the effects of radiation risk perception are diverse. It includes traumatic reactions (Fukasawa et al., 2020; Shigemura et al., 2018), depression (Ito et al., 2018), well-being (Murakami, Hirosaki, et al., 2018), food avoidance (Shirai et al., 2019; Suzuki et al., 2019), alcohol drinking (Orui et al., 2017), visiting Fukushima Prefecture for sightseeing (Shirai et al., 2019), safety views on the environment (Suzuki, Murakami, et al., 2018), pregnancy intention (Goto et al., 2019; Ito et al., 2018), self-esteem and self-efficacy (Ito et al., 2018), discrimination, and social division (Murakami et al., 2019; Sawano et al., 2018).

The mental health issues (a typically assessed effect) include psychological distress, traumatic reactions, depression, and well-being, or its related outcome, alcohol drinking. Others, including lack of intention to return or leave from work, declining safety views on the environment, hesitancy about visiting for sightseeing, and food avoidance, refer to the avoidance of radiation-related matter. In addition to pregnancy intention, there were also issues that had significant implications for personal development and life events, such as self-esteem, self-efficacy, discrimination, and social division.

In summary, radiation risk perception is governed by factors that are mostly difficult to modify (and often should not be modified) and works as a mediator to potentially affect mental health, avoidance of radiation-related matters, and the individual's life events and personality development.

Perspectives of radiation risk perception studies

This paper proposes that radiation risk perception is a natural human emotion, not right or wrong. Radiation risk perception is mainly governed by factors that are difficult to modify (or should not be modified). On the other hand, it is important to consider how to approach radiation risk perception issues given that it can affect mental health, discrimination, and social division.

There are three approaches to radiation risk perception. The first is to address it. Despite the limited effectiveness of knowledge delivery (Shirai et al., 2019), it is nonetheless important to communicate scientific knowledge regarding social and health issues, including genetic effects. Furthermore, authorities or experts have a moral obligation to provide it to those who need it. It would also be beneficial to foster a literacy of universal nature, such as health literacy (Kuroda et al., 2018; Nakayama et al., 2019). Given that experience associates radiation risk perception, one option may be to promote behavior that involves experiences (for example, enjoying foods produced in Fukushima with significant others) under guaranteed safety.

The second approach involves addressing outcomes rather than radiation risk perception. For example, psychological distress has been reported to be more strongly associated with general health anxiety than with radiation risk perception (Kashiwazaki et al., 2020) and mental health can be improved without changes in radiation risk perception (Imamura et al., 2016). Moreover, the formation of a positive image of the place of origin can encourage Fukushima food purchases (Shirai et al., 2019; Suzuki et al., 2019). Solving the problem through a different approach, rather than reducing radiation risk perception, is promising.

The third approach is to dispel the association between risk perception and outcomes via the mitigating mediators between them. For example, consider social division as an outcome. As mentioned, radiation risk perception is governed by individual attributes, identity, trust, and norms. These differences in values, norms, and positions often precede differences in opinion and decision-making, which may result in conflict and eventually social division. In addition, these differences in values, norms, and positions may be a source of divergence in radiation risk perception in terms of different opinions and decision-making during major events. In this case, conflicting opinions or decisions are regarded as mediators between risk perception and social division. On the other hand, previous studies reported that increasing empathy and the awareness of interdependence between groups may reduce discrimination and increase interaction between them (Kobayashi & Oikawa, 2015; Zaki, 2019). Thus, respecting diverse norms, values, and radiation risk perceptions may mitigate conflict and reduce the association between radiation risk perception and discrimination or social division.

In summary, the approach chosen depends on the goals: if it is to reduce the perceived radiation risk itself, the first approach is important. However, if the priority is to reduce psychological distress or social division, then the second and third approaches are more suitable. Therefore, it is important to deliberate both the problems at hand and the shared goals. Generally, research on radiation risk perception essentially requires us to debate the kind of world we want to live in (Select Committee on Science and Technology, 2000), rather than simply focusing on reducing risk perception or the extent of its impact.

Acknowledgments

This survey was partly supported by the National Health Fund for Children and Adults Affected by the Nuclear Incident. The findings and conclusions of this article are solely the responsibility of the authors and do not represent the official views of the Fukushima Prefecture government.

References

Bromet, E. J. (2012). Mental health consequences of the Chernobyl disaster. *Journal of Radiological Protection*, *32*(1), N71–N75. https://doi.org/10.1088/0952-4746/32/1/n71.

Bronfman, N. C., & Cifuentes, L. A. (2003). Risk perception in a developing country: The case of Chile. *Risk Analysis*, *23*(6), 1271–1285. https://doi.org/10.1111/j.0272-4332.2003.00400.x.

Fukasawa, M., Kawakami, N., Nakayama, C., & Yasumura, S. (2019). Relationship between use of media and radiation anxiety among the residents of Fukushima 5.5 years after the nuclear power plant accident. *Disaster Medicine and Public Health Preparedness*, 1–8. https://doi.org/10.1017/dmp.2019.132.

Fukasawa, M., Kawakami, N., Umeda, M., Akiyama, T., Horikoshi, N., Yasumura, S., et al. (2020). Longitudinal associations of radiation risk perceptions and mental health among non-evacuee residents of Fukushima prefecture seven years after the nuclear power plant disaster. *SSM - Population Health*, *10*. https://doi.org/10.1016/j.ssmph.2019.100523, 100523.

Fukasawa, M., Kawakami, N., Umeda, M., Miyamoto, K., Akiyama, T., Horikoshi, N., et al. (2017). Environmental radiation level, radiation anxiety, and psychological distress of non-evacuee residents in Fukushima five years after the Great East Japan earthquake: Multilevel analyses. *SSM - Population Health*, *3*, 740–748. https://doi.org/10.1016/j.ssmph.2017.09.002.

Fukushima Prefecture. (2020a). *Fukushima Health Management Survey: Pregnancy and birth survey*. Retrieved from https://www.pref.fukushima.lg.jp/site/portal/kenkocyosa-ninsanpu.html. (Accessed 4 November 2020).

Fukushima Prefecture. (2020b). *Fukushima Health Management Survey: Mental health and lifestyle survey*. Retrieved from https://www.pref.fukushima.lg.jp/site/portal/kenkocyosa-kokoro.html. (Accessed 4 November 2020).

Fukushima Prefecture. (2020c). *Public opinion survey on prefectural policies*. Retrieved from https://www.pref.fukushima.lg.jp/sec/01010e/koucho1-439.html. (Accessed 4 November 2020).

Goto, A., Tsugawa, Y., Fujimori, K., & Pregnancy Birth Survey Group of the Fukushima Health Management, S. (2019). Factors associated with intention of future pregnancy among women affected by the Fukushima nuclear accident: Analysis of Fukushima Health Management Survey data from 2012 to 2014. *Journal of Epidemiology*, *29*(8), 308–314. https://doi.org/10.2188/jea.JE20180015.

Hayashi, F., Sanpei, M., Ohira, T., Nakano, H., Okazaki, K., Yasumura, S., et al. (2020). Changes in the mental health status of adolescents following the Fukushima Daiichi nuclear accident and related factors: Fukushima Health Management Survey. *Journal of Affective Disorders*, *260*, 432–439. https://doi.org/10.1016/j.jad.2019.09.045.

Hidaka, T., Kakamu, T., Endo, S., Kasuga, H., Masuishi, Y., Kumagai, T., et al. (2019). Association of anxiety over radiation exposure and acquisition of knowledge regarding occupational health management in operation leader candidates of radioactivity decontamination workers in Fukushima, Japan: A cross-sectional study. *International Journal of Environmental Research and Public Health*, *17*(1), 228. https://doi.org/10.3390/ijerph17010228.

Hino, Y., Murakami, M., Midorikawa, S., Ohtsuru, A., Suzuki, S., Tsuboi, K., et al. (2016). Explanatory meetings on thyroid examination for the "Fukushima Health Management Survey" after the Great East Japan earthquake: Reduction of anxiety and improvement of comprehension, and satisfaction. *The Tohoku Journal of Experimental Medicine*, *239*(4), 333–343. https://doi.org/10.1620/tjem.239.333.

Hori, H., Orita, M., Taira, Y., Kudo, T., & Takamura, N. (2019). Risk perceptions regarding radiation exposure among Japanese schoolteachers living around the Sendai nuclear power plant after the Fukushima accident. *PLoS One*, *14*(3). https://doi.org/10.1371/journal.pone.0212917, e0212917.

Imamura, K., Sekiya, Y., Asai, Y., Umeda, M., Horikoshi, N., Yasumura, S., et al. (2016). The effect of a behavioral activation program on improving mental and physical health complaints associated with radiation stress among mothers in Fukushima: A randomized controlled trial. *BMC Public Health*, *16*(1), 1144. https://doi.org/10.1186/s12889-016-3819-x.

Ito, S., Sasaki, M., Okabe, S., Konno, N., & Goto, A. (2018). Depressive symptoms and associated factors in female students in Fukushima four years after the Fukushima nuclear power plant disaster. *International Journal of Environmental Research and Public Health*, *15*(11), 2411. https://doi.org/10.3390/ijerph15112411.

Kashiwazaki, Y., Takebayashi, Y., & Murakami, M. (2020). Relationships between radiation risk perception and health anxiety, and contribution of mindfulness to alleviating psychological distress after the Fukushima accident: Cross-sectional study using a path model. *PLoS One*, *15*(7). https://doi.org/10.1371/journal.pone.0235517, e0235517.

Kleinhesselink, R. R., & Rosa, E. A. (1991). Cognitive representation of risk perceptions: A comparison of Japan and the United States. *Journal of Cross-Cultural Psychology*, *22*(1), 11–28. https://doi.org/10.1177/0022022191221004.

Kobayashi, T., & Oikawa, M. (2015). The effects of meta-stereotypes and beliefs in group interdependence on responses toward outgroup. *The Japanese Journal of Psychology*, *86*(5), 467–473. https://doi.org/10.4992/jjpsy.86.14327 [in Japanese].

Kuroda, Y., Iwasa, H., Orui, M., Moriyama, N., Nakayama, C., & Yasumura, S. (2018). Association between health literacy and radiation anxiety among residents after a nuclear accident: Comparison between evacuated and non-evacuated areas. *International Journal of Environmental Research and Public Health*, *15*(7), 1463. https://doi.org/10.3390/ijerph15071463.

Kusama, T., Aida, J., Tsuboya, T., Sugiyama, K., Yamamoto, T., Igarashi, A., et al. (2018). The association between socioeconomic status and reactions to radiation exposure: A cross-sectional study after the Fukushima Daiichi nuclear power station accident. *PLoS One*, *13*(10). https://doi.org/10.1371/journal.pone.0205531, e0205531.

Kusumi, T., Hirayama, R., & Kashima, Y. (2017). Risk perception and risk talk: The case of the Fukushima Daiichi nuclear radiation risk. *Risk Analysis*, *37*(12), 2305–2320. https://doi.org/10.1111/risa.12784.

Lindell, M. K., & Barnes, V. E. (1986). Protective response to technological emergency: Risk perception and behavioral intention. *Nuclear Safety*, *27*(4), 457–467. https://doi.org/10.1016/0022-4375(89)90021-2.

Machida, N., Murakami, M., Takebayashi, Y., Kumagai, A., & Yamaguchi, T. (2020). Perceived risk and demands for countermeasures against diabetes and radiation in Fukushima after the nuclear accident: Effects of self-rated risk trade-off view and provision of risk information. *International Journal of Disaster Risk Reduction*, *49*. https://doi.org/10.1016/j.ijdrr.2020.101671, 101671.

Matsunaga, H., Orita, M., Iyama, K., Sato, N., Aso, S., Tateishi, F., et al. (2019). Intention to return to the town of Tomioka in residents 7 years after the accident at Fukushima Daiichi Nuclear Power Station: A cross-sectional study. *Journal of Radiation Research*, *60*(1), 51–58. https://doi.org/10.1093/jrr/rry094.

Ministry of Health Labour and Welfare. (2020). Retrieved from https://www.mhlw.go.jp/toukei/list/81-1a.html (accessed 4 November 2020).

Miura, I., Nagai, M., Maeda, M., Harigane, M., Fujii, S., Oe, M., et al. (2017). Perception of radiation risk as a predictor of mid-term mental health after a nuclear disaster: The Fukushima Health Management Survey. *International Journal of Environmental Research and Public Health*, *14*(9), 1067. https://doi.org/10.3390/ijerph14091067.

Murakami, M., Hirosaki, M., Suzuki, Y., Maeda, M., Yabe, H., Yasumura, S., et al. (2018). Reduction of radiation-related anxiety promoted wellbeing after the 2011 disaster: "Fukushima Health Management Survey". *Journal of Radiological Protection*, *38*, 1428–1440. https://doi.org/10.1088/1361-6498/aae65d.

Murakami, M., Kumagai, A., & Ohtsuru, A. (2018). Building risk communication capabilities among professionals: Seven essential characteristics of risk communication. *Radiation Protection Dosimetry*, *182*(1), 120–127.

Murakami, M., Kumagai, A., Stojarov, A. N., & Tsubokura, M. (2019). Radiation is not a political tool. *Science*, *366*(6465), 581–582. https://doi.org/10.1126/science.aaz3408.

Murakami, M., Nakatani, J., & Oki, T. (2016). Evaluation of risk perception and risk-comparison information regarding dietary radionuclides after the 2011 Fukushima nuclear power plant accident. *PLoS One*, *11*(11). https://doi.org/10.1371/journal.pone.0165594, e0165594.

Murakami, M., Takebayashi, Y., Takeda, Y., Sato, A., Igarashi, Y., Sano, K., et al. (2018). Effect of radiological countermeasures on subjective well-being and radiation anxiety after the 2011 disaster: The Fukushima Health Management Survey. *International Journal of Environmental Research and Public Health*, *15*(1), 124. https://doi.org/10.3390/ijerph15010124.

Nakayama, C., Sato, O., Sugita, M., Nakayama, T., Kuroda, Y., Orui, M., et al. (2019). Lingering health-related anxiety about radiation among Fukushima residents as correlated with media information following the accident at Fukushima Daiichi Nuclear Power Plant. *PLoS One*, *14*(5). https://doi.org/10.1371/journal.pone.0217285, e0217285.

Orita, M., Mori, K., Taira, Y., Yamada, Y., Maeda, M., & Takamura, N. (2020). Psychological health status among former residents of Tomioka, Fukushima Prefecture and their intention to return 8 years after the disaster at Fukushima Daiichi Nuclear Power Plant. *Journal of Neural Transmission*, *127*(11), 1449–1454. https://doi.org/10.1007/s00702-020-02160-8.

Orui, M., Fukasawa, M., Horikoshi, N., Suzuki, Y., & Kawakami, N. (2020). Development and evaluation of a gatekeeper training program regarding anxiety about radiation health effects following a nuclear power plant accident: A single-arm intervention pilot trial. *International Journal of Environmental Research and Public Health*, *17*(12), 4594. https://doi.org/10.3390/ijerph17124594.

Orui, M., Nakayama, C., Kuroda, Y., Moriyama, N., Iwasa, H., Horiuchi, T., et al. (2020). The association between utilization of media information and current health anxiety among the Fukushima Daiichi nuclear disaster evacuees. *International Journal of Environmental Research and Public Health*, *17*(11), 3291. https://doi.org/10.3390/ijerph17113921.

Orui, M., Ueda, Y., Suzuki, Y., Maeda, M., Ohira, T., Yabe, H., et al. (2017). The relationship between starting to drink and psychological distress, sleep disturbance after the Great East Japan earthquake and nuclear disaster: The Fukushima Health Management Survey. *International Journal of Environmental Research and Public Health*, *14*(10), 1281. https://doi.org/10.3390/ijerph14101281.

Sawano, T., Nishikawa, Y., Ozaki, A., Leppold, C., & Tsubokura, M. (2018). The Fukushima Daiichi Nuclear Power Plant accident and school bullying of affected children and adolescents: The need for continuous radiation education. *Journal of Radiation Research*, *59*(3), 381–384. https://doi.org/10.1093/jrr/rry025.

Sekiya, N. (2018). *Paper presented at the risk communication frontier 3, Fukushima*. (14 February 2018).

Select Committee on Science and Technology. (2000). *Science and technology, third report*. Retrieved from House of Lords https://publications.parliament.uk/pa/ld199900/ldselect/ldsctech/38/3801.htm. (Accessed 13 September 2021).

Shigemura, J., Someda, H., Tokuno, S., Nagamine, M., Tanichi, M., Araki, Y., et al. (2018). Disaster victim identification: Psychological distress and posttraumatic stress in dentists after the 2011 Fukushima disaster. *Psychiatry*, *81*(1), 85–92. https://doi.org/10.1080/00332747.2017.1297667.

Shirai, K., Yoshizawa, N., Takebayashi, Y., & Murakami, M. (2019). Modeling reconstruction-related behavior and evaluation of influences of major information sources. *PLoS One*, *14*(8). https://doi.org/10.1371/journal.pone.0221561, e0221561.

Siegrist, M., Cvetkovich, G., & Roth, C. (2000). Salient value similarity, social trust, and risk/benefit perception. *Risk Analysis*, *20*(3), 353–362. https://doi.org/10.1111/0272-4332.203034.

Sjoberg, L. (2000). Factors in risk perception. *Risk Analysis*, *20*(1), 1–11.

Slovic, P. (1987). Perception of risk. *Science*, *236*(4799), 280–285. https://doi.org/10.1126/science.3563507.

Starr, C. (1969). Social benefit versus technological risk. *Science*, *165*(3899), 1232–1238. https://doi.org/10.1126/science.165.3899.1232.

Suzuki, S., Murakami, M., Nishikiori, T., & Harada, S. (2018). Annual changes in the Fukushima residents' views on the safety of water and air environments and their associations with the perception of radiation risks. *Journal of Radiation Research*, *59*(suppl_2), ii31–ii39. https://doi.org/10.1093/jrr/rrx096.

Suzuki, T., Oishi, T., Kurokura, H., & Yagi, N. (2019). Which aspects of food value promote consumer purchase intent after a disaster? A case study of salmon products in disaster-affected areas of the Great East Japan earthquake. *Food*, *8*(1), 14. https://doi.org/10.3390/foods8010014.

Suzuki, Y., Takebayashi, Y., Yasumura, S., Murakami, M., Harigane, M., Yabe, H., et al. (2018). Changes in risk perception of the health effects of radiation and mental health status: The Fukushima Health Management Survey. *International Journal of Environmental Research and Public Health*, *15*(6), 1219. https://doi.org/10.3390/ijerph15061219.

Suzuki, Y., Yabe, H., Yasumura, S., Ohira, T., Niwa, S., Ohtsuru, A., et al. (2015). Psychological distress and the perception of radiation risks: The Fukushima Health Management Survey. *Bulletin of the World Health Organization*, *93*(9), 598–605. https://doi.org/10.2471/blt.14.146498.

Takebayashi, Y., Lyamzina, Y., Suzuki, Y., & Murakami, M. (2017). Risk perception and anxiety regarding radiation after the 2011 Fukushima nuclear power plant accident: A systematic qualitative review. *International Journal of Environmental Research and Public Health*, *14*(11), 1306. https://doi.org/10.3390/ijerph14111306.

Takebayashi, Y., Murakami, M., Nomura, S., Oikawa, T., & Tsubokura, M. (2020). The trajectories of local food avoidance after the Fukushima Daiichi nuclear plant disaster: A five-year prospective cohort study. *International Journal of Disaster Risk Reduction*, *46*. https://doi.org/10.1016/j.ijdrr.2020.101513, 101513.

Tsubokura, M., Nabeshima, Y., Murakami, M., Nemoto, T., Kambe, T., Nonaka, S., et al. (2020). Usefulness of the whole-body counter for infants and small children (BABYSCAN) as a risk communication tool after the Fukushima Daiichi nuclear power plant incident. *Proceedings of the Japan Academy, Series B*, *96*(2), 70–78. https://doi.org/10.2183/pjab.96.006.

Umeda, M., Sekiya, Y., Kawakami, N., Miyamoto, K., Horikoshi, N., Yabe, H., et al. (2014). Reliability and validity of radiation anxiety scale developed for Fukushima community residents. In *Proceedings of the 24th annual scientific meeting of the japan epidemiological association, Sendai* (In Japanese).

United Nations Scientific Committee on the Effects of Atomic Radiation. (2014). *Sources, effects and risks of ionizing radiation. Retrieved from UNSCEAR 2013 Reports to the General Assembly with Scientific Annexes*. New York: United Nations.

Visschers, V. H. M., & Siegrist, M. (2013). How a nuclear power plant accident influences acceptance of nuclear power: Results of a longitudinal study before and after the Fukushima disaster. *Risk Analysis*, *33*(2), 333–347. https://doi.org/10.1111/j.1539-6924.2012.01861.x.

World Health Organization. (2006). *Health effects of the Chernobyl accident and special health care programmes: Report of the UN Chernobyl forum expert group "health"*. Retrieved from http://www.who.int/ionizing_radiation/chernobyl/WHO%20Report%20on%20Chernobyl%20Health%20Effects%20July%2006.pdf.

Zaki, J. (2019). *The war for kindness: Building empathy in a fractured world*. New York: Crown.

Chapter 15

Influence of different media, producing stigma

Masaharu Tsubokura
Department of Radiation Health Management, Fukushima Medical University School of Medicine, Fukushima, Japan

A nuclear power plant accident can cause radioactive contamination and have health effects associated with radiation exposure on residents of the surrounding area. Health effects due to radiation exposure are the most important concerns, and assessment is important. Dose measurement and environmental monitoring from the early stage of the accident, evacuation, decontamination, contamination control, and other measures to reduce exposure are essential (Nomura et al., 2020; Tsubokura, Gilmour, Takahashi, Oikawa, & Kanazawa, 2012; Tsubokura, Kato, Nomura, et al., 2014; Tsubokura, Murakami, Takebayashi, et al., 2019). However, the effects on local residents associated with radiation disasters are not limited to those solely caused by radiation exposure (Tsubokura, 2018).

In the short term, evacuation and resettlement pose risks, and infrastructure destruction, such as lack of human and material resources or closure of hospitals, will become an issue (Murakami, Ono, Tsubokura, et al., 2015; Nomura, Gilmour, Tsubokura, et al., 2013; Sawano, Shigetomi, Ozaki, et al., 2021). In the medium term, while evacuation can reduce exposure doses, it has been reported to cause lifestyle changes, stress, posttraumatic stress disorder, and effects associated with changes in the living environment, such as worsening of lifestyle-related diseases and depression (Morita, Nomura, Tsubokura, et al., 2017; Murakami, Tsubokura, Ono, Nomura, & Oikawa, 2017; Nomura, Murakami, Ozaki, et al., 2021). In the long term, this has implications for social and cultural transformation, and eventually for national energy policy. It is important to assess these various secondary health effects and prioritize them while implementing countermeasures.

When considering these various health effects, the ramifications of media and news reports, rumors, social networking services (SNS), and incorrect news cannot be overlooked. In this paper, I will focus on the effects that occurred after the Fukushima nuclear accident, and outline the medium- and long-term health effects, with examples.

Health Effects of the Fukushima Nuclear Disaster. https://doi.org/10.1016/B978-0-12-824098-4.00019-2

The rise of the Internet and SNS

One of the major differences between the Chernobyl and Fukushima nuclear accidents was the change in the distribution and diffusion of information with the development of the Internet and SNS. Although the medium used varies from country to country and from age to age, the situation of being strongly influenced by SNS is the same in all countries. In the past, only certain entities such as the government or a newspaper company could send information. Now, though, individuals can be the main source, and information can be easily spread and seen by many people very quickly. As a result, false information, or information that is accompanied by fear or emotion, is more likely to spread and be shared among residents than before. This situation is not unique to radiation disasters, but can be seen in elections, scandals, and other disasters as well. The impact of the rise of social networking sites on people's perception and understanding of risk is enormous.

There have been various reports on the effects of SNS after the Fukushima nuclear accident, but most have been analyzed from the perspective of short-term information distribution and flaming. Here, I would like to introduce a report on the diffusion of radiation information via Twitter conducted by the author and others in the months following the Fukushima nuclear accident and beyond (Tsubokura, Onoue, Torii, et al., 2018).

The authors conducted a detailed analysis of the 25 million radiation-related tweets and retweets (the action of forwarding the contents of others' tweets directly to their followers) during the 6 months following the Fukushima nuclear accident, to determine what kind of information was sent out and how it was spread. A detailed description will be provided, but the following two points are important.

First, the tweets by a small number of highly influential accounts (influencers) were responsible for most of the information circulated. In fact, about half the radiation-related tweets sent out in the 6 months following the Fukushima nuclear accident were retweets, and about 40% of those were sent out by the top 100 influencers. Although SNS is a medium to promote communication among individuals, specialized content such as radiation in SNS was shown to be strongly influenced by these accounts. The influencers had various backgrounds, such as celebrities who had nothing to do with radiation, and accounts that became famous for repeatedly criticizing the government. On the other hand, the impacts of accounts by scientists such as physicists and biologists as well as government accounts were relatively small, especially those that anonymously disseminated scientific information. Government transmissions had a low impact because they were often made anonymously, and were rarely able to convey information ahead of time. The dissemination of information by scientists was the same, and was sometimes done by individuals striving for accuracy, which led to diminished collaboration and a decline in their own influence. As a result, their transmissions could not be as influential, emotional, or incendiary. The dissemination of information about radiation by influencers, especially actors and entertainers, would have been difficult because it would have affected their brand. Influencers are surrounded by subinfluencers who support them, and there is no end to the sources of emotional and incendiary information. Many attempts have been made to provide accurate information as a fact check, or to counter inaccuracies, but the effect is limited.

Second, information about radiation is fragmented according to its content, and residents who use SNS are easily exposed to one-sided information. When visualizing the spread of information about the Fukushima radiation accident, the statements of "Fukushima after the accident is safe" and "Fukushima after the accident is dangerous" solidified independently of each other. This fragmentation phenomenon, in which influencers with similar ideas connect and retweet, resulted in the same type of information being spread repeatedly in one community. This occurred within a month of the Fukushima nuclear accident, and it has been reported that the subsequent distribution of information between the fragmented groups became less frequent. This situation is called an echo chamber, and it is a characteristic phenomenon of the Internet and many SNSs. People form their own opinions by repeatedly and preferentially being exposed to information they want to know and information that supports their opinions. As a result, the information obtained by residents is greatly influenced by the initial information they are exposed to by chance, and their own opinions are formed by repeated exposure to the same type of information. This results in a structure that fixes prejudices against the residents of the surrounding area and creates stigma.

While it is unknown to what extent the lessons learned in 2011 can be applied to the present, at the very least, the impact of SNS on the spread of information, prejudice, and discrimination is significant, and it remains necessary to take countermeasures. Search results on SNS and the Internet have been improved, and various measures have already been taken, such as introducing a system to present reliable information at the top, suspending malicious accounts, and preventing spam. In fact, several years after the Fukushima nuclear accident, it was confirmed that bots automatically spread information on Twitter. The distribution of information changed after measures against such bots were taken.

Two details should be considered regarding the status of Twitter in the years after the Fukushima nuclear accident. First, the Twitter world regarding radiation was occupied by some influencers, who altered their chosen subject over time, similar to the way in which music sales rankings change over time. Few accounts remain as influencers of radiation issues for more than a few years, and most are very influential for several months or a year, and then disappear from the Twitter world altogether, or reappear as influencers on completely different social issues. This is not unique to the subject of radiation, but is a fundamental characteristic of the Internet.

Second, incorrect news and scientifically uncertain information that circulated in the early days after the accident will remain on the Internet and be referred to repeatedly for a long time. Once the information is sent, it will remain in perpetuity unless the site is deleted or closed. In addition, because it is readily available and can be easily spread, it provides a source that continues to be referenced, even after several years and despite being wrong. The Internet provides information only to individuals who want it, not to individuals who do not or do not search for it. For this reason, in a survey of search engines in which the author was involved years after the accident, the characteristics of individuals who searched the Internet for information on radiation were as follows: (1) those who were thinking of moving to Fukushima Prefecture; (2) those who used the Internet for radiation work; (3) those involved in the fishing industry; (4) those interested in radiation treatment, because they were diagnosed with some physical abnormalities through cancer screening, etc.; and (5) those who were interested in elections, etc.

Because of echo chambers and filter bubbles, individuals are unknowingly exposed to biased information in their favor that they prefer. Because individuals do not passively update their information, it becomes necessary to intervene from the outside to convey information actively. This may be done through radiation education, or the Guguru Project led by the Ministry of the Environment.

Existing media, such as TV, newspapers, and magazines, also had a great impact on the aftermath of the nuclear accident. Media outlets reported on the nuclear accident extensively, but the more nationally oriented the media (such as national newspapers and television), the more they discussed the pros and cons of nuclear power plants themselves and the future direction of nuclear power in Japan, rather than the actual health effects of the accident on the affected people. While the typical information needs of local residents immediately after the nuclear accident were about the safety of local vegetables and water, and the pros and cons of outdoor play for children (Sugimoto, Krull, Nomura, Morita, & Tsubokura, 2012; Sugimoto, Nomura, Tsubokura, et al., 2013), the content of the national media coverage was more about the pros and cons of the linear nonthreshold (LNT) model and the future direction of nuclear power. Studies analyzing the relationship between the media used by the individual and radiation anxiety showed that people who used national newspapers and national media had higher radiation anxiety than those who mainly used local media. The existing media continued to provide information on Fukushima Prefecture, including TV, newspapers, and magazines, and the media was structured to only report on anniversaries, such as the anniversary of the nuclear accident or special events. Examples include when the government or Tokyo Electric Power Company (TEPCO) made a decision, or when an adverse event occurred. Therefore, from the residents' point of view, the story of the Fukushima nuclear power plant was linked to a bad event or special anniversary, and was provided in such a way that it could easily be interpreted negatively or happening repeatedly. This meant that there were limited ways to disseminate positive information about the situation, progress, and recovery to the rest of the prefecture and the nation. This structure is the reason many people outside the prefecture stopped updating their information immediately after the nuclear accident, and it was perceived that the situation was continuing in Fukushima.

Unfortunately, this media structure is not adequate to prevent the adherence of discrimination and prejudice against the residents affected by the nuclear accident. Rather, it tends to be a structure that uses the affected residents as a story, and earns viewership by providing this information. However, this situation is not unique to radiation. This is a limitation of the media, which prioritizes newsworthiness and topicality. It is difficult to provide balanced information about the health effects of radiation and nuclear power because each of the media has a certain attitude toward the government and TEPCO, and a certain attitude toward radiation and nuclear power.

The echo chamber is also not specific to SNSs. Each media outlet has its own direction of information dissemination, and it must be recognized that only one can provide biased information. However, the situation is less extreme in the media than that of the Internet and SNS.

Dissemination of incorrect news

One of the most difficult problems in disseminating accurate information about radiation and preventing prejudice against residents is dealing with incorrect news. While it is difficult to eliminate the spread in times of disaster, incorrect news repeatedly shakes residents emotionally and it creates prejudice and discrimination, leading to fixed risk perception. The impact of incorrect news cannot be underestimated, as it is easily spread because of the influence of SNSs mentioned above.

News spread by residents in the disaster area after the Fukushima nuclear accident in 2018 elicits an example of the spread of incorrect news (Sawano, Ozaki, Hori, & Tsubokura, 2019). The problem began when a city councilor in the disaster area released a leaflet stating that the number of cancer and leukemia patients in his area was increasing rapidly due to radiation exposure. The basis for this statement was the original calculation by the councilor based on information about the number of outpatients presented to the council by the city hospital. He found an increase in the number of cancer and leukemia patients visiting hospitals, and claimed that this was due to radiation exposure, which was entirely false. This is because the increase in the number of patients was due to the increase in the number of medical personnel for the reconstruction, which resulted in an increased ability to provide specialized treatment for cancer and leukemia in those areas compared to before the nuclear accident. In other words, the number of patients who could not be treated due to the absence of specialists before the accident had apparently increased because of the arrival of specialists after the accident. While caution in interpreting the data was natural for public health experts, it was difficult for the public to understand, and this information was disseminated to the public with a certain degree of confidence. The courts even tried to use this information as a basis for judicial decisions related to radiation effects. In response, the local medical professionals, in cooperation with the government, repeatedly and intensively publicized through SNS and the Internet that this information was incorrect news and it was quickly suppressed. Local medical professionals and the government were able to collaborate and disseminate information, so it did not lead to major prejudice, but such incorrect news can be spread repeatedly over a long period of time. To counter such, it is essential for residents to acquire general basic knowledge and understanding of radiation, but this is difficult to achieve because the existing media are likely to report sensational or emotional information. In addition, it is difficult to dispel entrenched prejudices and stigmas at an early stage because residents' interest in radiation tends to wane over time.

There are several such cases not only in the immediate aftermath of the accident, but even several years after the accident. In the uncertain time immediately after the accident, it was inevitable that a lot of incorrect news would be spread, but even several years after the accident, incorrect news was spread in connection with various political movements, major events that attracted attention, and elections. These were attempts to politicize the issue of radiation (Murakami, Kumagai, Stojarov, & Tsubokura, 2019). Examples of this are the articles and actions taken in Japan and neighboring countries to stir up radiation fears when the Tokyo 2020 Olympic and Paralympics Games were held. Several foreign journalists and individuals sent out incorrect information using YouTube and other media with strong influence. While countermeasures are important to prevent stigma, prejudice, and discrimination against residents, securing resources for rebuttals becomes more difficult as time passes.

Immediately after the radiation accident or in its early stages, risk communication on radiation was a matter of concern for many residents, and the content of the information provided was an important issue. Conversely, after a certain amount of time has passed since the accident, many residents are less interested in radiation. Therefore, the timing and target of the information are more important than what kind of information is conveyed. Rather than basic and scientific knowledge of radiation, it is important to find a way to connect the information with the current interests of the people and to create a place to convey it. A good example of this is the counseling of residents regarding internal contamination tests using the Babyscan, a whole-body counter specifically for infants (Tsubokura, Nabeshima, Murakami, et al., 2020). It is not scientifically appropriate to test residents' internal exposure levels for infants, who consume less food and have a smaller body size; however, parents of infants and toddlers are very concerned about radiation, and it is important to provide them with accurate information. Parents generally spend a lot of time taking care of their children and doing household chores, and do not have enough time to interpret radiation information. Therefore, by conducting internal exposure examinations in a manner similar to infant health examinations, it is possible to provide parents with sufficient information. As the timing of information acquisition is affected by life events such as moving, marriage, pregnancy, and childbirth, it has become important to provide information in accordance with the timing of these events. Radiation was a major concern immediately after the accident, but has become less of a concern as time goes by, and it has become important to maintain a place to provide information about it.

One of the examples that the author perceives as a phenomenon of incorrect news is the dissemination of academic papers on the health effects after the Fukushima nuclear accident. Academic papers are often published after peer review, and it may not be scientifically correct to claim that the contents are incorrect. However, from the perspective of a medical professional who works in the affected areas after the nuclear accident, there is a problem that even events that are clearly considered to be caused by changes in the social and living environment after the nuclear accident are transmitted as if they are direct effects of radiation exposure. Reports of an increase in the number of undescended testes, maternal deaths, and cardiac malformations after a nuclear accident are further good examples. These phenomena are unlikely to be caused by radiation exposure, based on the current knowledge of radiological science and the estimated radiation exposure of the population (Hirata, Shimizu, Kumamaru, et al., 2020; Kojima, Yokoya, Kurita, et al., 2019; Ochi et al., 2016; Ohira et al., 2021). Rather, it is reasonable to assume that the effects are due to changes in the social and living environment caused by the nuclear accident, changes in medical practices such as surgical methods, and differences in the methods used to collect data. When large amounts of data are statistically analyzed in various ways, it is possible to find statistically significant points. The author feels that writing academic papers that ignore the major social and living environment changes caused by the accident and report that increases in diseases are the effect of radiation exposure is unbecoming of a scientist, but there is no end to such cases. Such academic papers tend to be spread more easily, and remain in the memory of the residents more than the papers that showed no clear effects of radiation after the nuclear accident. It is important to refute them individually by publishing letters or original papers (Kobashi, Sawano, Crump, Kami, & Tsubokura, 2020; Kobashi, Sawano, Ozaki, & Tsubokura, 2020).

Adherence to misconceptions about heredity and radiation bullying

One of the misperceptions of radiation effects that may have led to discrimination after the Fukushima nuclear accident was the genetic effect on the next generation. In a survey on mental health and lifestyle conducted by Fukushima Medical University, the subjects were asked whether they thought there were any effects of radiation exposure on the next generation over time. While approximately 60% of the respondents answered that there was an effect in fiscal year (FY) 2011 immediately after the nuclear power plant accident, the percentage gradually decreased to less than 50% in FY2012 and FY2013, and to less than 40% in FY2014. However, since then, the percentage has stopped falling and remains at the 30% level. A survey of Tokyo residents conducted in 2020 by the Mitsubishi Research Institute in Tokyo about harmful rumors associated with the nuclear power plant accident showed similar results: approximately 40% of respondents were aware that there might be genetic effects on Fukushima residents.

As the United Nations Scientific Committee on the Effects of Atomic Radiation (UNSCEAR) and various studies on radiation exposure have shown, genetic effects on next-generation residents due to exposure from the Fukushima Daiichi Nuclear Power Plant accident are highly unlikely. First, radiation effects have not been observed in humans. For example, there have been no increases in cancer or other diseases in the generation of pregnant women born after the atomic bombings of Hiroshima and Nagasaki (Otake, Schull, & Neel, 1990; Yamada, Furukawa, Tatsukawa, et al., 2021). In addition, the percentage of children with chromosomal abnormalities has not changed between those whose parents were exposed to radiation near the hypocenter and those whose parents were not (Izumi, Koyama, Soda, & Suyama, 2003). In addition, thousands of offspring of children treated for childhood cancers have been compared to their siblings, and no change in the frequency of chromosomal abnormalities, inherited diseases, or malformations has been reported (Green, Sklar, Boice Jr, et al., 2009). In addition, in 2021, researchers conducted a genome analysis of the children of workers after the Chernobyl nuclear power plant accident, showing that genetic effects are not clear (Yeager, Machiela, Kothiyal, et al., 2021). Radiation education plays an important role in dealing with this issue, but in reality, radiation education outside Fukushima Prefecture is not sufficient and should be further discussed.

A typical case of discrimination related to radiation is bullying against affected people in schools. According to Sawano et al., several hundred cases of bullying in schools have been reported, and one of the causes is the disparity due to evacuation and compensation (Sawano et al., 2019). Such bullying is related to the lack of understanding about radiation, but it is also one of the expressions of social strain caused by disasters. On the other hand, there is criticism that such bullying is not linked to radiation exposure but rather that the people involved believe that radiation was the cause of the bullying. While there are reports of discrimination against Fukushima residents in marriage, childbirth, and genetic effects, there are also cases where radiation exposure is not the reason for the bullying.

The author does not have an opinion on which of these ideas is correct. However, in any case, we need to continue our efforts to minimize discrimination and prejudice caused by radiation exposure. Genetic effects are like a curse, and children who have been greatly socially affected by the nuclear accident should not be in a situation where their potential is diminished because of misperceptions about radiation. The assessment of the risk of health effects due to radiation exposure based on LNT is a regulatory science. The author recognizes that the LNT-based assessment of the risk of health effects from radiation exposure is a necessary concept for making some regulations or conducting some administrative operations, but such a concept has a great disadvantage in preventing discrimination and the formation of stigma, and in considering the health and future of the population (Murakami & Tsubokura, 2021). It is a paradigm that leaves a curse of not being able to say that there is an impact, and its use needs to be fully understood and carefully considered.

Changes in the way information is received and given over time

An important issue to consider is the change in residents' perceptions of radiation risks over time. In the immediate aftermath of the nuclear power plant accident, interest in radiation was very high. However, as time passes, this wanes, creating a situation where false perceptions become fixed and incorrect news can easily spread. As an example, there was a report on changes in the avoidance rate of the consumption of locally produced food after the nuclear accident (Takebayashi, Murakami, Nomura, Oikawa, & Tsubokura, 2020). In Minamisoma City, a disaster-stricken area close to the power plant, a questionnaire on the intake of locally produced food was sent to guardians at the time of the annual internal exposure check for elementary and junior high school students. Takebayashi et al. conducted a detailed analysis of the changes over time (Takebayashi et al., 2020). In 2013, approximately 70%–80% of parents of small children avoided eating locally produced food while the percentage decreased year by year to approximately 30%–40% 5 years later. The avoidance rate continued to decline. Avoidance of consumption of locally produced food for fear of radioactive contamination is an effective indicator for understanding the risk perception of radiation. There are two important findings from this study. First, there are three main groups of residents in terms of food avoidance: (1) low concern adaptor, low avoidance of locally produced foods (less than 50%) from the first year, with further improvement over time; (2) high concern adaptor, high avoidance of locally produced foods (more than 75%) from the first year to 2015, with further improvement over time; and (3) high concern maintainer, high avoidance of local foods (80% or more) from the first year, with maintained avoidance over time (Takebayashi et al., 2020). There are various groups of residents who have different perceptions of radiation risks, and providing specific information is not enough to deal with such diversity. A particular problem is that the government and experts perceive those who maintain high concern as having inadequate knowledge of radiation, and try to conduct risk communication based on a deficit model. Such actions create further disconnect and distrust. It is necessary to provide information tailored to the needs of each individual.

Second, in the group of people who initially avoided eating locally produced food but whose avoidance tendency improved (high concern adaptor), the reason for the change was apparently not because they had sufficient knowledge about radiation. In fact, various government questionnaires have shown that residents' basic knowledge of radiation has been declining as time passes since the nuclear accident. The change in perception of radiation risk is not based on the acquisition of knowledge, but rather on policies such as the resumption of farming and the way it is handled in businesses, such as whether it is sold in supermarkets. In terms of water safety, for example, there is no scientific risk of radiation exposure from consuming local tap water because the water is filtered and purified to remove most of the radioactive contamination; however, there was a very strong sense of avoidance of local tap water by residents. This perception was encouraged by businesses selling mineral water. Such businesses emphasize the safety of the mineral water they sell, even if they do not promote the dangers of the local tap water, and their promotional activities have contributed to the perception that the local tap water is dangerous. This may have resulted in prolonged avoidance of local tap water. It is undisputed that conveying knowledge about radiation is important to prevent stigma, prejudice, and discrimination, but it is equally important to improve risk perception and prevent stigma from becoming fixed by reopening lost sales channels and treating sales as normal. As reported by Murakami et al., the perception of radiation is eventually linked to daily life and health concerns unrelated to radiation, and the lack of knowledge about radiation is not the central issue (Murakami & Tsubokura, 2021).

This does not mean that there is no need to educate children about radiation in the future. Radiation education in elementary, junior high, and high schools plays a significant role (Kuroda, Tsubokura, Sasaki, et al., 2020; Tsubokura, Kitamura, & Yoshida, 2018). In Fukushima Prefecture, radiation education at elementary and junior high schools became compulsory in the early stages after the accident. Basic materials and examples for radiation education were shared among schools, and many efforts have been made to standardize and improve radiation education. A questionnaire survey conducted by the authors on the basic knowledge and behavior of radiation before and after radiation education showed that the group with basic knowledge of radiation tended to choose more scientifically appropriate behavior. Radiation education in Japan is mainly provided only in Fukushima Prefecture. The enthusiasm for radiation education depends on the school, and interest is gradually waning. However, the effectiveness and importance of the program have been shared among stakeholders.

On the other hand, there are various opinions on what radiation education should aim for, and it is difficult to unify them. The author has been working with teachers to develop a curriculum for radiation education, and has encountered several problems in the process. For example, conveying the risk of minute radioactive contamination of food after a nuclear accident is a problem.

It is known that although there was contamination of food with radioactive cesium after the nuclear accident in Fukushima, the contamination was unevenly distributed among certain foods, and most of the foods had no detectable radioactive materials. Some foods, such as mushrooms from open fields, are susceptible to radioactive contamination, and shipping restrictions have been imposed for a long time. Also, most residents are below the detection limit in internal exposure tests, and there are almost no health effects due to internal exposure. However, the way to convey these facts to students differs depending on whether they focus on the fact that most things are undetectable or the fact that only a few things can be detected. From the perspective of improving student perception of risks related to radiation and preventing prejudice and discrimination, the former way of communicating is better while the latter is preferred from the perspective of disaster prevention and future countermeasures. This difference in communication is seen not only in education, but also in the dissemination of information through various media, and it leads to a difference in risk perception about radiation.

The same difference in perception can be observed in the radiation inspection system. Various radiation-related examinations were started immediately after the accident to assess the situation, but there are various opinions on whether they should be continued in the long term. The dilemma is that the tests started to provide information for children in the future to dispel stigma, but now that the amount of radiation exposure has been found to be very low, the tests themselves are creating stigma. The same argument was made regarding whether to continue the installation of monitoring posts and the presentation of daily radiation doses on TV and in newspapers. Regarding internal exposure inspections, one municipality in Hamadori established an inspection system with different defaults. As a result, only those adults who wanted to be were inspected while all children were subject to inspection. Unfortunately, the debate regarding this test has not gone well, as it focused on the pros and cons of the testing and not on the central issue of how to convey knowledge about radiation and deepen understanding of the current situation. This is often neglected. Although no conclusion has been reached as to which method is the best, it is important to note that the way in which the tests are conducted and the results reported can also lead to the fixation of stigma.

Future measures

What we can do to address the apparent misconceptions about radiation and genetic effects that have become entrenched in the early days after the accident is to communicate the scientific facts that have become clear. This can be done through radiation education or by using the media, but the difficulties are apparent. These actions can be called risk communication. However, what we must be most careful about is that information has a season and interest in radiation fades rapidly, especially in areas outside Fukushima Prefecture that were hardly affected by the nuclear accident. This means that interest shifts to other events.

There is a belief that people can make better decisions if they are told what is scientifically correct, but interest in radiation can only be maintained for a short period. Now, more than 10 years after the accident, the important issue is how to attract the interest of the residents and create a place where they can learn about radiation. To put it another way, the capacity to communicate needs to be secured, rather than how or what to communicate.

The issue of radiation is not something that interests people all the time, but it is something that arouses interest whenever there are various events that attract their attention, such as the decision on the final disposal site, the release of treated water from the nuclear power plant, the lifting of evacuation orders for areas that are difficult to return to, the termination of various nuclear power plant-related budgets, and that the Reconstruction Agency is a time-limited ministry. Radiation is expected to come back into the full view of society at various times in the future. It is important that effective information be provided at such times, and that such timing is taken as an opportunity to change the perception of radiation for the better. In this context, the Ministry of the Environment launched the Guguru Project in July 2021 to promote understanding of radiation. The project aims to reduce the misperception of genetic effects on Fukushima residents in Tokyo. As it stands, 40% of respondents perceive that Fukushima residents will have future genetic effects. The Guguru Project aims to reduce this to 20% within 10 years.

The project consists of five major pillars. The first is the counter paper project, which encourages people to gain the ability to read and interpret the various papers on health effects that have been published so far by learning how to write and interpret them. The second is the radiation college project to promote the creation of more opportunities for students to learn about radiation. The third is to promote the provision of information to promote decision-making. In other words, to promote the provision of information on the purpose of various post-nuclear accident tests so that people can fully understand the purpose, advantages, and disadvantages of the tests before undergoing them. The fourth project is to expand the functions of the Radiation Counselor Support Center, especially in areas where evacuation orders have been lifted, to provide more accessible information on radiation. The fifth is a project to promote the dissemination of information on radiation on websites and portal sites, based on access analysis. These projects were extensively organized, and started in July 2021.

Because perceptions and stigmas about radiation become entrenched at an early stage, and many people lose interest in radiation at an early stage, the major issue was how to actively create a space to communicate radiation knowledge thereafter. Radiation education plays the most important role for this purpose, but due to the above-mentioned difficulties, it is necessary to deal with stigma in the long term by combining various measures in addition to radiation education. Again, the health effects after the Fukushima nuclear power plant accident are not limited to those caused by radiation exposure, but are also reported to be caused by various changes in the social living environment. While these changes have had a significant impact on many people, and considering the children of the future, the author believes that discrimination and prejudice against them as well as adherence to stigma are the most important health effects that should be considered after the Fukushima nuclear accident. It is necessary for us to continue our activities to convey information and prevent unnecessary discrimination (Normile, 2021; Ozaki, Sawano, & Tsubokura, 2020).

References

Green, D. M., Sklar, C. A., Boice, J. D., Jr., et al. (2009). Ovarian failure and reproductive outcomes after childhood cancer treatment: Results from the childhood cancer survivor study. *Journal of Clinical Oncology*, *27*(14), 2374–2381.

Hirata, Y., Shimizu, H., Kumamaru, H., et al. (2020). Congenital heart disease after the Fukushima nuclear accident: The Japan cardiovascular surgery database study. *Journal of the American Heart Association*, *9*(13), e014787.

Izumi, S., Koyama, K., Soda, M., & Suyama, A. (2003). Cancer incidence in children and young adults did not increase relative to parental exposure to atomic bombs. *British Journal of Cancer*, *89*(9), 1709–1713.

Kobashi, Y., Sawano, T., Crump, A., Kami, M., & Tsubokura, M. (2020). Unambiguous evidence is required to accurately understand the health impacts of nuclear accidents. *Journal of Radiation Research*, *61*(1), 90–91.

Kobashi, Y., Sawano, T., Ozaki, A., & Tsubokura, M. (2020). Misunderstanding of the purpose and intermediate results of the Fukushima health management survey could impede progress in the radiation protection field. *Journal of Radiological Protection*, *40*(3), 924–927.

Kojima, Y., Yokoya, S., Kurita, N., et al. (2019). Cryptorchidism after the Fukushima Daiichi nuclear power plant accident: Causation or coincidence? *Fukushima Journal of Medical Science*, *65*(3), 76–98.

Kuroda, Y., Tsubokura, M., Sasaki, K., et al. (2020). Development of radiation education in schools after the Fukushima Daiichi nuclear power plant accident—A study from the perspectives of regionality, multidisciplinarity and continuity. *Radioprotection*, *55*(4), 317–324.

Morita, T., Nomura, S., Tsubokura, M., et al. (2017). Excess mortality due to indirect health effects of the 2011 triple disaster in Fukushima, Japan: A retrospective observational study. *Journal of Epidemiology and Community Health*, *71*(10), 974–980.

Murakami, M., Kumagai, A., Stojarov, A. N., & Tsubokura, M. (2019). Radiation is not a political tool. *Science*, *366*(6465), 581–582.

Murakami, M., Ono, K., Tsubokura, M., et al. (2015). Was the risk from nursing-home evacuation after the Fukushima accident higher than the radiation risk? *PLoS One*, *10*(9), e0137906.

Murakami, M., & Tsubokura, M. (2021). Deepening community-aligned science in response to wavering trust in science. *Lancet*, *397*(10278), 969–970.

Murakami, M., Tsubokura, M., Ono, K., Nomura, S., & Oikawa, T. (2017). Additional risk of diabetes exceeds the increased risk of cancer caused by radiation exposure after the Fukushima disaster. *PLoS One*, *12*(9), e0185259.

Nomura, S., Gilmour, S., Tsubokura, M., et al. (2013). Mortality risk amongst nursing home residents evacuated after the Fukushima nuclear accident: A retrospective cohort study. *PLoS One*, *8*(3), e60192.

Nomura, S., Murakami, M., Naito, W., Yasutaka, T., Sawano, T., & Tsubokura, M. (2020). Low dose of external exposure among returnees to former evacuation areas: A cross-sectional all-municipality joint study following the 2011 Fukushima Daiichi nuclear power plant incident. *Journal of Radiological Protection*, *40*(1), 1–18.

Nomura, S., Murakami, M., Ozaki, A., et al. (2021). Comparative risk assessment of non-communicable diseases by evacuation scenario—A retrospective study in the 7 years following the Fukushima Daiichi nuclear power plant accident. *Global Health Action*, *14*(1), 1918886.

Normile, D. (2021). Nuclear medicine. *Science*, *371*(6533), 978–982.

Ochi, S., Kato, S., Tsubokura, M., Leppold, C., Kami, M., & Shibuya, K. (2016). Voice from Fukushima: Responsibility of epidemiologists to avoid irrational stigmatization of children in Fukushima. *Thyroid*, *26*(9), 1332–1333.

Ohira, T., Shimura, H., Yasumura, S., Yokoya, S., Ohto, H., & Kamiya, K. (2021). Response to the letter to the editor: "Absorbed radiation doses in the thyroid as estimated by UNSCEAR and subsequent risk of childhood thyroid cancer following the great East Japan earthquake", by Ohira et al. *Journal of Radiation Research*, *62*(3), 425–426.

Otake, M., Schull, W. J., & Neel, J. V. (1990). Congenital malformations, stillbirths, and early mortality among the children of atomic bomb survivors: A reanalysis. *Radiation Research*, *122*(1), 1–11.

Ozaki, A., Sawano, T., & Tsubokura, M. (2020). Transition of originally external healthcare providers into local researchers: A case study of support activities in So-So District, Fukushima after the 2011 triple disaster. *Radioprotection*, *55*(4), 263–270.

Sawano, T., Ozaki, A., Hori, A., & Tsubokura, M. (2019). Combating "fake news" and social stigma after the Fukushima Daiichi nuclear power plant incident-the importance of accurate longitudinal clinical data. *QJM*, *112*(7), 479–481.

Sawano, T., Shigetomi, S., Ozaki, A., et al. (2021). Successful emergency evacuation from a hospital within a 5-km radius of Fukushima Daiichi Nuclear Power Plant: the importance of cooperation with an external body. *Journal of Radiation Research*, *62*(Supplement_1), i122–i128.

Sugimoto, A., Krull, S., Nomura, S., Morita, T., & Tsubokura, M. (2012). The voice of the most vulnerable: Lessons from the nuclear crisis in Fukushima, Japan. *Bulletin of the World Health Organization*, *90*(8), 629–630.

Sugimoto, A., Nomura, S., Tsubokura, M., et al. (2013). The relationship between media consumption and health-related anxieties after the Fukushima Daiichi nuclear disaster. *PLoS One*, *8*(8), e65331.

Takebayashi, Y., Murakami, M., Nomura, S., Oikawa, T., & Tsubokura, M. (2020). The trajectories of local food avoidance after the Fukushima Daiichi nuclear plant disaster: A five-year prospective cohort study. *International Journal of Disaster Risk Reduction*, *46*, 101513.

Tsubokura, M. (2018). Secondary health issues associated with the Fukushima Daiichi nuclear accident, based on the experiences of Soma and Minamisoma cities. *Journal of the National Institute of Public Health*, *67*(1), 71–83.

Tsubokura, M., Gilmour, S., Takahashi, K., Oikawa, T., & Kanazawa, Y. (2012). Internal radiation exposure after the Fukushima nuclear power plant disaster. *JAMA*, *308*(7), 669–670.

Tsubokura, M., Kato, S., Nomura, S., et al. (2014). Reduction of high levels of internal radio-contamination by dietary intervention in residents of areas affected by the Fukushima Daiichi nuclear plant disaster: A case series. *PLoS One*, *9*(6), e100302.

Tsubokura, M., Kitamura, Y., & Yoshida, M. (2018). Post-Fukushima radiation education for Japanese high school students in affected areas and its positive effects on their radiation literacy. *Journal of Radiation Research*, *59*(suppl_2), ii65–74.

Tsubokura, M., Murakami, M., Takebayashi, Y., et al. (2019). Impact of decontamination on individual radiation doses from external exposure among residents of Minamisoma City after the

2011 Fukushima Daiichi nuclear power plant incident in Japan: A retrospective observational study. *Journal of Radiological Protection*, *39*(3), 854–871.

Tsubokura, M., Nabeshima, Y., Murakami, M., et al. (2020). Usefulness of the whole-body counter for infants and small children (BABYSCAN) as a risk communication tool after the Fukushima Daiichi nuclear power plant incident. *Proceedings of the Japan Academy. Series B, Physical and Biological Sciences*, *96*(2), 70–78.

Tsubokura, M., Onoue, Y., Torii, H. A., et al. (2018). Twitter use in scientific communication revealed by visualization of information spreading by influencers within half a year after the Fukushima Daiichi nuclear power plant accident. *PLoS One*, *13*(9), e0203594.

Yamada, M., Furukawa, K., Tatsukawa, Y., et al. (2021). Congenital malformations and perinatal deaths among the children of atomic bomb survivors: A reappraisal. *American Journal of Epidemiology*, *190*(11), 2323–2333.

Yeager, M., Machiela, M. J., Kothiyal, P., et al. (2021). Lack of transgenerational effects of ionizing radiation exposure from the Chernobyl accident. *Science*, *372*(6543), 725–729.

Chapter 16

Crises and resilience among local communities

Tomoyuki Kobayashi[a] **and Kazuki Yoshida**[b]
[a]*Department of Disaster Psychiatry, Fukushima Medical University School of Medicine, Fukushima, Japan,* [b]*Iryo Sosei University, Faculty of Nursing, Fukushima, Japan*

Community disruption and resilience after disasters

The 2011 Fukushima nuclear accident had a serious detrimental effect on the systems and social relationships of many communities. At first, residents were ordered or voluntarily evacuated from their hometowns to avoid the risk of radiation exposure due to accidents at the Fukushima Daiichi and Daini Nuclear Power Plants. During the chaos after the earthquake and tsunami, the information the residents had about the nuclear power plants was exceedingly lacking (National Diet of Japan Fukushima Nuclear Accident Independent Investigation Commission, 2012). While some residents evacuated with the assistance of local government officials, others evacuated on their own or with the assistance of family or friends. The local governments did not have enough transportation capacity to evacuate all their residents. As a result, people were dispersed to various evacuation shelters. According to the Earthquake Record of Okuma, there were more than 20 shelters with capacities ranging from a few dozen to more than 1000 people (Okuma Town, 2017).

After that, some in-kind support, such as temporary housing and public housing, was provided, but the nuclear evacuees were never again able to reunite with all the members of their original communities. Some residents could not relocate with their original community because they and their family members had to go to work and school. Others wanted to leave Fukushima due to the perceived risks of radiation. Furthermore, in public housing blocks where capacity was limited, tenants were decided by lottery, although in some cases consideration was given to maintaining the original neighborhoods before the nuclear accident (Takaki, 2018).

As a corollary to the physical loss of these local communities, their cultural systems were also damaged. During the evacuation, traditional festivals and events were canceled for a variety of reasons, such as a naturally restrained

Health Effects of the Fukushima Nuclear Disaster. https://doi.org/10.1016/B978-0-12-824098-4.00014-3

general mood, a lack of venues to hold the celebrations, and inadequate human resources to maintain the festivals because the young people who worked at the festivals typically evacuated farther away. Some evacuees were also confused by their drastically different living environments. For example, many of those who lived near the nuclear power plants lived in detached houses, but found themselves living in temporary housing or public housing, where neighbors were so close they could hear the everyday noises of life. This was both unfamiliar and stressful to them (Miura, Okayasu, & Miura, 2018). Furthermore, their community interactions, such as exchanging vegetables they had grown, were gone due to the loss of farmland.

Even though disaster experiences typically increase mutual assistance and cooperation (Drury, 2018), some evacuees had trouble forming relationships with their neighbors. Another remarkable issue was the discrimination faced by the evacuees following the nuclear accident. Imai (2017) surveyed the nuclear disaster evacuees and reported that 65.6% reported being victims of discrimination or bullying, or had known of it happening around them. They were mainly discriminated against regarding radiation exposure and nuclear damage compensation. Among their experiences in relocation areas, for example, their children were bullied at school, accused of infecting others with radiation poisoning. When job seeking, they were asked why they were working even though they had received compensation money. They were accused of using the compensation money to live lives of luxury. Furthermore, there were refusals to marry or get along with neighbors, graffiti and vandalism on cars and houses, and insidious bullying.

In addition to the above, some evacuees were bullied even after returning to their hometown. Takagi (2017) reported a case of those who voluntarily evacuated in areas where evacuation orders had not been given; they were stigmatized as "escapees." When they returned home, they were in a weaker position and had to engage in community work for "nonescapees" to be accepted again by the community.

In previous disaster studies, communities have displayed various social connections that contribute significantly to people's acquisition of stress resilience (Aldrich & Meyer, 2015) which is the process of successful adaptation under challenging or threatening circumstances following a disaster (Masten, Best, & Garmezy, 1990). Orui et al. (2018) conducted a study of Fukushima residents and found that those who maintained predisaster relationships with friends and who had socially conducive ties were more likely to acquire resilience. Similarly, Takebayashi et al. (2020) showed that socially conducive ties were included in the same factor as behavioral activation to engage in hobbies and new experiences, leading to a reduction in psychological distress. These results suggest that the resilience of affected people requires not only social connections that provide passive social support (e.g., between affected people and healthcare providers), but also social connections that settle mutual social support and allow evacuees themselves to participate in contributing to society (e.g., among neighbors in the same community) (Kobayashi, Yoshida, Takebayashi, et al., 2019; Orui et al., 2018). Furthermore, residents who had positive social relationships were able to maintain disaster preparedness and made use of health checkups (Hasegawa, Murakami, Takebayashi, Suzuki, & Ohto, 2018). Therefore, the survivors in Fukushima were caught up in the great disruption of their community, even though they had various potential benefits of the community at their disposal. As a result, the Fukushima population differs from other affected prefectures in terms of serious mental health problems and suicide rates (see Chapter 8 for a detailed discussion). Responders needed to plan and manage social connections for the resilience of the people affected in the postdisaster period.

Social connections for life after the Fukushima nuclear accident

Successful social connections, which include both formal and informal relationships, help people to acquire stress resilience after a disaster. Formal connections refer to the support and care provided mainly by the national and local governments through explicit, legal, and institutionalized authority over society (Szreter & Woolcock, 2004). The government can give people access to important public goods they would not otherwise have (Fraser & Aldrich, 2021). For example, in the Fukushima nuclear accident, national and prefectural governments provided services included in the Act on Compensation for Nuclear Damage: decontamination work, screenings, the provision of temporary housing and reconstruction public housing, medical treatment, and free highway use. The special act for nuclear evacuees guaranteed that evacuees would be able to receive medical, welfare, and educational services in their relocation areas (Ministry of Internal Affairs and Communications, n.d.). Moreover, the evacuated local governments kept in contact with residents who had evacuated both within and outside the prefecture, providing ongoing surveys and information regarding their hometown to members of the original community and transportation to gatherings in the original community. In the case of Hirono Town, located about 20 km from the plant, Fukushima Medical University collaborated with the local government for risk communication by providing information and holding consultation meetings (Kobayashi, Yoshida, Kumagai, et al., 2019). At these meetings, doctors, radiation counselors, and other specialists met with residents in small groups for roundtable discussions. In addition to radiation concerns, the topics included increased traffic, the increased number of outsiders (e.g., decontamination workers), and other everyday concerns (e.g., the safety of crops and water).

Informal connections are those formed among individuals within a community, such as with family members, friends, or neighbors. These connections lead to increased levels of practical social support, which in turn provide access to various resources in a disaster situation, including information, aid, financial resources, and child care, along with emotional and psychological support

(Aldrich & Meyer, 2015). According to our interviews conducted in a public disaster housing development, evacuees achieved improvements in their living environment after the disaster by cooperating with their neighbors, for which government support was not enough. This type of support is characterized by emotional connections, strong close relationships, and homophily in demographic characteristics, attitudes, and available information and resources (Aldrich & Meyer, 2015; McPherson, Smith-Lovin, & Cook, 2001).

Disaster studies have also shown that informal connections are temporarily strengthened and expanded after a disaster occurs (Fritz, 1996; Jencson, 2001; Jong, Whitehouse, Kavanagh, & Lane, 2015; Rich, Edelstein, Hallman, & Wandersman, 1995). Drury (2018) noted that mass emergencies, such as disasters and pandemics, can create a sense of "us in relation to the emergency" among those who share the same experiences. This sense is noted for typically increasing group cohesion and improving social interactions and relationships among those affected by an emergency. In fact, after the Great East Japan Earthquake, the slogan "Ganbarou Nippon! (Hang in there Japan!)" became popular in Japan, suggesting that the Japanese people, including those both directly and indirectly affected by the disaster, tried to cooperate with each other to achieve recovery. Therefore, public health practitioners may encourage more effective informal social support by attempting to use this psychological response to build messages and networks.

Our study revealed that the maintenance and formation of informal social connections were among the most difficult aspects of the Fukushima nuclear accident. There are three important points that may explain why, despite the benefits of the psychological response to a shared emergency, it did not work well. The first is the unsuitable social environment for the formation of social connections. Many evacuees left without sufficient information and repeatedly moved to different cities. Many were also separated from their families and friends during these repeated evacuations (Yabe et al., 2014). Some lost their family members due to health problems during the evacuation. Living in temporary housing was always economically stressful, and it was difficult to secure satisfactory privacy, causing numerous interpersonal problems. In such an environment, it was difficult to foster sufficient social networks, and so many evacuees were under interpersonal stress.

The second point is discrimination against evacuees, as mentioned above. The Fukushima nuclear accident caused the general population to experience health anxiety related to radiation exposure. Social psychologists have pointed out that many prejudices can be understood as responses to threats against human welfare, such as one's health (Schaller & Neuberg, 2012). The evacuees were not exposed to high doses of radiation, and radiation exposure is not an infectious disease, yet many people were not knowledgeable about radiation and had a vague perception of it as an invisible fore that causes severe damage to the body. Information on whether a person came from Fukushima was more accessible to them than information on whether the person had been exposed to radiation, thus they associated this image with the evacuees and felt disgusted by them in response to perceived threats to their own health (Kwesell & Jung, 2019). This prejudice based on health anxiety was observed against in-group members in the community (Schaller & Neuberg, 2012).

Furthermore, some nuclear evacuees were stereotyped for having received some sort of compensation. At the time, the issue of compensation from the Tokyo Electric Power Company (TEPCO) for the nuclear accident became one of the hot topics of discussion throughout Japan. A false rumor was spread that many evacuees were receiving high amounts of compensation. Although people generally had a supportive attitude toward evacuees, this stereotype could suppress such attitudes. In addition, the evacuees' outings and dining out were sometimes perceived as luxuries stemming from their large amounts of compensation. According to our interviews, some evacuees deliberately chose to wear more modest clothing when they went out.

The final point is the intercommunity conflict that emerged between the evacuee community and the communities who received them. Even within the Fukushima Prefecture, each community has a different culture in terms of language, customs, values, and climate. For example, Aizuwakamatsu City, where Okuma residents evacuated to, has a snow country culture while it rarely snows in Okuma. Therefore, the reception of evacuees was an event of cross-cultural contact. Continuous first-hand contact between groups of individuals from different cultures can cause stress (Redfield, Linton, & Herskovits, 1936) due to feelings of depression based on the experiences of cultural loss and anxiety stemming from uncertainty in their new location; this is known as "acculturation stress" (Berry, 2006a). This stress can be experienced not only by residents of the evacuee community, but also by residents of the host community (Kobayashi, Yoshida, Takebayashi, et al., 2019). The elderly, women, and people with little social support are more likely to suffer from acculturation stress than others (Sam & Berry, 2010). Similar to reactions to other common stressors (Lazarus & Folkman, 1974), people with acculturation stress can achieve adjustment through the assessment of and coping with the cultural problems they face. If the problems faced are difficult to control, their stress will be greater (Berry, 2006a).

Iwaki City has the largest number of public housing facilities in the Fukushima Prefecture (Fukushima Revitalization Station, 2019). It was preferred as an evacuation destination because of its similar climate and geographical proximity to the evacuation area (Soso District). However, in Iwaki City, there were several cases of intergroup conflict, such as slanderous and offensive graffiti against evacuees and vandalism of evacuees' cars (Kawafuku, 2013). Iwaki City and the Soso District have subtle differences in local rules (e.g., categories into which garbage is sorted for disposal). Moreover, Iwaki City did not have a large enough welfare and economic infrastructure to receive evacuee communities with different cultures. Some host residents in Iwaki City lost their homes in the tsunami of 2011, and the influx of evacuees from the Soso District caused housing prices to soar. While some evacuees who received compensation from TEPCO could afford to buy a home, those affected by the tsunami in Iwaki City had a hard time finding new homes. The rapid population growth also raised real estate prices and put pressure on the housing market, health care, and administrative services, and increased the number of people who were not following local customs, such as participating in neighborhood associations, organizing local festivals, and staging community cleaning efforts. These circumstances may have created acculturation stress for both host residents and evacuees. Among the host residents who felt a sense of identity with Iwaki City, more psychological stress was felt by those willing to accept evacuees from the Soso District (Kobayashi, Yoshida, Takebayashi, et al., 2019).

Resilient communities with evacuees

In postdisaster societies, experts and policymakers must consider creating communities that encourage the affected people to acquire resilience. For evacuees to form sufficient social connections, especially informal social ones, a community that integrates both, rather than just some of, the host and evacuee communities is required. Forming adaptive communities after a disaster may require both material resources and social functions (Norris, Stevens, Pfefferbaum, Wyche, & Pfefferbaum, 2008).

First, the material resources of economic development and information are essential as the basis for running a community (Norris et al., 2008). Economic development includes land, physical capital, accessible housing, health services, schools, and employment opportunities (Godschalk, 2003). As mentioned in the previous section, it may be relatively easy to receive evacuees if there is enough housing and if health services are guaranteed in the community, but a lack of economic security within the community can create economic disparities among the residents, which in turn can create conflicts. Lower socioeconomic status is associated with adverse mental health after disasters (Norris et al., 2002; Orui et al., 2018). In the case of the Fukushima accident, changes in economic conditions such as those brought on by new living conditions, unemployment, and lower incomes after the disaster were also reported as significantly related to poor subjective health (Nagai et al., 2017; Sawano et al., 2019).

In situations of high uncertainty due to disasters, both individuals and communities are required to make decisions based on analyses and evaluations of information. However, information should not be merely communicated; it needs to become usable knowledge for decision-makers. In Fukushima, risk communication was recommended as a way to treat information and knowledge. Communication with the residents is important for deciding how to live with uncertain risks such as radiation, rather than just one-way communication instigated by experts (Kobayashi, Yoshida, Kumagai, et al., 2019). Moreover, community decision-makers are required to survey their members continuously about their condition and opinions. This also helps to make organizational networks within and outside the community function flexibly (Norris et al., 2008). However, it is necessary to be cautious about excessive numbers of surveys, as they impose a burden on the residents (Kobayashi, Takebayashi, & Murakami, 2020).

Second, the social functions needed to run an adaptive community include organizing, perceived support, mutual trust, cooperation, shared goals, a shared willingness to work for the common good of the neighborhood, the effectiveness of organized community actions, and efforts to improve the community damaged by the crisis (Norris et al., 2008; Rich et al., 1995; Sampson, Raudenbush, & Earls, 1997). The social identity approach states that sharing a social identity with others, which is a part of one's self-concept defined by the social groups to which they belong (Tajfel & Turner, 1979), includes psychological resources such as an feeling of interconnectedness to others, a sense of community, social support, and efficacy, all of which promote health and well-being (Haslam, Jetten, Cruwys, Dingle, & Haslam, 2018; Van Dick, Ketturat, Häusser, & Mojzisch, 2017). However, the promotion of a shared social identity may also be a source of intergroup conflict if the community is not a single community, but instead a complex community that may include both host and evacuee communities (Kobayashi, Yoshida, Takebayashi, et al., 2019).

We introduce two approaches as the solution. The first approach is for residents to identify a superordinate in-group. Gaertner and Dovidio's common in-group identity model explains that inducing members of potentially conflicted groups to recategorize themselves as members of a more inclusive superordinate group can reduce intergroup conflicts (Gaertner & Dovidio, 2000). Intergroup conflicts are caused by four types of threats: realistic threats (i.e., those related to the economy and resources), symbolic threats (i.e., those related to values and norms), intergroup anxiety, and negative stereotypes (Stephan & Stephan, 2000). A common in-group identity can reduce these four threats and promote a friendly attitude toward out-group members (Riek, Mania, Gaertner, McDonald, & Lamoreaux, 2010). For example, Adam-Troian, Çelebi, Bonetto, Taşdemir, and Yurtbakan (2020) used a flag-priming paradigm to activate a common national identity and found that it improved intergroup attitudes between Turks and Syrians and attitudes toward minority groups (Maghrebis) in France. There is no empirical research on this model in Fukushima, but as one of the grassroots activities, picture book author Teppei Ikumasa has created a kamishibai, which means storytelling with illustrated posters, related to this. He described the importance of cooperating as related to Fukushima residents' ability to achieve recovery in the situation amid conflicts between host residents and evacuees in Iwaki City.

Another approach is to integrate the host community with the evacuee community. The acculturation model suggests that people adapt to acculturation stress using four different strategies: integration, assimilation, separation, and marginalization (Berry, 2006b). These are strategies for determining which norms and social relationships are more important or self-identifying in the process of cultural adaptation with a different community, the original community, or the new community. The separation strategy calls for remaining attached to the original community while the assimilation strategy aims to adapt completely to the new community. Some people succeed with the integration strategy, that is, they internalize the original and new communities' identities. This strategy allows for dual cultural competence and access to dual resources (Sam & Berry, 2010), leading to better adaptation outcomes, as individuals can switch between the two identities depending on the context (Benet-Martinez & Haritatos, 2005); therefore, the most commonly preferred approach is psychological and sociocultural adaptation. Avoiding involvement in either community is called the marginalization strategy and is generally known to have the worst mental health outcomes (Sam & Berry, 2010). Han, Gui, Zheng, and Berry (2015) reported that evacuees who used the integration strategy were more successful in acquiring resilience than those who used other strategies following the 2008 China earthquake.

In the context of intergroup conflicts between host residents and nuclear evacuees in Iwaki City, Kobayashi et al. (2021) showed that adopting an integration strategy is facilitated by a belief in group interdependence—that is, the extent to which an individual believes that group function is realized through interdependence with related other groups. Communities often have connections to other communities in the form of interdependence in the economic, political, and technological spheres for group performance and development (Barbieri, 1996; Ertur & Koch, 2007). Or, connections among different communities, as forms of interdependence in social spheres, can be described as loose social ties among acquaintances and individuals that span communities (Aldrich & Meyer, 2015). It can be argued that these facilitate communication and the beneficial flow of information and resources among communities and are more likely to provide social resources such as employment opportunities (Granovetter, 1983). In such intercommunity relationships, people can favor their own group while perceiving the outgroup as part of the management and development of their group and adopt the integration strategy in the context of acculturation. In fact, nuclear evacuees and resilient hosts with a firm belief in group interdependence were likely to have integrated the evacuee and host communities' social identities and were more likely to interact with both members of the in- and out-group communities (Kobayashi et al., 2021).

Health promotion activities for reconstruction in Iwaki City

Here, we discuss the recovery activities in a local community affected by the Great East Japan Earthquake, focusing on the activities of public health nurses at the Community General Support Center, which supports the elderly in all areas of Iwaki City, Fukushima. The Center employs public health nurses, social workers, and chief care managers. Public health nurses in Japan work with local residents to maintain and improve their health. The specific activities of public health nurses include community health data synthesis, identifying any issues, and then effectively resolving them as well as conducting general health consultations, leading health education efforts, and conducting home visits. They were one of the major sources of aid during the chaos and recovery period following the earthquake.

Iwaki City is on Japan's Pacific coast, about 200 km (124 miles) northeast of Tokyo and about 30 km (19 miles) from the Fukushima Daiichi Nuclear Power Plant. This is a disaster-stricken area affected by the near-simultaneous earthquake, tsunami, and nuclear accident in 2011. At the same time, the city received many evacuees from the nuclear power plant's evacuation zone. These crises disrupted communications and resources, making the usual collaborations among the responders from various professions and the residents problematic. Public health nurses had a hard time providing care for the residents in the midst of the confusion and unprecedented experiences, including an overwhelming lack of credible information, unknown local leaders and organizations, too many residents to care for, and delayed organization-based crisis responses. They had to respond using trial and error.

To solve local problems that were greatly exacerbated by the disaster, it was obvious that economic development, information administration, and the reconstruction of social networks were important. However, public health nurses were also tasked with gathering information about the communities, analyzing their issues, and considering how to intervene, given their actual specific conditions. For example, after the disaster, the approach depended on whether the community had collapsed or survived. In collapsed communities, nurses first gathered information about the area, assessed the residents' safety, and then considered how they could reach the contactable residents to begin the rebuilding process. They did this by sending out information on activities to reunite separated residents. By contrast, in surviving communities, one activity aimed to support those elderly and disabled residents who could not be evacuated. According to one public health worker, the public health nurses' preexisting personal relationships with local leaders and other organizations made it possible for them to respond quickly to the disaster, setting up evacuation shelters and confirming the safety of elderly residents. One particularly valuable resource was the Community Information Notebook, which was compiled and regularly updated. Public health nurses at the Community General Support Center shared the notebook with members of the center's staff so they could efficiently respond to the disaster, even without a designated local leader.

We have been evaluating activities in Iwaki City based on the network with two foci: improving community health activity specialists' skills and creating an evacuee support system. Under those circumstances, the Community General Support Center's public health nurses faced a lack of skills for providing postdisaster community redevelopment activities, the need for an established organization system for recovery, and the health issues of residents and nuclear zone evacuees. After the disaster, the national and municipal governments provided community support in the form of financial and welfare service resources, but it was actually the members of the community—those on the ground—who communicated with the residents about government resources. Public health nurses had to be knowledgeable enough to educate residents about radiation, health and welfare services, and the operations of the medical institutions. They also had to be proactive in acquiring new skills because they had numerous day-to-day tasks to perform, such as maintaining residents' good health to reduce the demand for nursing care and holding meetings with residents to resolve problems in the community. They improved their skills by using existing in-service education for public health nurses and incorporating these programs into their work framework as well as by holding independent study groups.

Our first activity was skills improvement; we targeted a study group of public health nurses. Along with public health nurses from the Community General Support Center, we trained them in problem solving and how to statistically process community data. Their increased capabilities accelerated the situation analysis. This also helped us to establish a symbiotic relationship between the elderly and the study group's human resource developers. In the future, we expect requests from chief care managers to improve their long-term care management skills, and we believe that this activity will have a ripple effect by effectively and efficiently alleviating the community's problems. One of these effects is building trust. To date, it has taken 3 years of repeated visits and interactions to begin to see results, but we believe that the accumulation of these activities and the rooting of these activities in the community will lead to community reconstruction and the resolution of the community's health issues.

Second, the health issues found among Iwaki City's blended original and evacuee communities required the creation of a system that would draw on the strengths of both groups and enable residents to resolve their own life and health issues. This required addressing some major problems. When municipal officials from the evacuated towns of Okuma and Tomioka tried to meet with their former residents, the public health nurses were unable to locate them all. This inability to locate and efficiently inform the evacuees delayed the provision of financial and other service support by their home municipalities and the creation of the disaster recovery plan.

Our second activity focus, building a postdisaster community support system, has entailed our visiting public housing sites reserved for disaster and recovery, discussing health issues with evacuees and related organizations, and considering measures to address evacuees' health issues. Residents' mutual relationships have been changing, and the evacuees often feel isolated and in conflict with their neighbors. This necessitated a community development plan. However, it must be noted that the historical data and theories were often unsuited to the extraordinary postdisaster context. For example, the large number of evacuees of an advanced age is increasingly overloading the community's nursing care capabilities, negatively impacting both their functionality and well-being. Therefore, community development efforts required not only the formation of a resident's network but also the strengthening of nursing care provision so evacuees could maintain independent lifestyles. This system has been possible because it is being built on a strong foundation of trust between the residents and public health nurses.

In short, the postdisaster situation in Iwaki City clarified three requirements for improving the Community General Support Center's public health nursing efforts: clarifying community health issues; collaborating with both types of residents (original residents and new evacuees) to improve those issues; and improving the skills of individual public health nurses. Scientific findings can be useful in local public health activities. It is important for local public health nurses to rely on both the evidence before them and their own experiences in their work. However, to make valuable findings useful for the local community, flexible collaboration between scientists and the frontline stakeholders is necessary. In our activities, the relationships with stakeholders have deepened through building and expanding a network by introductions via acquaintances from predisaster times or by frequent site visits. We believe that it is important for researchers and research institutes to stay connected to the field, especially proactively, through everyday, precrisis community conditions.

Summary

This chapter focused on the community crises that followed the Fukushima nuclear accident. We first discussed community resilience based on a review of previous studies and then discussed implementing the research findings in real-world activities. The findings pointed out that elements of community resilience include economic and welfare development, information administration, and community-building efforts that care for both the host and evacuee communities. In actuality, however, such community resilience can be achieved by having discussions and coordinating activities with people who know the community well (e.g., residents and practitioners in the area). As for economic and welfare development and information administration, the Community General Support Center in Iwaki City that we worked with has faced the problems of not having enough access or time to acquire the necessary knowledge and skills for providing services and collecting and communicating information effectively and efficiently. It was important for the center not only to plan services and surveys, but also to ensure that the public health workers could acquire the skills to handle those services and data adequately. Furthermore, resolving intergroup conflicts between the host and evacuee communities required cultural competence; at the same time, these workers were expected to recognize and address other issues (e.g., the aging of the evacuees) from the standpoint of disaster recovery. During a disaster, it is necessary to coordinate the work of academics, who accumulate data and build theories, with the work of practitioners in the field, who transfer these theories into practice. In particular, activities with the affected community could involve more people, such as residents and practitioners, instead of only focusing on those individuals directly affected by the disaster. Therefore, activities can be affected by differences in individuals' values and preferences, and often they may fail in unexpected manners. If that happens, the residents in the community could be losing out on the potential benefits of the community for building stress resilience. To achieve successful community resilience, it is important for researchers, practitioners, and residents to build trust daily, regardless of the circumstances of the disaster.

Acknowledgments

The authors are grateful to Drs. Hitoshi Oto and Masaharu Maeda from Fukushima Medical University for their valuable discussions of the manuscript for this chapter.

References

Adam-Troian, J., Çelebi, E., Bonetto, E., Taşdemir, N., & Yurtbakan, T. (2020). Together we stand? Belonging motive moderates the effect of national ingroup salience on attitudes towards ethnic minorities. *International Journal of Intercultural Relations*, *77*, 95–109. https://doi.org/10.1016/j.ijintrel.2020.05.002.

Aldrich, D. P., & Meyer, M. A. (2015). Social capital and community resilience. *American Behavioral Scientist*, *59*, 254–269. https://doi.org/10.1177/0002764214550299.

Barbieri, K. (1996). Economic interdependence: A path to peace or a source of interstate conflict? *Journal of Peace Research*, *33*, 29–49. https://doi.org/10.1177/0022343396033001003.

Benet-Martinez, V., & Haritatos, J. (2005). Bicultural identity integration (BII): Components and psychosocial antecedents. *Journal of Personality*, *73*, 1015–1050. https://doi.org/10.1111/j.1467-6494.2005.00337.x.

Berry, J. W. (2006a). Stress perspectives on acculturation. In D. L. Sam, & J. W. Berry (Eds.), *The Cambridge handbook of acculturation psychology*. Cambridge, United Kingdom: Cambridge University Press.

Berry, J. W. (2006b). Contexts of acculturation. In D. L. Sam, & J. W. Berry (Eds.), *The Cambridge handbook of acculturation psychology*. Cambridge, United Kingdom: Cambridge University Press.

Drury, J. (2018). The role of social identity processes in mass emergency behaviour: An integrative review. *European Review of Social Psychology*, *29*, 38–81. https://doi.org/10.1080/10463283.2018.1471948.

Ertur, C., & Koch, W. (2007). Growth, technological interdependence and spatial externalities: Theory and evidence. *Journal of Applied Econometrics*, *22*, 1033–1062. https://doi.org/10.1002/jae.963.

Fraser, T., & Aldrich, D. P. (2021). The dual effect of social ties on COVID-19 spread in Japan. *Scientific Reports*, *11*, 1596. https://doi.org/10.1038/s41598-021-81001-4.

Fritz, C. E. (1996). Disasters and mental health: Therapeutic principles drawn from disaster studies. *Historical and comparative disaster series*. Disaster Research Center.

Fukushima Revitalization Station. (2019). *Fukkoukoueijuutaku: chikugoto no kouteihyou to sinchokujoukyou* [*The public revitalization housing: Schedules in the areas*]. (WWW document) URL http://www.pref.fukushima.lg.jp/site/portal/ps-fukkoukouei009.html. (Accessed 17 March 2019).

Gaertner, S. L., & Dovidio, J. F. (2000). *Reducing intergroup bias: The common ingroup identity model*. Philadelphia: Psychology Press.

Godschalk, D. R. (2003). Urban hazard mitigation: Creating resilient cities. *Natural Hazards Review*, *4*, 136–143. https://doi.org/10.1061/(ASCE)1527-6988(2003)4:3(136).

Granovetter, M. (1983). The strength of weak ties: A network theory revisited. *Sociological Theory*, *1*, 201–233. https://doi.org/10.2307/202051.

Han, L., Gui, Y., Zheng, Y., & Berry, J. W. (2015). Differences in resilience by acculturation strategies: A study with Qiang nationality following 2008 Chinese earthquake. *International Journal of Emergency Mental Health and Human*, *17*(2), 573–580.

Hasegawa, M., Murakami, M., Takebayashi, Y., Suzuki, S., & Ohto, H. (2018). Social capital enhanced disaster preparedness and health consultations after the 2011 Great East Japan Earthquake and Nuclear Power Station accident. *International Journal of Environmental Research and Public Health*, *15*, 516. https://doi.org/10.3390/ijerph15030516.

Haslam, C., Jetten, J., Cruwys, T., Dingle, G. A., & Haslam, S. A. (2018). *The new psychology of health: Unlocking the social cure* (1st ed.). London: Routledge.

Imai, A. (2017). Genpatsusaigai hinansya no jittaityousa: Roku ji [The states of nuclear disaster evacuees: The sixth survey]. *Jichisoken*, *462*, 1–34 (in Japanese).

Jencson, L. (2001). Disastrous rites: Liminality and communitas in a flood crisis. *Anthropology and Humanism*, *26*, 46–58. https://doi.org/10.1525/ahu.2001.26.1.46.

Jong, J., Whitehouse, H., Kavanagh, C., & Lane, J. (2015). Shared negative experiences lead to identity fusion via personal reflection. *PLoS One*, *10*. https://doi.org/10.1371/journal.pone.0145611, e0145611.

Kawafuku, S. (2013). Circumstances affecting the acceptance of nuclear evacuees: The case of Iwaki City. *Research on Environmental Disruption*, *42*, 37–41.

Kobayashi, T., Takebayashi, Y., & Murakami, M. (2020). Disaster-zone research: Participants should benefit too. *Nature*, *579*, 193. https://doi.org/10.1038/d41586-020-00696-z.

Kobayashi, T., Yoshida, K., Kumagai, A., Yasui, K., Goto, A., Takebayashi, Y., et al. (2019). Saigai kanren kenkou risuku ni taisuru komyunikesyon to kyoudou [Communication and collaboration toward disaster-related health risks]. *Journal of Japan Society for Safety Enginieering*, *58*, 387–393. https://doi.org/10.18943/safety.58.6_387.

Kobayashi, T., Yoshida, K., Takebayashi, Y., Goto, A., Kumagai, A., & Murakami, M. (2019). Social identity threats following the Fukushima nuclear accident and its influence on psychological distress. *International Journal of Disaster Risk Reduction*, *37*. https://doi.org/10.1016/j.ijdrr.2019.101171, 101171.

Kobayashi, T., Yoshida, K., Takebayashi, Y., Goto, A., Kumagai, A., & Murakami, M. (2021). Belief in group interdependence: Facilitating evacuee–host interactions after the Fukushima nuclear accident. *Journal of Applied Social Psychology*, *51*, 513–521. https://doi.org/10.1111/jasp.12753.

Kwesell, A., & Jung, J.-Y. (2019). A multidimensional analysis of stigma: Findings from a qualitative study of Fukushima residents following Japan's 2011 nuclear disaster. *Journal of International Crisis & Risk Communication Research*, *2*, 233–258. https://doi.org/10.30658/jicrcr.2.2.4.

Lazarus, R., & Folkman, S. (1974). *Stress, appraisal, and coping*. New York: Springer.

Masten, A. S., Best, K. M., & Garmezy, N. (1990). Resilience and development: Contributions from the study of children who overcome adversity. *Development and Psychopathology*, *2*, 425–444. https://doi.org/10.1017/S0954579400005812.

McPherson, M., Smith-Lovin, L., & Cook, J. M. (2001). Birds of a feather: Homophily in social networks. *Annual Review of Sociology*, *27*, 415–444. https://doi.org/10.1146/annurev.soc.27.1.415.

Ministry of Internal Affairs and Communications. (n.d.). *Genpatsu hinansya tokureihou ni motoduku shiteishichouson oyobi tokureijimu no kokuchi tou ni tsuite* [*Notification of the municipalities and special affairs based on the special act for nuclear evacuees*]. Ministry of Internal Affairs and Communications. (WWW document) URL https://www.soumu.go.jp/menu_kyotsuu/important/48479.html. (Accessed 14 June 2021).

Miura, A., Okayasu, T., & Miura, M. (2018). Mental health among children relocating to temporary housing following the Fukushima nuclear accident. *Shinrigaku Kenkyu*, *89*, 104–110. https://doi.org/10.4992/jjpsy.89.16333.

Nagai, M., Ohira, T., Zhang, W., Nakano, H., Maeda, M., Yasumura, S., et al. (2017). Lifestyle-related factors that explain disaster-induced changes in socioeconomic status and poor subjective health: A cross-sectional study from the Fukushima health management survey. *BMC Public Health*, *17*, 340. https://doi.org/10.1186/s12889-017-4247-2.

National Diet of Japan Fukushima Nuclear Accident Independent Investigation Commission. (2012). Overview of the damage and how it spread. In *The official report of the Fukushima nuclear accident independent investigation commission* National Diet of Japan.

Norris, F. H., Friedman, M. J., Watson, P. J., Byrne, C. M., Diaz, E., & Kaniasty, K. (2002). 60,000 disaster victims speak: Part I. An empirical review of the empirical literature, 1981–2001. *Psychiatry: Interpersonal and Biological Processes*, *65*, 207–239. https://doi.org/10.1521/psyc.65.3.207.20173.

Norris, F., Stevens, S., Pfefferbaum, B., Wyche, K., & Pfefferbaum, R. (2008). Community resilience as a metaphor, theory, set of capacities, and strategy for disaster readiness. *American Journal of Community Psychology*, *41*, 127–150.

Okuma Town. (2017). *Earthquake record of Okuma: From the town where Fukushima Daiichi nuclear power station is located.* Okuma Town.

Orui, M., Nakajima, S., Takebayashi, Y., Ito, A., Momoi, M., Maeda, M., et al. (2018). Mental health recovery of evacuees and residents from the Fukushima Daiichi Nuclear Power Plant accident after seven years-contribution of social network and a desirable lifestyle. *International Journal of Environmental Research and Public Health*, *15*, 2381. https://doi.org/10.3390/ijerph15112381.

Redfield, R., Linton, R., & Herskovits, M. (1936). Memorandum for the study of acculturation. *American Anthropologist*, *38*, 149–152. https://doi.org/10.1525/aa.1936.38.1.02a00330.

Rich, R. C., Edelstein, M., Hallman, W. K., & Wandersman, A. H. (1995). Citizen participation and empowerment: The case of local environmental hazards. *American Journal of Community Psychology*, *23*, 657–676. https://doi.org/10.1007/BF02506986.

Riek, B. M., Mania, E. W., Gaertner, S. L., McDonald, S. A., & Lamoreaux, M. J. (2010). Does a common ingroup identity reduce intergroup threat? *Group Processes & Intergroup Relations*, *13*, 403–423. https://doi.org/10.1177/1368430209346701.

Sam, D. L., & Berry, J. W. (2010). Acculturation: When individuals and groups of different cultural backgrounds meet. *Perspectives on Psychological Science*, *5*, 472–481. https://doi.org/10.1177/1745691610373075.

Sampson, R. J., Raudenbush, S. W., & Earls, F. (1997). Neighborhoods and violent crime: A multilevel study of collective efficacy. *Science*, *277*, 918. https://doi.org/10.1126/science.277.5328.918.

Sawano, T., Kambe, T., Seno, Y., Konoe, R., Nishikawa, Y., Ozaki, A., et al. (2019). High internal radiation exposure associated with low socio-economic status six years after the Fukushima nuclear disaster: A case report. *Medicine (Baltimore)*, *98*. https://doi.org/10.1097/MD.0000000000017989, e17989.

Schaller, M., & Neuberg, S. L. (2012). Danger, disease, and the nature of prejudice(s). In J. M. Olson, & M. P. Zanna (Eds.), *Advances in experimental social psychology* (pp. 1–54). Academic Press. https://doi.org/10.1016/B978-0-12-394281-4.00001-5.

Stephan, W. G., & Stephan, C. W. (2000). An integrated threat theory of prejudice. In S. Oskamp (Ed.), *Reducing prejudice and discrimination, the Claremont symposium on applied social psychology* (pp. 23–45). Mahwah, NJ, US: Lawrence Erlbaum Associates Publishers.

Szreter, S., & Woolcock, M. (2004). Health by association? Social capital, social theory, and the political economy of public health. *International Journal of Epidemiology*, *33*, 650–667. https://doi.org/10.1093/ije/dyh013.

Tajfel, H., & Turner, J. (1979). An integrative theory of intergroup conflict. In W. G. Austin, & S. Worchel (Eds.), *The social psychology of intergroup relations* (pp. 33–47). Monterey, CA: Brooks/Cole.

Takagi, A. (2017). Evacuation and community issues caused by nuclear disaster in Fukushima Japan. *Journal of Asian Development*, *3*, 40–50. https://doi.org/10.5296/jad.v3i1.10578.

Takaki, R. (2018). Problem of the community in the public housing for evacuees from the Fukushima nuclear accidents. *Annual Reports of the Tohoku Sociological Society*, *47*, 11–23. https://doi.org/10.11271/tss.47.11.

Takebayashi, Y., Maeda, M., Orui, M., Nakajima, S., Momoi, M., Ito, A., et al. (2020). Resilience factors contributing to mental health among people affected by the Fukushima disaster: Development of Fukushima Resilience Scale. *Frontiers in Public Health*, *8*, 159. https://doi.org/10.3389/fpubh.2020.00159.

Van Dick, R., Ketturat, C., Häusser, J. A., & Mojzisch, A. (2017). Two sides of the same coin and two routes for improvement: Integrating resilience and the social identity approach to well-being and ill-health. *Health Psychology Open*, *4*. https://doi.org/10.1177/2055102917719564, 2055102917719564.

Yabe, H., Suzuki, Y., Mashiko, H., Nakayama, Y., Hisata, M., Niwa, S.-I., et al. (2014). Psychological distress after the Great East Japan Earthquake and Fukushima Daiichi Nuclear Power Plant accident: Results of a mental health and lifestyle survey through the Fukushima Health Management Survey in FY2011 and FY2012. *Fukushima Journal of Medical Science*, *60*, 57–67. https://doi.org/10.5387/fms.2014-1.

Chapter 17

A comparative study of psychological reactions of the former Soviet Union, Scandinavia, and Fukushima people to nuclear disasters

Tatsuo Ujiie[a] and Yuji Tsutsui[b]
[a]*The Open University of Japan, Nagoya, Japan,* [b]*Fukushima University, Fukushima, Japan*

Introduction

The Chernobyl nuclear explosion in the former Soviet Union on April 26, 1986, and the meltdown of the TEPCO Fukushima Daiichi Nuclear Power Plant caused by the Great East Japan Earthquake on March 11, 2011, are classified as major accidents corresponding to Level 7 of the International Nuclear and Radiological Event Scale. These incidents had enormous psychological and social impacts on vast groups of people.

Not only has the impact of the Chernobyl disaster affected the three Soviet republics of Russia, Belarus, and Ukraine, but it has also spread over Europe. Scandinavian countries were the most affected by the disaster after the three Soviet republics. Various research institutions such as IAEA, WHO, and EC conducted many studies on the impact of the disasters on the three Soviet republics. They concluded that "the mental health impact of Chernobyl is the largest public health problem caused by the accident to date" (WHO, 2006, p. 95). While their effects were not as long-lived as those in the three Soviet republics, the Scandinavian people also feared the health effects of radiation exposure and experienced severe anxiety.

Despite the differences in the psychological effects of the Chernobyl disaster between the Soviet Union and Scandinavia, there are commonalities in the primary causes. There are also differences and commonalities between these regions and Fukushima. Unique features of radiation accidents also contributed to these psychological impacts. In this chapter, we attempt to clarify the effects

Health Effects of the Fukushima Nuclear Disaster. https://doi.org/10.1016/B978-0-12-824098-4.00009-X

of nuclear disasters on both people and society by comparing the cause of psychological effects in the Soviet Union, Scandinavia, and Fukushima, and learning lessons from these experiences.

Comparative studies of the people's reaction

Reactions of the people in the Soviet Union

Reactions immediately after the accident

Owing to a lack of information about the disaster, the people in the Soviet Union did not have a strong negative reaction immediately afterward. At the time of the disaster, the fire could be seen from the buildings of Pripyat. Alexievich (2011) has published several testimonies of Pripyat citizens who witnessed the accident. The people watched the fire from their balconies without fear. Some people even rushed dozens of kilometers by car or bike to catch a glimpse upon hearing rumors of a fire. The next morning, children rode close to the power plant on their bikes to watch.

On April 27, the 45,000 residents of Pripyat were quickly evacuated. They were informed that it would be a 3-day evacuation. According to one child evacuated from Pripyat to Kharkiv, the evacuation was like the quick preparation and departure for a Pioneer Camp, and this child firmly believed that they would return home (Pidkorytov, Shestopalova, & Kukuruza, 2000). They noted that the nuclear power plant explosion and fire were fairly localized, and it did not become a major man-made disaster. The threat to life and health was not very real. Awareness of the threat arose gradually as the people acquired information through official and unofficial channels.

Studies of psychogenic issues in the Soviet Union

Aleksandrovskij (1989) was dispatched by the Soviet Union Ministry of Public Health to a severely contaminated region to conduct a year-long study of psychogenic issues beginning on April 29. The study was divided into three periods: the acute period of the disaster from April 29 to May 5, the period of early effects from May 6 to the end of October, and the period of late effects from November to April of the next year. Evacuation of those within a 30-km radius was largely completed in the acute period of the disaster. In most citizens, what was observed was not a psychotic level, but the first-stage stress reaction, such as a stand-still response or an escape response. Entering the period of early effects, the people began to experience chronic stress due to the second calamity caused by evacuation (loss of home or assets, separation from relatives, a difficult everyday life). Six months after the disaster, many evacuees began to exhibit psychiatric disorders. They developed depression with physical symptoms, persistent radiophobia (note: this term was used at the time but is not used now), and psychosomatic symptoms. The chronic stress of lacking information on the effects of radiation exposure and countermeasures against it, the difficulty of adapting to life in their new locations, and becoming the targets of stigma lowered the people's psychological well-being.

Delayed-onset psychological issues

Several years after the disaster, the people experienced the onset of stress disorders, psychosomatic symptoms, and learned helplessness (Lee, 1996). These were accompanied by irrational coping strategies such as avoidance, fatalism, and passive refusal of protective measures. Victims were disturbed by the extent of anxiety and fear and the widespread social disruption caused by the exodus from the area.

Such issues were both caused and exacerbated by the government's secretiveness with respect to the information about the disaster and contamination. Contamination maps and radiation levels were considered highly classified information until 1989. However, according to a report published on the 25th anniversary of the disaster by the government of Ukraine, the scale of the accident made concealing information impossible. Information on resettling the residents of Pripyat and Chernobyl spread promptly among the population of Ukraine, Belarus, and Russia. "The hiding of the Chernobyl accident information led to the formation and spread of unbelievable rumors about the possible consequences of the accident. This, in turn, caused a very large social and psychological tension among the population and distrust of official information" (Ministry of Ukraine of Emergencies, 2011, p. 22).

Local doctors have played a key role in the outbreak of psychosomatic symptoms (Bromet, 2011; International Advisory Committee, 1991). Upon hearing that radioactivity would cause several illnesses, victims visited doctors for an examination of various symptoms. Despite a lack of expert knowledge on radiation, the doctors ultimately attributed the symptoms to radiation exposure.

Pidkorytov et al. (2000) explained that inconsistent information about regular health check-ups and the effects of radiation led parents to believe that radiation exposure caused their children's illnesses. These regular health check-ups were conducted on the premise that a child had been exposed to radiation and, as such, bred distrust in the official information that radiation exposure had no health effects for many parents. These health check-ups did detect several issues, which many parents attributed to radiation exposure.

The nuclear disaster also impacted children's development and adjustment. According to Pidkorytov et al. (2000), many of the children evacuated from Pripyat and Chernobyl to Kharkiv between the ages of 4–7 developed borderline disorders in 10 years. Numerous psychosocial factors contributed to the development of these disorders. For example, the children were led to become hypersensitive to declines in their health due to how they were treated by parents and grandparents who were worried about the health effects of radiation on their children. Conflict in the community and with peers was also related to the children's psychological issues. They perceived themselves as victims of

Chernobyl, but at the same time tried to hide that fact during everyday conversations with local peers. While their preferential treatment as victims was a kind of special privilege, it could lead to the envy of the local people and was also associated with perceptions about their health being poor. Interactions with family members strengthened their desire to return home, but there was a higher chance of developing intellectually and professionally in a big city.

Lagutin et al. (Lagutin, 1998; Lagutin & Rogozin, 1999) found that children with physical illnesses caused by low-level radiation exposure were more likely to exhibit asthenia, anxiety, and depressive symptoms. The authors suggested that the negative psychosocial factors experienced by children from the Chernobyl disaster may have increased the onset rate and severity of borderline disorders in those with physical illnesses.

Victim syndrome

Surveys periodically administered by Ukraine's Institute of Sociology starting in 1992 revealed that those affected by the disaster were developing a strong victim mentality and a passive attitude toward life. Each year in Ukraine, an increasing number of people viewed themselves and their children as victims whose entire lives had been ruined by the disaster (referred to as "victim syndrome"). According to Garnetz (2000), victim syndrome increased from 19% in 1992 to 35% in 1999 and persisted even in 2003. Soboleva (2003) found that 34% of residents of Zone 2 (the compulsory evacuation zone) and the *samosely* citizens who illegally returned to live in Zone 1 (the exclusion zone) responded that Chernobyl had destroyed their health and lot in life.

The effects of the disaster cast a dark shadow across the Ukrainians. According to Sayenko (1996), almost none of the Ukrainian people viewed the problems of Chernobyl as fully resolved in 1994. Moreover, one in three responded that they could not count on anyone to resolve it. Soboleva (2015) reported that in 1998, 70% responded that they could not hope for their lives to get better soon. By 2005, this number had dropped to just under 20%, and 40% of respondents stated that it would likely get better soon. However, in response to the question "How well are you adapting to your current circumstances?," 34% still responded that they were passively waiting for their situation to get better in 2006.

Chernobyl's Legacy (Chernobyl Forum: 2003-2005, 2006) states that victims had renounced the responsibility of caring for their own health; for example, they were eating mushrooms and berries picked from contaminated forests. Yet, the results of a survey by the Institute of Sociology revealed victims who did not fit the conclusions made in Chernobyl's Legacy. According to Sayenko (1996), in a 1994 survey, residents of Zone 2 estimated less radioactive contamination in their neighborhoods than residents in Zone 3 (the voluntary migration zone). For example, while 86% of residents of Zone 3 thought that foodstuffs were contaminated, this rate was only 45% for residents of Zone 2. However, concerning what they were eating, more than 60% of residents of both zones responded that they eat food even if it is contaminated. These results can be interpreted as the result of passive adaptation by residents of Zones 2 and 3 to a reality in which they could not find uncontaminated food, rather than indifference to contamination. According to Drottz-Sjöberg (2016), when she visited the Russian city of Novozybkov 5 years after the disaster, the store shelves were nearly empty. Material economic conditions took a drastic downturn after Ukraine seceded from the Soviet Union and gained independence in 1991 (Lavrinenko, 2000). Sayenko (1996) made several interpretations of the surprisingly low levels of environmental radioactive contamination estimated by residents of the highest risk area, Zone 2: It may be adaptive, a defense mechanism, or fear of migrating to a new place far from home that is unknown, unfamiliar, and probably even hostile.

Based on the Institute of Sociology, victim syndrome and passive social attitude were in part caused by the sudden obliteration of their living environment and everyday habits due to the disaster as well as political and economic disruption and the paternalism seen in the citizens of a totalitarian state. The people had been forced to abandon thinking autonomously and taking the initiative for centuries (Sayenko, 1996).

Psychological reactions of the people of Scandinavia and Fukushima

Here, we will explore the psychological reactions by comparing Scandinavia and Fukushima, as the responses of the people of these two regions can be seen from a common perspective.

Survey results

Research directly measuring the people's responses to the effects of the Chernobyl disaster has been conducted in Sweden and Norway. Drottz-Sjöberg and Sjöberg (1990) surveyed residents of three regions of Sweden (the highly contaminated Gävle, the mildly contaminated Stockholm, and the uncontaminated Bohus) about the risk perception and perplexity about radiation exposure in September of 1986. They found that 72% of respondents selected responses of "very worried" or "rather worried." Parents, women, and farmers, in particular, responded that they are worried. In residents of the highly contaminated Gävle, these rates were roughly double those in residents of the other two regions. In Norway, a survey with a nationwide sample was conducted from May to June 1986 (Weisæth & Tønnessen, 1995). According to the results, about one of five respondents experienced very unpleasant thoughts and feelings after the accident.

Tsutsui, Ujiie, Takaya, and Tominaga (2020) of Fukushima University surveyed parents of infants in Fukushima Prefecture, areas outside the evacuation zone, from June to July 2011. The authors measured stress related to the earthquake and nuclear disaster and found higher stress scores than parents in prefectures that were not affected.

According to a survey of mental health and everyday habits conducted with evacuees by Fukushima Prefecture roughly 1 year after the disaster (Radiation Medical Science Center for the Fukushima Health Management Survey, 2019), approximately 15% of evacuees aged 16 or older had depressed moods or severe anxiety, prompting the need for support. As the nationwide average is 3%, this demonstrates that psychological health issues persisted in those who had evacuated even 1 year after the disaster. Further, the results concerning children's psychological health using the Strength and Difficulties Questionnaire (SDQ) revealed that total scores exceeded the cutoff value for a high percentage of children, with rates surpassing 20% in early childhood and elementary school and reaching 16% in middle school.

A telephone storm

We can understand people's confusion, anxiety, and fear of their behaviors. In Scandinavia, expert institutions, the media, fire departments, emergency services, and the police were hit with a telephone storm from citizens (Norges Offentlige Utredninger, 1986). In a press conference, the director of the Swedish Radiation Protection Institute (Statens strålskyddsinstitut: SSI) appealed to the people not to let their anxiety drive them to call emergency services, fire departments, or the police (Löfveberg, 1986). The phone lines at the emergency task force and SSI were jammed starting on April 28, the day the first radioactive material released from Chernobyl was detected in Sweden. Even after 9 days, on May 6, they were still receiving 1000 calls.

On March 17, Fukushima Prefecture established the Support Services for Questions about Radiation. In its first month, the service dealt with a total of 8800 inquiries. Fukushima Prefecture published the radiation dose for each region on their website on May 16, but an article published on May 17 in *Fukushima Minpo*, a local newspaper, reported that getting through was difficult due to the massive influx of visitors. The Fukushima Daiichi disaster occurred in the era of widespread access to the Internet and social networking services. As such, the people's primary source of information was not phone lines set up by the prefecture or the media. Rather, the people sought and shared information online.

Changes in diet and everyday behaviors

Fahlen and Hammarström (1990) studied diet and everyday habits in the highly contaminated cities of Kramfors and Sollefteå in northern Sweden. Of those surveyed, 84% responded that their eating habits had changed. More than half the respondents went beyond the recommendations of the county authorities and stopped or reduced berry picking and mushroom foraging.

The parents in Fukushima City changed their daily activities after the accident (Tsutsui et al., 2020); they restricted their children's outdoor play. Less than 10% of parents allowed their children to play outside while 60% of people stopped hanging their laundry outside to dry after the disaster. Only 22% hung their laundry outside as they had before. In surveys conducted in other prefectures in 2013 and 2015, almost all parents allowed their children to play outside and less than 10% did not hang their laundry outside to dry. Fukushima City was not an evacuation zone. *Fukushima Minpo* published information about avoiding outdoor radiation exposure several times after the disaster, but this information was intended for regions told to shelter indoors and did not apply to Fukushima City. Public relations bulletins for Fukushima City ran two question-and-answer articles about radiation, but neither recommended such countermeasures in any way.

Voluntary evacuation

At the time of the Three Mile Island accident on March 28, 1979, the governor of Pennsylvania recommended evacuation for all pregnant women and preschool-age children within 8 km of the NPP. The governor instructed other citizens within 16 km to shelter indoors. While these evacuation recommendations applied to roughly 3500 individuals, 200,000 people evacuated (Erikson, 1990), thus voluntarily traveling an average of 160 km away from the NPP.

It is unclear how many people voluntarily evacuated from the vicinity of the NPP immediately after the Chernobyl disaster. The actual status of voluntary evacuees in Scandinavia is also unknown. However, it seems that residents of highly contaminated regions in Sweden voluntarily evacuated or sent their children to the uncontaminated areas. In Höijer's interview study, one participant said that many people tried to evacuate their children away from the city of Gävle. That participant evacuated her 7-year-old daughter to a safe city (Höijer, 1987). On May 21, the tabloid newspaper *Aftonbladet* ran an article with a headline saying residents of Gävle evacuate (Löfveberg, 1986). It seems that regardless of how much government agencies and experts explained that evacuation was unnecessary, some families were not convinced and decided to evacuate temporarily (Löfveberg, 1986).

The Fukushima Prefectural Headquarters for Disaster Control confirmed that more than 40,000 had evacuated voluntarily within the prefecture by March 15. There was also a considerable number of evacuees who left the prefecture. On March 12, the Japanese government ordered residents within a 20-km radius of the NPP to evacuate. Many of the towns and villages in the south of the Soso area, where the nuclear power plant is located, were subject to evacuation orders. Residents of other municipalities in Soso who were not subject to evacuation orders voluntarily evacuated. By March 15, the number of voluntary evacuees from the Soso region had reached 12,205.

Those who voluntarily evacuated experienced a type of hardship different from those who evacuated due to the evacuation order. This situation brought about not only psychological hardship, but financial difficulties as well. Many mothers evacuated with their children and were required from the start to evacuate at their own risk. Moreover, the government and TEPCO's financial compensation and support were insufficient (Institute of Disaster Area Revitalization, Regrowth and Governance, Kwansei Gakuin University, and Japan Civil Network, and Save Fukushima Children Lawyer's Network, 2015).

Subsequent people's reaction

Scandinavia

In Sweden, the value of news related to Chernobyl declined over time (Findahl & Lindblad, 1987). Chernobyl was mentioned on public broadcasting TV an average of 6–7 times per day in the first week, but this dropped to two or fewer times by week four. In unaffected regions, memories of the disaster quickly faded. Nonetheless, the people of regions strongly impacted by the disaster distinctly remembered the event's moments more than 10 years later. Memories of the disaster remained vivid and embellished with strong emotion (Enander, 2000).

The fallout from Chernobyl extensively contaminated the region inhabited by the Sami people. Reindeer meat was highly contaminated, and most of the meat was discarded when many reindeer were culled. Consumption and the price of the meat dropped after the news that reindeer meat was contaminated spread throughout the country. According to a 2018 interview survey conducted by the authors with the Sami people living in the mountain areas of Norway and Sweden, it took 5–10 years for the economic downturn to recover.

The Norwegian government tracked the radiation dose in both the Sami people and reindeer meat continuously until 1993. However, according to Skuterud of the Norwegian Radiation Protection Authority (Statens strålevern), who had been involved with the Sami people for 20 years beginning in 1996, the Sami people in 1996 suffered increasing stress due to simultaneous distrust in experts and the government as well as anxiety about their health and that of their children (Skuterud, 2016).

Fukushima

According to a survey of mental health and everyday habits in evacuees conducted by Fukushima Prefecture (Radiation Medical Science Center for the Fukushima Health Management Survey, 2019), the percentage of evacuees 16 years or older thought to require support for depressed mood or anxiety dropped from 14.6% in 2011 to 5.7% in 2018. However, considering this rate is 3% in the general population, evacuees continued to exhibit higher depressive tendencies even 7 years after the disaster.

A survey of parents living outside evacuation zones (Tsutsui et al., 2020) reported that anxiety in mothers decreased over time. The percentage of parents who did not allow their children to play outside had nearly reached zero by 2014. However, even in 2016, only 54% of parents in Fukushima allowed outside activities of children, compared to 76% of parents in other prefectures. In 2011, only 22% of parents responded that they always hang their laundry outside to dry, but this number had increased to more than 40% in 2016. Nonetheless, nearly 70% of parents in other prefectures responded that they always hang their laundry out to dry, revealing that parents in Fukushima continue to take precautions against radiation even in 2016. These results demonstrate that parents in Fukushima continue to have long-term anxiety about radioactivity.

Parental anxiety has negative impacts on the psychological health and development of young children in particular. The authors studied correlations of parental risk perception and parental behavior with indicators of child development at 3.5 years old in mothers and their children born in 2011, most of whom were pregnant during the earthquake and nuclear disaster. This research revealed that perceiving the effects of the disaster as risk lowered mothers' sensitivity toward their children and reduced their children's effortful control, a factor thought to be associated with the development of executive function and school adjustment (Sánchez-Pérez, Fuentes, Eisenberg, & González-Salinas, 2018; Zhou, Chen, & Main, 2012). The cognitive and socioemotional development of children born in 2011 after they had reached school age was investigated based on these results. A preliminary analysis of SDQ scores suggested that children born in 2011 may face unique issues. The percentage of children in the northern Soso region, northern Fukushima Prefecture, and neighboring prefectures born in 2011 who had a total SDQ score exceeding the cutoff value was greater than the national standard for Japan. In contrast, this rate was no different from the Japanese national standard for children born in 2012. The results for children born in 2012 are comparable to the results of a survey of mental health and everyday habits in Fukushima Prefecture. According to this survey, after the disaster, the percentage of children above the cutoff value for total score was higher than the national standard for Japan in all age groups from early childhood to middle school, but this percentage decreased over time, reaching the standard level of 9% by 2018 in all age groups.

On March 3, 2018, *Asahi Shimbun* published an article on the results of a survey conducted by the *Asahi Shimbun* and *Fukushima Hoso*, a local TV station in Fukushima, on 1000 Fukushima residents in February 2018 (Shimbun, 2018 March 3, p. 7). According to this article, 66% of people had anxiety about radioactive material from the disaster on themselves and their families. As this number was about 80% when the same question was asked in 2012, it is clear that the percentage of people with anxiety is decreasing. Nonetheless, more than half of Fukushima Prefecture residents have been unable to shake their anxiety.

At peak, more than 8000 people had evacuated to Niigata Prefecture, which is adjacent to Fukushima Prefecture. This figure includes those who evacuated voluntarily from outside the evacuation zone. According to a review by the Niigataken seishinhukushi kyoukai (2019), most of Niigata Prefecture's evacuees were people living in areas to which they had no ties and no relatives or acquaintances. The number of people living as evacuees in Niigata Prefecture had decreased to just under 3000 by 2017, but during the long-term duration of their evacuation, evacuees' challenges became more complex, more serious, and latent. In 2017, 24.8% were thought to require support for depressed mood or anxiety. Comparing this to the rate of 6.4% from the Fukushima Prefecture Health Survey conducted in the same year demonstrates that evacuees have feeble psychological health. This is true for all people who are continuing to live anywhere as evacuees.

Analysis of people's psychological reactions

The role of information in radioactive contamination disasters

Nuclear disasters and the contamination caused by their radioactive materials cause severe fear and health-related anxiety. Following the Three Mile Island accident, the people felt a deep and profound dread toward radioactivity, and many voluntarily evacuated (Erikson, 1990). Fear of radioactivity is fundamentally different from the fear of things that can be perceived with the senses. As human senses cannot detect radioactivity, the dread of radioactivity is thought to arise when we find out that there is unnatural radioactivity around us and that our food and drink have been contaminated (this happens by being informed, as it is difficult for ordinary citizens to find this out on their own). The question here is when, how, and from whom citizens are notified of this information.

In the case of the radioactivity leak from the Fernald Preserve in the Ohio in the United States, people did not know for a long time that the factory produced nuclear fuel. Although the factory had been leaking (discarding) radioactive material long before the people knew, the government intentionally kept information on contamination hidden for quite some time. People learned about the leakage through media reports and discovered the government's cover-up at a congressional public hearing. Even after the leakage came to light, the factory denied radiation exposure and attempted to minimize the situation. As the facts gradually became clear, the people began to exhibit a condition referred to by Green, Lindy, and Grace (1994) as the "Informed of Radioactive Contamination Syndrome." The time lag in which information was revealed became a stressor that negatively impacted the people's psychological well-being.

In the Soviet Union, there was no information on the disaster or the risk of contamination from the start. However, there was no shortage of unofficial information warning of the dangers and rumors of sick or dying young due to radiation. Unbelievable rumors were born and spread soon. When the Soviet Union government publicly announced the amount of contamination and radiation exposure and their risk levels, the affected citizens already firmly believed that something "fatal" had happened. The safety news after that only intensified their anger and distrust. The policies for evacuation and compensation enacted by the government of the Soviet Union further strengthened rumors that a lethal event was taking place.

However, it seems the affected people were unexpectedly tough. The testimony of one *samosely* in Alexhevic's book that the annoyance is their work and that our work is silent is highly representative (Alexievich, 2011). The *samosely* is a group of people who returned to the exclusion zone of their own accord and lived by growing crops in the contaminated soil. Information was not particularly necessary for the *samosely* because the fear of migrating to unknown, unfamiliar, and probably even hostile areas far from home (Sayenko, 1996) was likely stronger. Also, to survive, they had to stay and eat contaminated food.

After the disaster, scientists such as Nesterenko took measurements in various locations and warned people about the dangers. Nevertheless, people would not be easily persuaded by such unofficial claims from scientists (Alexievich, 2011).

Sweden faced a different problem. The world learned of the Chernobyl disaster when radioactive material was detected at the Forsmark NPP. At first, Forsmark suspected a leak within their power plant. They promptly conveyed facts such as the suspected leak, extent of contamination, and no special measures to the people via radio. Upon learning that the contamination came from Chernobyl, the SSI informed people through the media. However, the information provided by the power plant and the SSI intensified people's anxiety and spurred them to what could be called excessive action. People were unable to understand the information they received. In their media briefings, SSI used complex numbers without sufficient explanation. Gradually, it became clear that the contamination level was much worse than it had initially been thought, but the SSI repeatedly stated that special countermeasures were unnecessary. A local newspaper published a complaint from a father that he did not know how to utilize all the information provided by SSI in his daily life (Engholm, 1987). A report by Enander (2000) found that this was not just excessive criticism from a small group of people. Even Löfveberg, an SSI employee, acknowledges the media criticism that the information provided by the SSI was lightyears away from what the people needed (Löfveberg, 1986). The issue is how it was conveyed (Löfveberg, 1986; Nohrstedt & Lekare, 1987). Many people felt that the SSI was hiding the truth or that they were being deceived because of how the SSI reported the information. Anér (1986), a renowned anti-nuclear power activist, argued that the information was hidden in becquerel, gray, sievert, micro x-ray, and new technology terminology, and people could not be understood. He insisted SSI intentionally made it incomprehensible to people. While this is an exaggeration, it is likely an accurate representation of the people's sentiment. The kind of chaos that arose in Sweden is known as an "information crisis." The information released by the SSI not only failed to give the people peace of mind but unintentionally invoked an intensely negative reaction. According to Engholm (1987), SSI chief Gunnar Bengtsson said to *Gefle Dagblad*, a local newspaper, that radiation was 80% an information problem and 20% a radiation protection problem.

The Fukushima Daiichi NPP disaster was caused by the Great East Japan Earthquake. This megadisaster threw the information system into disorder. The Headquarters for Disaster and Accident Control was located in the prime minister's official residence and was thus quite cut off from information from the NPP and the contaminated region (Fukuyama, 2012; Hosono & Torigoe, 2012; Kimura, 2012). Information from the official residence tended to be delayed and was often unclear. Even so, people had no choice but to use the information from the government as a guide for judging the situation and deciding what action to take, as it is impossible for members of the general public to determine radioactive contamination on their own.

Following the disaster, the information from the government conveyed the message that although the status of the NPP was critical, the government would be able to overcome it. The government issued an evacuation order for residents close to the power plant but publicly stated that it was a "precautionary" measure. Nonetheless, the situation deteriorated beyond the government's expectations. Several nuclear reactors exploded, and the evacuation zone was expanded. The government's decision on evacuation from shelter-in-place zones wavered, and their instructions were vague. Trust in the government's response and the information they provided was already being lost less than a month after the disaster (Fukuyama, 2012). On April 19, the Ministry of Education, Culture, Sports, Science, and Technology (MEXT) issued guidelines concerning the duration of outdoor activities in schools. These guidelines were met with widespread criticism. Government experts strongly criticized the guidelines at a protest resignation conference on April 30, causing people to distrust government information and measures (Shibata, 2012). This became a new crisis for the people. Unfortunately, by this time, people had already been exposed to radiation. The anger, fear, and anxiety of people about the future were intensified by the information that radiation exposure is dangerous.

Alternative information

In Sweden, independent contamination information from unofficial experts and criticism and suspicion toward the risk ratings and countermeasures of the SSI fueled the information crisis. In one example, a researcher from Uppsala University appeared on the Uppland radio program on the afternoon of April 29 (Nordlund, 1987). The piercing sound of the dosimeter played from the radio as he measured the radioactivity in the surrounding area using a Geiger counter. He stated that there was an extremely high level of radiation, contrary to the amount reported by the SSI that morning. He went on to say that it is best to stay indoors if possible. When asked what he would do if this level was measured in a laboratory, he replied that it would be contaminated with radioactivity and need to be decontaminated. The local radio station that broadcast this statement was inundated with calls from anxious citizens.

In the Soviet Union, there was nothing but unofficial information. Unless there was formal and reliable information, even unbelievable rumors were information for people. There were allegations that some firefighters had died on duty and others had developed severe radiation damage and were transported to Moscow. These rumors were common, many of which were perpetrated by the district's officials and soldiers who were desperately attempting to force the affected citizens to evacuate. The officials ordering evacuation threatened the people, stating that the soil and food were contaminated and that they would die if they continued to stay and eat the food from the region. Alexievich (2011) recorded testimony from farmers, which is a vivid reminder of this kind of attempt to drive them out.

People were convinced that they would develop all sorts of diseases; hence, they consulted the doctor for even slight symptoms. This was spurred on by the fact that medical fees were waived as compensation to victims. Further, local doctors did not necessarily possess any specialized knowledge of radiation, and their diagnoses triggered fear and anxiety (Bromet, 2011).

In Fukushima, people did not have such a high distrust for the information reported by the prime minister's official residence immediately after the disaster. Nonetheless, feelings of distrust toward such information were already growing among Japanese citizens by the end of March (Fukuyama, 2012). Various factors were involved in this, including confusion about evacuation orders, voluntary evacuation from shelters at home regions, uniform restricted access, and prohibition of distribution of farm goods produced in Fukushima Prefecture. As a follow-up blow to this already dismal situation, a government expert on radiation protection held a public resignation interview protesting the government's establishment of a radiation dose criterion for school playgrounds of 20 mSv per year. He appealed to the audience that the standard must be 1 mSv for children. His interview had a powerful impact on the nation's people. This not only made distrust toward official experts and the government's official residence definitive, but also forced those who had previously believed and continued living normally in nonevacuation zones, particularly parents, to become aware that they had allowed their children to receive irreparable exposure. The parents of Fukushima directed their concern toward past radiation exposure. Yamashita, a risk communicator in various cities of Fukushima Prefecture, called the resignation interview the delayed agitation of danger after the true crisis had already passed (Shibata, 2012).

The extent of the damage

The extent of the damage includes not only the level of physical damage (i.e., the amount of radioactive material fallout and radiation exposure) but also cultural and socioeconomic damage. The people of the Gästrike region, the most highly contaminated region of Sweden, felt that nature had been contaminated and no longer found joy or pleasure in spending time outdoors (Enander, 2000). They stopped picking berries, foraging for mushrooms, and fishing, the activities characterizing the Swedish people's culture.

The Sami people were forced to stop grazing reindeer and to limit the consumption of reindeer meat. Grazing reindeer was the foundation of the lifestyle of the Sami people. Eating reindeer meat was their culture and formed a portion of their identity. The economic damage caused by the meat contamination was massive, but the cultural effects were also substantial.

Both Soviet and Fukushima enacted evacuation policies. As a result of the evacuation, many people lost assets such as land and homes, their hometown, a lifestyle surrounded by nature, and their community ties. These losses led to extreme stress.

In Fukushima, radioactive material was detected from agricultural products even outside the evacuation zone, resulting in the suspension of shipping and planting for the entire prefecture. While farmers were compensated, the perception that farm goods produced in Fukushima are contaminated has not easily faded, either domestically or abroad. The effects of this tattered reputation persist even today. The nuclear disaster continued to cause major long-term economic damage not only to farmers but to many residents of the prefecture.

Mechanisms evoking memories

Anniversaries of the disaster recall the event. According to Dew and Bromet (1993), symptoms in those strongly impacted by the Three Mile Island disaster were exacerbated by major news of NPPs. Symptoms began to worsen 81 months after the disaster when the undamaged nuclear reactor was restarted and were exacerbated further on the 10th anniversary of the accident.

Events that evoke memories in Fukushima are periodically reported in the media. Recently, a council publicly announced their decision to use a method in which water contaminated with tritium would be released into the ocean. Regardless of how many times experts explained that this method is used globally with almost no impact on the environment, they could not convince fishers who feared the reputational damage of misinformation. Each time people see or hear this kind of news, it reaffirms the status of Fukushima as an area impacted by a nuclear disaster.

Pidkorytov et al. (2000) found that for many parents and children, regular health checkups not only have the effect of convincing them that the exposure is pathogenic, but also have the effect of reminding them of various things involved in the accident. In a survey by Enander (2000) focusing on citizens who had been undergoing whole-body counting, participants stated that although they acknowledged the benefits of undergoing the procedure, they recall the disaster events each time they go in for measurement. Radiation levels in the Sami people are being monitored to this day, and the contamination level of reindeer meat in the region fluctuates each year. The Fukushima Prefecture Health Survey is also ongoing. This testing specifically includes a repeated examination of children's thyroid glands. Many parents must worry each time their children undergo the examination.

Disaster victims are often the subjects of many surveys. Each time a questionnaire is sent, those who participate must recall their memories from the event. One of the participants in the above-mentioned survey by Enander (2000) said that he felt like a guinea pig. However, according to Seong (2015), there were as many positive opinions toward being targeted for a survey as negative opinions. Many people want their circumstances to be accurately understood.

Discrimination and feelings of separation

Evacuees from Chernobyl were discriminated against and stigmatized. Alexievich (2011) includes examples of children being called "fireflies" and fear that radioactivity was contagious. Discrimination and bullying of evacuees occurred in Fukushima as well.

According to Pidkorytov et al. (2000), relations with citizens in regions to which evacuees traveled were not so simple. Eighty-two percent of evacuees, including children, experienced negative attitudes or sentiments from the citizens of their evacuation destinations. Because evacuees received several benefits, citizens of the host communities saw them as potential rivals that would place a financial burden on the government and regions that were already strained and would likely preferentially steal limited resources. To some extent, this was likely true in Fukushima. Evacuees becoming permanent residents in their evacuation destinations increased land prices and resulted in a shortage of residential lots.

As their evacuee lifestyle drew on, evacuees adapted to the community of their evacuation destination gradually. In Ukraine, the number of victims increased to the extent that the victims felt that they were part of the natives. According to Pidkorytov et al. (2000), this also meant less social interest in the problems evacuees experienced, ending their suffering. The people undoubtedly felt ambivalent emotions. Because the prices of agricultural products produced in Fukushima Prefecture have fallen due to reputational damage, for the people of Fukushima Prefecture, it must be economically desirable for many people to forget about the nuclear accident. However, they also felt strong resistance and dissatisfaction toward the idea of the massive burden they were forced to bear.

There is a historical background for the psychological issues faced by the Sami people. According to Skuterud, the Sami people were historically excluded from the central government. Even before the Chernobyl disaster, the Sami people had been contaminated by the atmospheric effects of nuclear weapons testing. This is because they were eating reindeer meat containing radioactive materials. The government was aware of the risks but did not inform them. After the disaster, the government continued measuring radiation until 1993, at which time the Norwegian Radiation Protection Authority considered the effects of Chernobyl in Norway to be over. The Sami people continued to have anxiety about their health and that of their children because of the massive economic impact and a lack of sufficient information. When Skuterud was dispatched to the lands of the Sami people 10 years after the disaster, he found that they felt their people had been abandoned by the government of Norway and were living in a state of increasing stress (Skuterud, 2016).

The people of Fukushima felt that they had been excluded from other regions of Japan and the rest of the world. The Olympic inviting speech given by the then prime minister in Buenos Aires in 2013 is a good example. At the start of his speech, the prime minister stated, "Some may have concerns about Fukushima. Let me assure you, the situation is under control. It has never done and will never do any damage to Tokyo (Prime Minister of Japan and his Cabinet, 2013)." At that time, the nuclear disaster did not seem to be under control. Further, the Prime Minister's emphasis that Tokyo was safe and had not been damaged seemed like an exclusion for the people of Fukushima. They felt that their prefecture had been excluded from the rest of the country by the government.

There are divisions in Fukushima, even within families. Conflicts occur between generations and spouses concerning whether to evacuate or how far to go with countermeasures. Within communities, there were different ways of feeling and thinking. For example, those who did not evacuate likely respond coldly to those who voluntarily evacuated. Resentment about having evacuated may at times come to the surface when people return home after evacuating. Various rumors concerning financial compensation have also developed. These events are deeply traumatic for many people.

Conclusion

In a nuclear disaster, the government and expert institutions must provide accurate and prompt information. However, based on the experiences of Chernobyl and the Fukushima Daiichi disaster, the information provided by the government and expert institutions has not satisfied either of these conditions. Scientific information will not necessarily be understood as accurate by the general public. For instance, when a scientist says that there is no risk, this means that there is not enough data currently to demonstrate risk, not that there is no risk altogether. This determination will change if new data are obtained. Such a predicament is difficult for the general public to accept. Because by the time it was later discovered that exposure had serious negative effects on health, the exposure had already occurred.

The information provided by the SSI was scientifically correct, yet it was misunderstood. Citizens have access to information beyond what is officially announced and may draw various conclusions. When the SSI declares no risk, people accuse it of misleading them. Thus, governments and experts must be more familiar with the general public's psychology and respect their perspective. Governments and experts tend to question people's low scientific literacy and insist on the need for education for people when they do not understand the information provided, forgetting that it could be scientists who need psychological education. These experts do not seem to have the ability to explain things in ways that the general public and the media can understand (Löfveberg, 1986).

Information crises involve many different opinions, and this is essential to democracy. Governments and companies are notorious for hiding the truth and trying to minimize the damage. Scientists and others who declare different opinions do so after investigating official information and countermeasures from their conviction and sense of responsibility. However, it is also true that diverse opinions increase feelings of anxiety in people. "Agitations after the crisis" likely only serve to trap people in fear. To avoid such situations, scientists from different backgrounds and vulnerable citizens should assess the situation and establish countermeasures early. They should also consider multiple possibilities. In Finland, methods of providing information and establishing countermeasures at the time of the Chernobyl disaster were chaotic; consequently, the country established a system in which multiple stakeholders, including citizens, participate in determining countermeasures (Aaltonen, 2016).

Although there is not much evidence, various psychosocial factors associated with the nuclear disasters (both Chernobyl and Fukushima) and evacuation led to negative impacts on children's development and psychological health. Young children are particularly susceptible to increased stress from a major disaster and parental anxiety about the health effects of radiation exposure. Our findings reveal that mothers who were pregnant at the time of the disaster or who became pregnant immediately after the disaster could have had severe anxiety and stress. These effects were not easy to resolve after that. Moreover, the effects of parental anxiety and stress during early childhood are likely to have a long-term impact on children's development and psychological health. Providing support to pregnant women and mothers in any major disaster can protect both mothers and children (Masten & Osofsky, 2010).

The effects of nuclear disasters are long-lasting. This chapter identified three commonalities by comparing three different sites: the extent of socioeconomic and cultural damage, the existence of mechanisms that evoke memories, and discrimination and feelings of separation. These factors are relevant to all citizens and should be considered when establishing countermeasures for nuclear disasters. The effects of nuclear disasters are physical but also infringe on people's social and mental lives. Natural science methodologies and knowledge alone are not sufficient to effectively cope with nuclear disaster. We believe that practices and understanding from the humanities and social sciences are indispensable.

Acknowledgments

This chapter is based on studies supported by 16H05721 and 18H01082 from Japan Society for the Promotion of Science. We thank Dr. Britt-Marie Drottz Sjöberg and Dr. Ghulbarsyn Chepurko for their supports with our collecting papers and reports published Swedish, Norwegian, and Ukrainian. We would like to especially thank Dr. Yoshihiro Sato, because he provided us with a chronological summary of what happened in Sweden after the Chernobyl accident.

References

Aaltonen, H. (2016). Towards recovery after a nuclear or radiological emergency; A Finnish approach. In *Presentation at the seminar on "The psychological effects of Chernobyl in Nordic countries: Lessons for Fukushima" at Helsinki.*

Aleksandrovskij, Y. A. (1989). *Pseuchoneurotic disorders associated with the Chernobyl accident.* IAEA-TECDOC-516 (pp. 283–291).

Alexievich, S. (2011). *Chernobyl's prayer.* Iwanami (Japanese paperback edition).

Anér, S. (1986). *Tjernobyl och Sverige (Chernobyl and Sweden).* Elanders (in Swedish).

Bromet, E. J. (2011). Lessons learned from radiation disasters. *World Psychiatry, 10*, 83–84.

Chernobyl Forum: 2003-2005. (2006). *Chernobyl's legacy: Health, environmental and socio-economic impacts and recommendations to the governments of Belarus, the Russian Federation and Ukraine.* IAEA.

Dew, M. A., & Bromet, E. J. (1993). Predictors of temporal patterns of psychiatric distress during 10 years following the nuclear accident at Three Mile Island. *Social Psychiatry and Psychiatric Epidemiology, 1993*, 49–55.

Drottz-Sjöberg, B.-M. (2016). The JSP2 project: Social and psychological consequences of the Chernobyl accident. In *Presentation at the seminar on "The psychological effects of Chernobyl in Nordic countries: Lessons for Fukushima" at Trondheim.*

Drottz-Sjöberg, B.-M., & Sjöberg, L. (1990). Risk perception and worries after the Chernobyl accident. *Journal of Environmental Psychology, 10*, 135–149.

Enander, A. (2000). *Psykologiska reaktioner vid radioaktivt nedfall fran en karnenergiolycka: Ett svenskt beredskapsperspektirt.* Forsvarshogskolan Ledarskapsinstitutionen (in Swedish).

Engholm, M. (1987). *Nar det osannolika blev sant: En studie av fyra lokaltidningars rapportering efter Tiernobyl 12-24 maj 1986.* Styrelsen för Psykologiskt Försvar (in Swedish).

Erikson, K. (1990). *Toxic reckoning: Business faces a new kind of fear* (pp. 118–126). Harvard Business Review.

Fahlen, G., & Hammarström, L. (1990). *Hur påverkas våra levnadsvanor i Kramfors och Sollefteå tre år efter Tjernobylolyckan?.* Länsstyrelsen Västernorrlands län meddelar (in Swedish).

Findahl, O., & Lindblad, I.-B. (1987). 40 dagar med Tjernobylnyheter i radio och TV: En i nnehallsanalys pa tre nivaer. In O. Findahl, G. Ganson, B. Höijer, & I. B. Lindbrd (Eds.), *"Va va de dom sa"?: I radio och tv om Tjernobyl och hur upplevde vi nyhetsinformationen?* Statens Strålskyddsinstitute (in Swedish).

Fukuyama, T. (2012). *Genpatsu kiki: Kantei karano shougen.* Chikuma (in Japanese).

Garnetz, O. (2000). Система соціально-психологічної допомоги чорнобильським постраждалим (System of social and psychological aid for victims of the Chernobyl catastrophe). *Чорнобиль і соціум, 6*, 99–110 (in Ukrainian).

Green, B. L., Lindy, J. D., & Grace, M. C. (1994). Psychological effects of toxic contamination. In R. J. Ursano, B. G. McCsughey, & C. S. Fullerton (Eds.), *Individual and community responses to trauma and disaster: The structure of human chaos* (pp. 154–176). Cambridge University Press.

Höijer, B. (1987). Tjernobylolyckan i människors medvetande: En studi av informationsinhamtande och uppleveler. In O. Findahl, G. Ganson, B. Höijer, & I.-B. Lindbrd (Eds.), *"Va va de dom sa" ? I radio och tv om Tjernobyl och hur upplevde vi nyhetsinformationen?* Statens Strålskyddsinstitute (in Swedish).

Hosono, G., & Torigoe, S. (2012). *Shougen: Hosono Gohshi.* Kodansha (in Japanese).

Institute of Disaster Area Revitalization, Regrowth and Governance, Kwansei Gakuin University, & Japan Civil Network, & Save Fukushima Children Lawyer's Network. (2015). *Genpatsu hinan hakusho.* Jin-bun Shoin (in Japanese).

International Advisory Committee. (1991). *The international Chernobyl project: Technical report.* IAEA.

Kimura, H. (2012). *Kensho Fukushima-genpatsu jiko: Kantei no 100 jikan.* Iwanami (in Japanese).

Lagutin, A. (1998). Патофізіологічн підходи до діагностування психосоматичної патології у дітей, опромінених внаслок Чорнобильської катастстофи (Pathological approaches to diagnostics of psychosomatic pathology at children irradiated as a result of the Chernobyl catastrophe). *Чорнобиль і соціум, 4*, 216–221 (in Ukrainian).

Lagutin, A., & Rogozin, O. (1999). Патофізіологічні механізми постчорнобильської психосоматичної патології у дітей м. Києва і області (Patho-physiologic mechanisms of after-Chernobyl psychosomatic pathology among the children born in Kyiv town and region). *Чорнобиль і соціум, 5*, 294–3008 (in Ukrainian).

Lavrinenko, N. (2000). Адаптація інституту сім'ї до постчорнобильської ситуації (Adaptation of the institution of family to post-Chernobyl situation). *Чорнобиль і соціум, 6*, 81–98 (in Ukraine).

Lee, T. R. (1996). Environmental stress reactions following the Chernobyl accident. In *One decade after Chernobyl. Summing up the consequences of the accident* (pp. 283–310). IAEA.

Löfveberg, S. (1986). *En strålande vår: Dagbok om Tjernobyl.* Utbildningrsproduktion AB (in Swedish).

Masten, A., & Osofsky, J. D. (2010). Disasters and their impact on child development: Introduction to the special section. *Child Development, 81*, 1029–1039.

Ministry of Ukraine of Emergencies. (2011). *Twenty-five years after Chornobyl accident: Safety for the future.* KIM.

Niigataken seishinhukushi kyoukai. (2019). *Changes in support needs for evacuees outside the prefecture and issues of support (Kengai hinansha shien niokeru shien niizu no henka to shien no kadai).* (in Japanese).

Nohrstedt, S. A., & Lekare, K. (1987). *Att rapportera det oförutsedda: En studie av lokaltidningarnas Tjernobylnyheter i Uppsala och Gävleborgs län under maj och iuni 1986.* Styrelsen för Psykologiskt Försvar (in Swedish).

Nordlund, R. (1987). *"Ovanligt hög lokal radioaktiviet" : En studie av Radio Upplands Tjernobylbevakning de 29 april-30 juni 1986.* Styrelsen för Psykologiskt Försvar (in Swedish).

Norges Offentlige Utredninger. (1986). *Informasjonskriser.* Universitetsfortaget A/S (in Norwagean).

Pidkorytov, V., Shestopalova, L., & Kukuruza, A. (2000). Особливості соціально-психологічної адаптації дітей і підлітків, евакуйованих із 30-кілометровоїзони (Peculiarities of social and psychological adaptation of children and adolescents evacuated from the 30-km zone). *Чорнобиль і соціум, 6*, 111–123 (in Ukrainina).

Prime Minister of Japan and his Cabinet. (2013). *Presentation by prime minister Shinzo Abe at the 125th Session of the International Olympic Committee (IOC).* https://japan.kantei.go.jp/96_abe/statement/201309/07ioc_presentation_e.html.

Radiation Medical Science Center for the Fukushima Health Management Survey. (2019). *Fukushima Medical University Report of the Fukushima health management survey*. Fukushima Medical University. http://kenko-kanri.jp/img/report_r1.pdf.

Sánchez-Pérez, N., Fuentes, L. J., Eisenberg, N., & González-Salinas, C. (2018). Effortful control is associated with children's school functioning via learning-related behaviors. *Learning and Individual Differences*, *63*, 78–88.

Sayenko, Y.(.С.,.Ю.). (1996). Вісім з половиною років після катастрофи. In *Соціальні наслідки Чорнобильської катастрофи* (pp. 155–164). Харкіа: Фоліо (in Ukraininan).

Seong, W. (2015). *Owaranai hisai no jikan*. Sekihusha (in Japanese).

Shibata, Y. (2012). *Fukushima-genpatsu jiko: Naibu hibaku no sinjitsu*. Nagasaki Shinbun (in Japanese).

Shimbun, A. (2018). *Houshaseibussitsu "Huan" 66%: Fukushima kenmin kyoudo yoronchousa (in Japanese)* (p. 7).

Skuterud, L. (2016). Long term follow-up of population groups. In *Presentation at the seminar on "The psychological effects of Chernobyl in Nordic countries: Lessons for Fukushima" at Trondheim.*

Soboleva, N. (2003). Відлуння лиха: Відображення соціально-психолгічних наслідків природних і техногенних катастроф у масовій свідомоті (Echo of the disaster: Reflection of social and psychological consequences of natural and technological catastrophes in the mass conscience). *Чорнобиль і соціум*, *9*, 82–91 (in Ukrainian).

Soboleva, N. (2015). The long echo of Chernobyl: The psychological consequences of the Chernobyl disaster. In *Presentation at the International Seminar on the Psychological Impact of Nuclear Disaster at Fukushima.*

Tsutsui, Y., Ujiie, T., Takaya, R., & Tominaga, M. (2020). Five-year post-disaster mental changes: Mothers and children living in low-dose contaminated Fukushima regions. *PLoS One*, *15*(12), e0243367.

Weisæth, L., & Tønnessen, A. (1995). Public reactions on Norway to radioactive fallout. *Radiation Protection Dosimetry*, *62*, 101–106.

WHO. (2006). *Health effects of the Chernobyl accident and special health care programmes: Report of the UN Chernobyl Forum expert group "Health"*. WHO.

Zhou, Q., Chen, S. H., & Main, A. (2012). Commonalities and differences in the research on children's effortful control and executive function: A call for an integrated model of self-regulation. *Child Development Perspectives*, *6*, 112–121.

Section IV

Long-term support

Chapter 18

A mother's work is to worry: Accepting maternal concerns during a health crisis

Aya Goto[a], Kohta Suzuki[b], and Chihaya Koriyama[c]
[a]*Center for Integrated Science and Humanities, Fukushima Medical University, Fukushima, Japan,* [b]*Department of Health and Psychosocial Medicine, Aichi Medical University School of Medicine, Nagakute, Japan,* [c]*Department of Epidemiology and Preventive Medicine, Graduate School of Medical and Dental Sciences, Kagoshima University, Kagoshima, Japan*

Background

Natural and human-made disasters can have a large impact on people's lives, starting from the reproductive cycle. A review of studies on fertility after disasters in the United States found fluctuations in birth rate trends (Zotti, Williams, Robertson, Horney, & Hsia, 2013). After Hurricane Hugo in 1989, a time-series analysis indicated that marriage, birth, and divorce rates increased in the following year in affected areas compared with control areas (Cohan & Cole, 2002). In contrast, after Hurricane Katrina, the number of births was considerably lower in severely affected counties (Hamilton, Sutton, Mathews, Martin, & Ventura, 2009). These results suggest that, depending on the severity, a life-threatening event might either motivate or discourage significant life decisions. Disasters have long-lasting effects on child rearing, and mothers of young children have been identified as one of the groups at the greatest risk of negative mental health consequences (Bromet, Havenaar, & Guey, 2011; Harville, Xiong, & Buekens, 2010). The severity of exposure predicts mental health outcomes in pregnant and postpartum women, which further influences child health outcomes (Harville et al., 2010).

The Fukushima nuclear accident in 2011 was no exception. A study analyzing government statistics from Fukushima City, the capital of the prefecture located outside the forced evacuation zone, revealed a significant reduction in the birth rate during the 2 years after the accident (Kurita, 2019). Moreover, a large migration, especially of mothers and children, occurred even outside the designated evacuation zones. The proportion of the Fukushima population aged

Health Effects of the Fukushima Nuclear Disaster. https://doi.org/10.1016/B978-0-12-824098-4.00010-6

5 years or under was 8% lower in the year after the accident than before. Fukushima City conducted a survey of 5000 residents and 500 evacuees in 2014 and found that health concerns regarding radiation exposure were more severe among the evacuees (Fukushima City, 2017). Another study analyzing data from a prefecture-wide survey (later explained as the Fukushima Health Management Survey) of about 7000 mothers who registered their pregnancies between August 2011 and July 2012 and gave live births reported that nearly half had evacuated and that their rate of depression was higher compared with those who did not evacuate (Ota et al., 2019) (Table 1).

Ms. Otsuki, a local resident whose two children were toddlers at the time of the Fukushima nuclear accident, shared her experiences at several conferences and in an academic journal (Otsuki, 2016). Her story began with considerable feelings of anxiety and helplessness about her ability to protect her children. "The situation became worse as time passed, and I was panic-stricken. I wanted to just run away, but I couldn't," she said. "I remember feeling desperate, knowing that I had to protect my children at all costs." After evacuating, she decided to return home. "Although I didn't want to breathe the air in Fukushima, I thought it was best to talk everything over with my family to determine if Fukushima was actually so bad that we couldn't live there." She gathered radiation-related scientific information, started measuring radiation levels,

TABLE 1 Evacuation and mental health of mothers in 2011 after the Fukushima nuclear accident.

		Depression tendency *N* (%)		
		Positive	*Negative*	*P* value
Residential area (*N*=6911)	Outside the evacuation zone (*N*=6256)	1538 (24.6)	4718 (75.4)	<0.001
	Inside the evacuation zone (*N*=655)	207 (31.6)	448 (68.4)	
Evacuation status (*N*=6866)	Never evacuated (*N*=3643)	870 (23.9)	2773 (76.1)	<0.001
	Evacuated and returned (*N*=2691)	674 (25.0)	2017 (75.0)	
	Currently evacuated (*N*=532)	192 (36.1)	340 (63.9)	

A Chi-square test was used.
Data source: Ota, M., Hattori, S., Arai, M., Shimizukawa, Y., Ishii, K., Goto, A., Yasumura, S., & Fujimori, K. (2019). Lives and depressive tendencies of pregnant women and nursing mothers who evacuated after the 2011 Great East Japan Earthquake: An investigation based on the Pregnancy and Birth Survey of the FY2012 Fukushima Health Management Survey. *Journal of the Japan Maternal and Infant Care Association, 12*(2), 21–31.

and enlisted her family to help her decontaminate their neighborhood. "I was once again reminded of my responsibility as a mother," she remarked. "I was sure of what it would mean to raise my children in Fukushima." She also said, "I was freed from the feelings of insecurity and fear." Her story is representative of the experience many mothers in Fukushima had, transitioning from confusion and uncertainty to increasing confidence.

A question we address in this chapter is whether mothers' concerns are indicative of their vulnerability or their resilience. One study in the United States reported that mothers' concern about their daughters' risk of breast cancer was a significant predictor of their intent to teach their daughters how to lead a healthy life (Neuberger, Silk, Yun, Bowman, & Anderson, 2011). The study used the protection motivation theory as its analytical framework and tested the extent to which self-efficacy, response efficacy, and threat appraisal influenced protective behavioral intent. In this theory, coping appraisal (the first two items of the breast cancer study) and threat appraisal (the last item) are posited to motivate people to change their behaviors. Parenting concerns per se are often regarded as a negative predictor or outcome (Combs-Orme, Cain, & Wilson, 2004; DeVet & Ireys, 1998). However, as in the protection motivation theory, we argue that having such concerns as a natural parental response, especially when facing a major crisis, is positive because it leads to positive parenting behaviors. This perspective might lead to recommending a parenting support strategy that aims to enhance intrinsic maternal resilience.

In this chapter, we review mothers' pregnancy intention, parenting concerns, and the support provided to them to explore their resiliency after the Fukushima nuclear accident. We further tested our hypothesis of maternal concern as a motivator for health promotion behaviors using data collected during the ongoing COVID-19 pandemic.

Decline and recovery of pregnancy intention

Fukushima Medical University launched the Fukushima Health Management Survey (FHMS) to assess residents' level of radiation exposure as well as their mental and physical health, and to provide counseling services to distressed residents. As a part of the program, the Pregnancy and Birth Survey has been conducted each year among women who registered their pregnancies after the nuclear accident (Fujimori et al., 2014). In Japan, pregnant women are required to register their pregnancy, after which they receive a maternal and child health handbook that is coupled with free access to prenatal care and well-child visits.

Using data from this prefecture-wide survey, we found that pregnancy intention among primiparas continued to decline during the 3 years after the disaster in 2011 but increased among multiparas immediately after 2012 (Goto, Tsugawa, & Fujimori, 2019). Consistent with the findings among multiparas, the government's vital statistics in Fukushima showed a steep decline in the number of births in 2012 followed by an increase in 2013. Concern about radiation contamination was significantly associated with the lower pregnancy intention among primiparas. It was a rational decision for younger mothers to wait to make sure that the regional radiation contamination would not affect the health of their babies.

Maternal depression and confidence

Using the 2012 and 2013 survey data from the FHMS, we further explored mothers' negative and positive parenting status (Goto et al., 2017). We assessed maternal depression tendency as the negative indicator and maternal confidence as the positive indicator. Similar proportions of mothers expressed decreased maternal confidence and depressive symptoms during the 2 years after the accident: 53.2% lacked confidence and 25.1% screened positive for depressive symptoms in 2012, and these percentages were 54.5% and 24.4%, respectively, in 2013. It is noteworthy that disaster-related stressors (evacuation and concern about radiation) were not associated with maternal confidence but showed a significant association with depressive symptoms (Table 2). The study did not merely show a decline in maternal mental health but suggested positive adaptation and resilience among mothers. Similarly, a mixed-method study of disadvantaged mothers affected by Hurricane Katrina in the United States found that more than 60% were categorized as resilient; that is, their psychological distress increased during the first year after the disaster but decreased in the following years (Lowe, Rhodes, & Waters, 2015). Qualitative analysis of these resilient cases revealed their tendency to utilize available resources, positively evaluate their experiences, and grow through adversity.

Listening to concerns

As part of the FHMS program, public health nurses and midwives provided telephone counseling to women who had depressive symptoms or who expressed concerns in open-ended items on the questionnaire (Ishii, Goto, Ota, Yasumura, & Fujimori, 2017a). The proportion of those who required support was about 15% during the first 3 years after the nuclear accident (Ishii, Goto, Ota, Yasumura, & Fujimori, 2017b). In most cases, FHMS program staff provided support by actively listening to the concerns of the women (Table 3). It has been documented that women with postpartum depression often recover after talking with health workers who provide therapeutic listening. For example, Britain's National Institute for Clinical Excellence recommends a program called Listening Visits as an evidence-based treatment for mild-to-moderate postnatal depression (Holden, Sagovsky, & Cox, 1989). In this program, a series of home visits is conducted by trained nurses known as "health visitors" who carefully listen to the mother's problems and engage in collaborative problem solving. The program has been adapted to the United States and has been shown to improve mothers' level of depression and quality of life (Segre, Brock, & O'Hara, 2015).

TABLE 2 Association of disaster stressors with depressive symptoms and maternal confidence in 2012 and 2013.

	Maternal confidence *N* (%)			
	N	*Not confident*	*Confident*	P *value*
Currently evacuated				
Yes	857	484 (56.5)	373 (43.5)	0.11
No	12,252	6576 (53.7)	5676 (46.3)	
Concerned about radiation				
Yes	340	198 (58.2)	142 (41.8)	0.10
No	12,769	6862 (53.7)	5907 (46.3)	
	Depression tendency *N* (%)			
	N	*Positive*	*Negative*	P *value*
Currently evacuated				
Yes	857	289 (33.7)	568 (66.3)	<0.001
No	12,251	2958 (24.1)	9293 (75.9)	
Concerned about radiation				
Yes	340	162 (47.6)	178 (52.4)	<0.001
No	12,769	3085 (24.2)	9684 (75.8)	

A Chi-square test was used.
Data source: Goto, A., Bromet E. J., Ota M., Ohtsuru, A., Yasumura, S., Fujimori, K., & The Pregnancy and Birth Survey Group of the Fukushima Health Management Survey (2017). The Fukushima nuclear accident affected mothers' depression but not maternal confidence. *Asian Pacific Journal of Public Health,29*(S2), S139–150.

Facing another health crisis

Both radiation and viruses are invisible and nonsensible risk factors that affect people's health. Immediately after the first wave of COVID-19 infections in Japan, researchers (the authors) from Aichi Medical University, Fukushima Medical University, and Kagoshima University conducted a survey in May 2020 to observe changes in the behavior of the general public in four regions. Among the target regions, Tokyo and Aichi Prefectures had the highest number of reported COVID-19 cases. Fukushima and Kagoshima Prefectures had a much lower number of reported cases. A survey questionnaire was sent to randomly selected residents aged 16–89, and their responses were collected by the postal method. The response rate was as high as 65% without sending out reminders.

TABLE 3 Telephone counseling for respondents (*N* = 1251) of the Fukushima Health Management Survey's Pregnancy and Birth Survey in 2011.

	N (%)
Counseling content	
Concerns about radiation and its effects	376 (30.1)
Mother's mental and physical health	251 (20.1)
Child rearing issues (e.g., weaning, colic)	188 (15.0)
Child's mental and physical health	129 (10.3)
Evacuation	124 (9.9)
Family life	63 (5.0)
Support provided	
Active listening	968 (77.4)
Information provision	198 (15.8)
Recommending medical care	187 (14.9)
Referral (e.g., municipal health centers, doctors)	14 (1.1)
Did not answer or refused support	186 (14.9)

Multiple choice. The proportions in each item do not add up to 100%.
Data source: Ishii, K., Goto, A., Ota, M., Yasumura, S., & Fujimori, K. (2017a). Content and characteristics of mothers receiving telephone counseling following the Fukushima Nuclear Power Plant accident: Data from the Pregnancy and Birth Survey of the Fukushima Health Management Survey 2011. *Japanese Journal of Maternal Health, 57*(4), 652–659.

For this study, in accordance with our research focus on mothers, we extracted and analyzed data of 483 women in their 20s to 40s who lived with their children. Among them, 60% were in their 40s, 44% lived in Aichi Prefecture, 26% were not employed, and 60% had a level of education higher than high school (Table 4). We found that 70% of mothers were clearly worried about COVID-19 (Fig. 1), which was not associated with their age, residential region, employment status, or education level. When comparing infection prevention practices among those with and without anxiety toward COVID-19, we found that those with anxiety were more likely to wash their hands, wear masks, and avoid eating out and taking trips (Table 5). As a subanalysis of attending events, eating out, and taking business and personal trips, we eliminated those who answered "not as often as before," then repeated the analyses and obtained similar results. The proportion of decreased frequency of event attendance was 94% for the mild anxiety group and 99% for the anxiety group ($P = 0.03$). The data for eating out were 94% and 98% ($P = 0.03$) and for personal travel were 95% and 100% ($P < 0.001$), respectively.

TABLE 4 Characteristics of women ($N=483$) in their 20s to 40s who have children, extracted from a COVID-19 survey

	N (%)
Age	
20s	12 (3)
30s	179 (37)
40s	292 (60)
Region	
Aichi	210 (44)
Tokyo	190 (39)
Fukushima	42 (9)
Kagoshima	41 (8)
Employment	
Employed	358 (74)
Not employed	125 (26)
Education	
High school or lower	149 (31)
Higher than high school	332 (69)
Past medical history	
Present	21 (4)
Absent	457 (96)

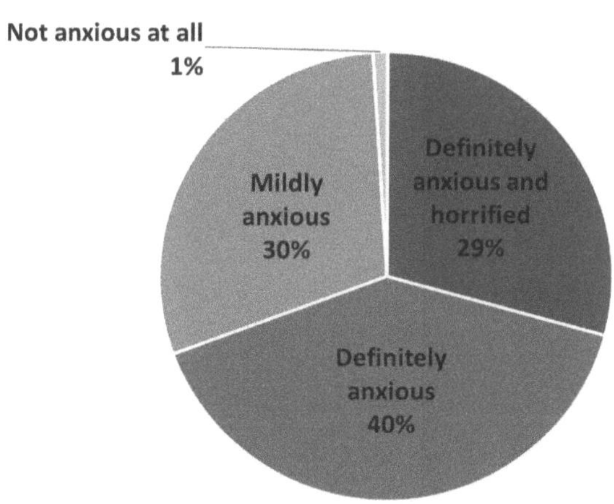

FIG. 1 Level of anxiety about COVID-19 among women in their 20s to 40s who have children.

TABLE 5 Differences in infection prevention practices according to the level of anxiety about COVID-19.

	Anxiety about COVID-19 *N* (%)		
	No or mild anxiety N = *148*	*Definite anxiety (with fear)* N = *335*	*P* value
Handwashing			
Not always	28 (19)	22 (7)	<0.001
Always	120 (81)	312 (93)	
Wearing a mask			
Not always	36 (24)	39 (12)	<0.001
Always	111 (76)	295 (88)	
Influenza vaccination in the last year			
No	82 (55)	165 (49)	0.22
Yes	66 (45)	169 (51)	
Event attendance			
Increased or did not change	4 (3)	2 (1)	0.05
Decreased, not as much as before	144 (97)	333 (99)	
Eating out			
Increased or did not change	8 (5)	6 (2)	0.03
Decreased, not as much as before	140 (95)	329 (98)	
Business trips			
Increased or did not change	7 (5)	7 (2)	0.11
Decreased, rarely as before	141 (95)	327 (98)	
Personal trips			
Increased or did not change	6 (4)	0 (0)	<0.001
Decreased, not as much as before	142 (96)	335 (99)	

Most respondents shared their opinions in the space provided at the end of the questionnaire. Although not statistically significant, the proportion of mothers who shared their opinions in this manner was higher among those with a higher level of anxiety: 57% among the mildly anxious group, 61% among the definitely anxious group, and 66% among the horrified group. We entered a total of 797 sentences into a text-mining program called KH Coder version 3. Correspondence analysis was performed to examine differences in word usage according to the level of anxiety about COVID-19. Anxiety level labels and words frequently used by each group were visualized in two dimensions. Words that appear on the opposite side of the contrasting group and close to the anxiety level label are characteristic of that group. Group-specific words for the mildly anxious group were "worry" and "school closure." Examples of opinions using these words included "I *worry* about my child's mental health. It is not only about their education, but a total lack of communication for 2 months," and "I agree with the *school closure* for preventing infections. But after the school reopens, I *worry* about my children falling behind in their education as well as their favorite activities being canceled." This group was particularly worried about their children's schooling.

As for the definitely anxious group, the group-specific words were "self-quarantine," "television," and "situation." Examples of opinions using these words include "In a time when we can get information from *television* and the Internet, I am worried that I won't be able to distinguish which information is true and which is not," "I've never experienced a *situation* like this before, so I'm very worried about my health and the economy," and "We will do our best to continue to *self-quarantine* with the aim of ending the coronavirus pandemic as soon as possible." This was the only group to mention media literacy and social uncertainty as well as the need for individual prevention efforts.

The most worried group was horrified with the COVID-19 situation and was characterized by the words "horrified" and "outside." Examples of opinions using these words include "I am *horrified* because it is caused by an invisible virus" and "I feel sorry because I want to let my child run *outside* in the park but I can't because I'm afraid they'll catch the virus." This group stated that their fears were due to an invisible risk and mixed feelings about keeping their children at home.

All the above opinions show that mothers were worried but tried their best for the benefit of their children. There is no doubt that such a global health crisis would affect the daily lives and health of parents and children. However, more in-depth investigation is needed to explore how some mothers are able to turn a negative experience into a positive gain. Posttraumatic growth is a positive adaptation process in which people reevaluate a traumatic experience. This phenomenon has been observed since the Fukushima nuclear accident (Iwasa et al., 2019). Relevant research is also warranted to investigate the potential positive impacts of coping with the COVID-19 outbreak (Tamiolaki & Kalaitzaki, 2020).

Conclusion

Mothers in Fukushima made rational reproductive decisions after the nuclear accident. The disaster stressors (evacuation and radiation anxiety) affected their mental health, but they were resilient in maintaining maternal confidence. Active listening was a key component of the support provided to those with depressive symptoms and parenting concerns. Our exploratory analysis of COVID-19 survey data further suggested parental concerns as a motivator for practicing infection prevention measures. The free-written opinions of mothers about COVID-19 showed that they struggled but also tried their best to protect their children's health. Such evidence supports the notion that maternal concern is a natural part of the recovery process following a health crisis, and accepting their concerns and applying a participatory approach in solving their problems may enhance their resilience to recover. When facing a health crisis, healthcare providers are recommended to hold positive, receptive attitudes toward mothers who express concerns and to seek ways to work with them to turn these concerns into positive preventive actions.

References

Bromet, E. J., Havenaar, J. M., & Guey, L. T. (2011). A 25 year retrospective review of the psychological consequences of the Chernobyl accident. *Clinical Oncology*, *23*(4), 297–305. https://doi.org/10.1016/j.clon.2011.01.501.

Cohan, C. L., & Cole, S. W. (2002). Life course transitions and natural disaster: Marriage, birth, and divorce following Hurricane Hugo. *Journal of Family Psychology*, *16*(1), 14–25. https://doi.org/10.1037/0893-3200.16.1.14.

Combs-Orme, T., Cain, D. S., & Wilson, E. E. (2004). Do maternal concerns at delivery predict parenting stress during infancy? *Child Abuse and Neglect*, *28*(4), 377–392. https://doi.org/10.1016/j.chiabu.2003.09.019.

DeVet, K. A., & Ireys, H. T. (1998). Psychometric properties of the maternal worry scale for children with chronic illness. *Journal of Pediatric Psychology*, *23*(4), 257–266. https://doi.org/10.1093/jpepsy/23.4.257.

Fujimori, K., Kyozuka, H., Yasuda, S., Goto, A., Yasumura, S., Ota, M., et al. (2014). Pregnancy and birth survey after the Great East Japan Earthquake and Fukushima Daiichi Nuclear Power Plant accident in Fukushima prefecture. *Fukushima Journal of Medical Science*, *60*(1), 75–81. https://doi.org/10.5387/fms.2014-9.

Fukushima City. (2017). *Second report of citizen survey about radiation*. Fukushima City Website. https://www.city.fukushima.fukushima.jp/kohoka-koho/shise/kocho/anketo/kouchou12090501.html.

Goto, A., Bromet, E. J., Ota, M., Ohtsuru, A., Yasumura, S., & Fujimori, K. (2017). The Fukushima Nuclear accident affected mothers' depression but not maternal confidence. *Asia-Pacific Journal of Public Health*, *29*(2), 139–150. https://doi.org/10.1177/1010539516684945.

Goto, A., Tsugawa, Y., & Fujimori, K. (2019). Factors associated with intention of future pregnancy among women affected by the Fukushima nuclear accident: Analysis of Fukushima health management survey data from 2012 to 2014. *Journal of Epidemiology*, *29*(8), 308–314. https://doi.org/10.2188/jea.JE20180015.

Hamilton, B. E., Sutton, P. D., Mathews, T. J., Martin, J. A., & Ventura, S. J. (2009). The effect of Hurricane Katrina: Births in the U.S. Gulf Coast region, before and after the storm. *National Vital Statistics Reports : From the Centers for Disease Control and Prevention, National Center for Health Statistics, National Vital Statistics System*, *58*(2), 1–32.

Harville, E., Xiong, X., & Buekens, P. (2010). Disasters and perinatal health: A systematic review. *Obstetrical and Gynecological Survey*, *65*(11), 713–728. https://doi.org/10.1097/OGX.0b013e31820eddbe.

Holden, J. M., Sagovsky, R., & Cox, J. L. (1989). Counselling in a general practice setting: Controlled study of heath visitor intervention in treatment of postnatal depression. *British Medical Journal*, *298*(6668), 223–226.

Ishii, K., Goto, A., Ota, M., Yasumura, S., & Fujimori, K. (2017a). Content and characteristics of mothers receiving telephone counseling following the Fukushima Nuclear Power Plant accident: Data from the Pregnancy and Birth Survey of the Fukushima Health Management Survey. *Japanese Journal of Maternal Health*, *57*(4), 652–659.

Ishii, K., Goto, A., Ota, M., Yasumura, S., & Fujimori, K. (2017b). Pregnancy and Birth Survey of the Fukushima Health Management Survey. *Asia-Pacific Journal of Public Health*, 56S–62S. https://doi.org/10.1177/1010539516684534.

Iwasa, H., Moriyama, N., Kuroda, Y., Nakayama, C., Orui, M., Horiuchi, T., et al. (2019). Recovery from radiation anxiety and posttraumatic growth among community dwellers after the nuclear disaster in Fukushima. *Cogent Psychology*, *6*(1). https://doi.org/10.1080/23311908.2019.1602970.

Kurita, N. (2019). Association of the Great East Japan Earthquake and the Daiichi Nuclear Disaster in Fukushima City, Japan, with birth rates. *JAMA Network Open*, *2*(1). https://doi.org/10.1001/jamanetworkopen.2018.7455, e187455.

Lowe, S. R., Rhodes, J. E., & Waters, M. C. (2015). Understanding resilience and other trajectories of psychological distress: A mixed-methods study of low-income mothers who survived Hurricane Katrina. *Current Psychology*, *34*(3), 537–550. https://doi.org/10.1007/s12144-015-9362-6.

Neuberger, L., Silk, K. J., Yun, D., Bowman, N. D., & Anderson, J. (2011). Concern as motivation for protection: An investigation of mothers' concern about daughters' breast cancer risk. *Journal of Health Communication*, *16*(10), 1055–1071. https://doi.org/10.1080/10810730.2011.571339.

Ota, M., Hattori, S., Arai, M., Shimizukawa, Y., Ishii, K., Goto, A., et al. (2019). Lives and depressive tendencies of pregnant women and nursing mothers evacuated due to the 2011 Great East Japan Earthquake: An investigation based on the Pregnancy and Birth Survey of the FY2012 Fukushima Health Management Survey. *Journal of the Japan Maternal and Infant Care Association*, *12*(2), 21–31.

Otsuki, M. (2016). Raising children in Fukushima. *Annals of the ICRP*, *45*(2), 65–70. https://doi.org/10.1177/0146645316666495.

Segre, L. S., Brock, R. L., & O'Hara, M. W. (2015). Depression treatment for impoverished mothers by point-of-care providers: A randomized controlled trial. *Journal of Consulting and Clinical Psychology*, *83*(2), 314–324. https://doi.org/10.1037/a0038495.

Tamiolaki, A., & Kalaitzaki, A. E. (2020). That which does not kill us, makes us stronger: COVID-19 and post-traumatic growth. *Psychiatry Research*, *289*.

Zotti, M. E., Williams, A. M., Robertson, M., Horney, J., & Hsia, J. (2013). Post-disaster reproductive health outcomes. *Maternal and Child Health Journal*, *17*(5), 783–796. https://doi.org/10.1007/s10995-012-1068-x.

Chapter 19

Telephone support

Naoko Horikoshi[a], Masaharu Maeda[b], and Maho Momoi[a,b]
[a]*Radiation Medical Science Center for the Fukushima Health Management Survey, Fukushima Medical University, Fukushima, Japan,* [b]*Department of Disaster Psychiatry, Fukushima Medical University School of Medicine, Fukushima, Japan*

Introduction

Since the accident at the Fukushima Daiichi Nuclear Power Plant caused by the Great East Japan Earthquake, the Radiation Medical Science Center (RMSC) for the Fukushima Health Management Survey (FHMS) at Fukushima Medical University has provided telephone support to evacuated residents responding to the Mental Health and Lifestyle Survey (MHLS) since financial year (FY) 2011 on commission from the Fukushima Prefecture (Yasumura et al., 2012). This disaster caused many residents to be separated from their communities and families (Yabe et al., 2014), and even now, 10 years later, the situation has not been resolved. The telephone support service provided by the RMSC for FHMS is an attempt to provide long-term care for the physical and mental health of evacuees who have been physically and psychologically separated from their communities and families, and to connect them with local medical and social resources as needed. A variety of telephone support services are provided during disasters, but most of them are hotlines (incoming calls), where the user makes the call and a support staff member on standby provides advice. In the Fukushima Daiichi disaster, many hotline services were established in the three affected prefectures. However, the telephone support provided by the RMSC for FHMS uses a more proactive method in which target residents are screened in advance using a questionnaire, and the supporters call the residents considered to be at high risk. We refer to this telephone support as "outreach telephone support" in analogy to outreach support, where supporters visit the homes of residents. This section introduces a large-scale outreach telephone support program. Projects of this kind have rarely been conducted in Japan in the past.

Health Effects of the Fukushima Nuclear Disaster. https://doi.org/10.1016/B978-0-12-824098-4.00022-2

Structure

Purpose of telephone support

We started providing telephone support in February 2012, about a year after the Great East Japan Earthquake. The target population every year is approximately 210,000 people living in the evacuation zone. In the confusion of the postdisaster period, because of limited human resources, we started providing outreach telephone support as a rapid service to many high-risk individuals who had been evacuated across the country. The purpose of telephone support is to listen to the individual's worries and ascertain their living situation, assess them for potential mental or physical crises (e.g., suicide), and provide crisis intervention as necessary.

Structure of telephone support

The process of telephone support is as follows. First, we send the MHLS questionnaire to the target residents annually by mail, asking about their physical and mental health condition. Next, we select high-risk residents from the completed questionnaires based on our support criteria. The Mental Health Support Team subsequently calls the telephone number provided on the questionnaire to assess the mental and physical health of high-risk residents and provides advice and psychological education. Finally, in urgent cases such as suicidal thoughts, domestic violence, or suspected abuse, we provide information to the health staff of the predisaster municipalities, request support from the Fukushima Center for Disaster Mental Health through the municipality, and refer the respondent to registered doctors and relevant institutions (Fig. 1). If we consider the postal questionnaire as primary screening, telephone support serves as secondary screening.

Target population for the Mental Health and Lifestyle Survey (MHLS)

The target population was approximately 210,000 people per year, including residents of all ages living in the evacuation zone specified by the government, namely Hirono Town, Naraha Town, Tomioka Town, Kawauchi Village, Okuma Town, Futaba Town, Namie Town, Katsurao Village, Iitate Village, Minamisoma City, Tamura City, Kawamata Town, and parts of Date City (containing specific spots recommended for evacuation).

Primary study variables

The MHLS questionnaire is divided into five categories according to age: 0–3 years, 4–6 years, primary school age, junior high school age, and adults (16 years and older).

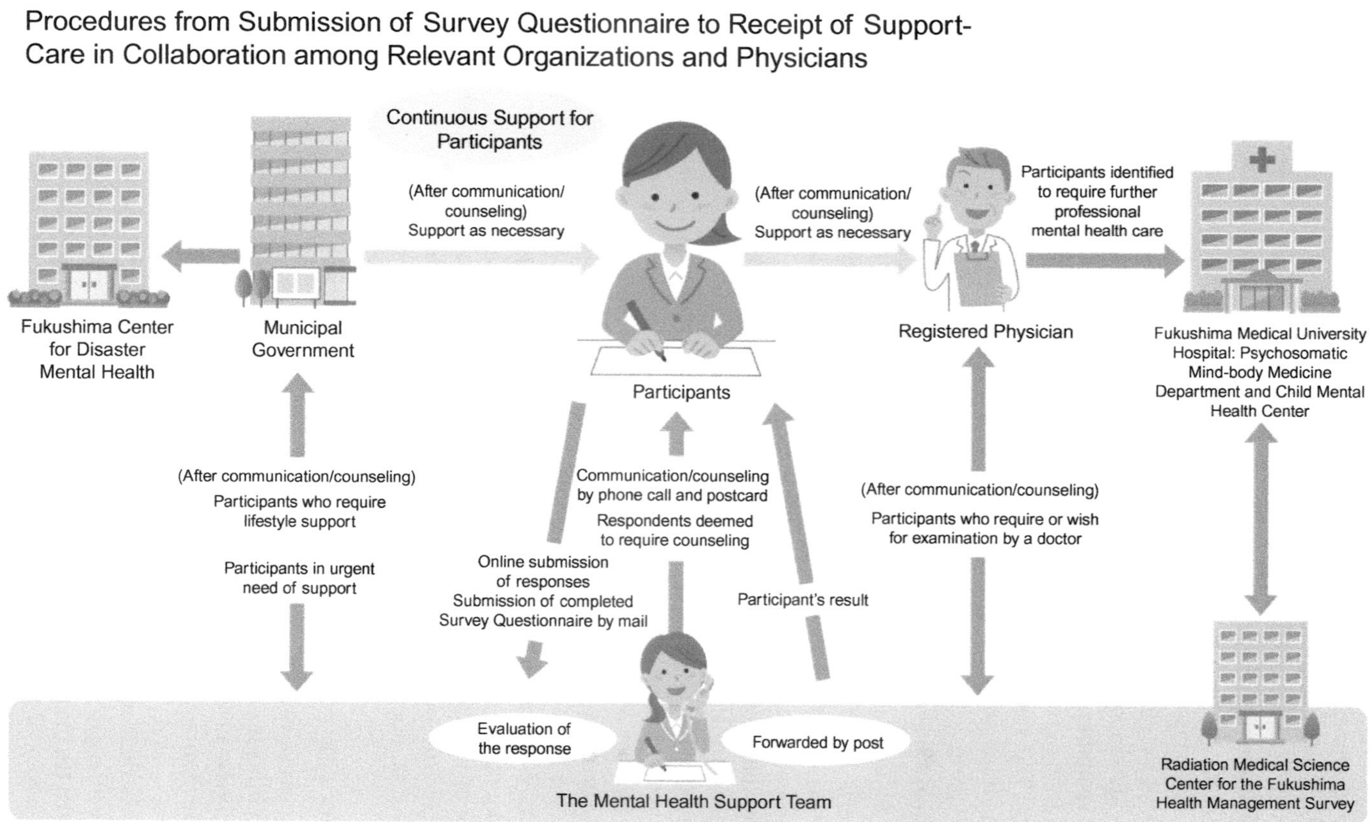

FIG. 1 Processes from submission of the survey questionnaire to receipt of support care in collaboration between relevant organizations and physicians. *(From Radiation Medical Science Center for the Fukushima Health Management Survey, Fukushima Medical University. (2020). Report of the Fukushima Health Management Survey FY2019, 32.)*

The survey items included subjective health perceptions, mental health-related measures such as Kessler's K6 (Furukawa et al., 2008; Kessler et al., 2002)/Posttraumatic Stress Disorder (PTSD) Checklist Stressor-Specific Version (PCL-S; Blanchard, Jones-Alexander, Buckley, & Forneris, 1996) for adults and the Strengths and Difficulties Questionnaire (SDQ; Goodman, 1997) for children as well as exercise habits, sleep satisfaction, eating habits, alcohol consumption, smoking, and medical history. By including questions about lifestyle and physical health as well as mental health, we are able to provide support for both aspects of health. In addition, we can infer psychological symptoms that the person may not be aware of based on physical symptoms and living conditions, and use this information to prevent more serious problems. The survey forms are self-administered or administered by guardians. In addition, we have enabled questionnaire completion online since 2016.

Criteria for telephone support

To facilitate the implementation of high-risk support, we set detailed support criteria according to questionnaire responses. Based on previous studies, we set the cutoff point for adults as a score of 13 or higher on the K6 (Furukawa et al., 2008; Kessler et al., 2002) or 44 or higher on the PCL (Blanchard et al., 1996), and for children as 16 or higher on the SDQ (Goodman, 1997). However, it is practically impossible to provide support to all respondents who comply with these criteria. Thus, although it would be ideal to support as many residents as possible, we set our own reference points so that the number of targeted residents would be between approximately 3,000 and 4,000 per year, based on the limited number of support staff and the expectation that the first contact be made within 3 months from the arrival of the questionnaire. Physical criteria include items related to sleep disturbances, alcohol consumption and alcohol dependence (CAGE) scores, hypertension, and history of diabetes. We also check the open-ended fields of each questionnaire and provide more intensive support to those who indicated urgency, such as suicide risk, domestic violence, and abuse, as needed.

Interview survey to optimize screening efficacy

In disaster situations, support criteria based on mental health measures require flexibility to meet mental health requirements. Therefore, we conducted an interview survey to optimize the screening efficacy of the MHLS in 2012. More than 20 survey staff with enough experience in healthcare, mainly public health nurses, were gathered throughout Japan and visited affected individuals to assess their mental health status and needs as correctly as possible (Fig. 2). According to the results of the survey, we were able to establish criteria optimized for the provision of interventions, which were based on a combination of the results of the K6 and PCL (Member of the Mental Health Survey Working Group in FY 2012 (Chair; Yabe, H.), 2013).

FIG. 2 Meeting with interviewers conducted in a temporary housing complex.

Structure of support

Support team

When the program started, it was conducted mainly by a few clinical psychologists, public health nurses, and nurses, with support from many outside sources. Currently, support is provided by a dedicated Mental Health Support Team, which comprises more than a dozen multidisciplinary professionals, including doctors, clinical psychologists, nurses, public health nurses, and social workers, all well-trained and experienced staff (Figs. 3 and 4).

FIG. 3 Support room at the Radiation Medical Science Center for the Fukushima Health Management Survey.

FIG. 4 Mental Health Support Team of the Radiation Medical Science Center for the Fukushima Health Management Survey.

Method of support

Telephone support is generally provided on weekdays from 9 a.m. to 5 p.m. According to the records of the first year of the project, the average time required for support was about 10–15 min per person, with about 60% of the consultations taking less than 10 min and 5% taking more than 30 min (Kashiwazaki et al., 2016). The telephone support staff first thank the respondents for completing the survey. Next, they confirm their physical condition, sleep status, depression, medical visits, living with family, support resources such as relatives in the neighborhood, and social interactions. Based on this information, they provide psychological education, lifestyle guidance, recommendations for medical consultations, and information on social resources (Fig. 5).

In addition, in the case of radiation-related issues that cannot be handled by the support staff, respondents are referred to the Radiation Health Consultation Team at Fukushima Medical University. Furthermore, we provide information to predisaster municipalities in urgent cases, such as suicide risk, domestic violence, and suspected abuse. We are also collaborating with the Fukushima Center for Disaster Mental Health (FCDMH), and at the discretion of the local government, the staff of the municipality or the FCDMH may provide visiting support to evacuees.

FIG. 5 Mental Health Support Team member wearing a headset to provide support.

If we are unable to reach respondents by telephone, we send them a postcard confirming whether they would like to receive telephone support, informing them of our dedicated number, and including educational pamphlets. When the target person calls the dedicated phone number for consultation, the Mental Health Support Team also gives the same advice as in outreach-type support. The actual method of providing such telephone support is described in detail in a book (Maeda, Momoi, & Takebayashi, 2020), which we published urgently due to COVID-19 issues. We recommend that interested readers refer to this book.

Educational training for mental health support team

To provide support in the long term, it is important to ensure not only an adequate number of staff members but also the service quality and care provided by each of them. In the first year, in the confusion after the disaster, it was extremely difficult to find supporters with sufficient professional skills. Many individuals gathered as supporters were themselves evacuees of the disaster and may have experienced difficulties in their role as supporters. However, for residents who were anxious, scared, and angry, it was very meaningful that supporters who had experienced the same disaster could provide assistance in a familiar dialect. On the other hand, our telephone support service had the cooperation of many professionals outside Fukushima Prefecture. Without their assistance, we would not have been able to provide outreach telephone support to 210,000 people. In addition, support staffs are often exposed to the uncontrollable anger of evacuees (Maeda, Ueda, Nagai, Fujii, & Oe, 2016), and it is essential to have a support system to prevent staff burnout. At our center, we have gradually developed a system that includes (1) regular supervision to provide guidance and advice from a professional standpoint, (2) training programs to improve qualifications, and (3) personal interviews to prevent burnout.

Collaboration with the community

In cases of high urgency (e.g., suicide risk, obvious depressive symptoms, need for medical care but not being able to see a doctor for some reason, or isolated without family care despite having serious difficulties), we provide information to the public health nurses of the predisaster municipalities with the consent of the respondents. The information is provided in writing, and we aim to be as clear as possible about the purpose of providing this information (i.e., what kind of support we are seeking). The information provided includes basic data such as address, name, age, gender, and telephone number as well as mental health scale scores, medical consultation status, sleep problems, and drinking and smoking status. In addition, we are careful to include specific information (e.g., suicidal thoughts, significant weight loss). As needed, public health nurses of the predisaster municipalities and FCDMH staff may provide visiting support to evacuees. We also aim to share the results of these activities as much as possible.

Results

Trends in the number of residents receiving telephone support

In financial year (FY) 2011, we provided telephone support to about 9000 residents with the help of many professionals living in Fukushima Prefecture as well as outside the prefecture. From the following year, we aimed to support 3000–5000 people per year, taking into account the capacity of our team. Consequently, we have supported a total of more than 30,000 residents over the 8 years from FY 2011 to FY 2018 (Figs. 6 and 7).

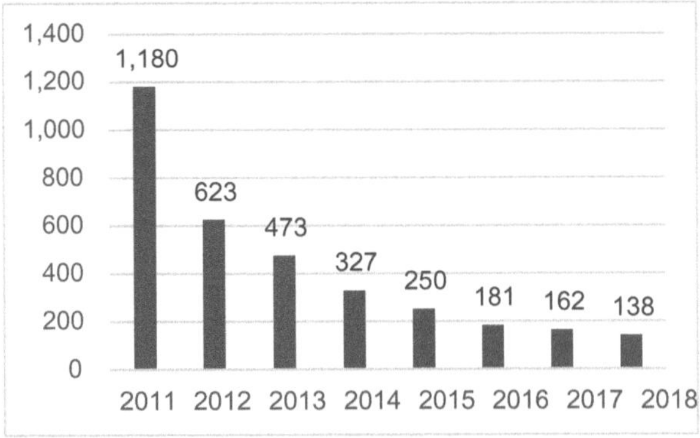

FIG. 6 Trends in the number of cases of telephone support (children). *(Figure developed by the author with permission from Kamiya, K., Ohto, H., Yasumura, S. (2021). Report of the Fukushima health management survey FY2011-2020. Radiation Medical Science Center for the Fukushima Health Management Survey, Fukushima Medical University, 133–134.)*

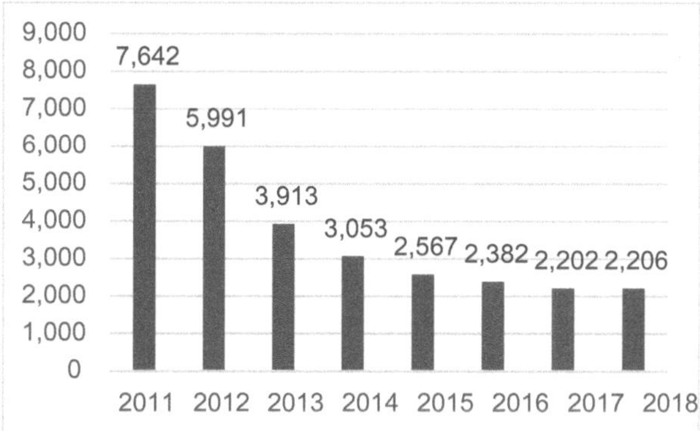

FIG. 7 Trends in the number of cases of telephone support (adults). *(Figure developed by the author with permission from Kamiya, K., Ohto, H., Yasumura, S. (2021). Report of the Fukushima health management survey FY2011-2020. Radiation Medical Science Center for the Fukushima Health Management Survey, Fukushima Medical University, 133–134.)*

Contents of consultations

In the consultations for children, the most frequently consulted topic was "anxiety" in FY 2012, and "school-related matters." On the other hand, for adults, "physical health," "sleep," and "depression" accounted for the highest proportion of consultations in that order from FY 2012 to FY 2018. In addition, in the first few years after the disaster, "concerns about the living environment" were common due to the effects of evacuation, but since FY 2014, "anxiety about the future" has appeared (Mental Health and Lifestyle Survey for the Fukushima Health Management Survey, 2020).

In this way, the content of consultations has changed over time since the disaster. In FY 2011 and FY 2012, immediately after the disaster, there was a strong concern about the health effects of radiation, and many residents asked questions such as "Can I drink the tap water?" and "Can I let my children play outside?" In addition, there were many residents reporting anger and frustration at having to leave their hometown because of the accident, physical problems due to fatigue and frustration caused by repeated evacuations, and many cases of intense thoughts of death.

In addition, we heard appeals of "stigma" and "self-stigma" from FY 2013 to FY 2016. For example, some people experienced prejudice in their workplace and neighborhood regarding compensation (such as the idea that "evacuees receive high compensation") or had anxiety and fear about being discriminated against. For these reasons, some residents "hid the fact that they had been evacuated" or "cut off interaction with their neighbors," which was a serious situation that led to isolation at the evacuation site.

Since 2017, more than 5 years after the earthquake, more municipalities have lifted their evacuations while others have no prospect of lifting them; thus, consultations about returning home following the lifting of evacuations have become more diverse. Residents whose evacuations were lifted had a variety of concerns, such as "wondering whether they should return to their homes," "being forced to live separately due to differences of opinion within the family," and "regretting the decision to settle in the evacuation area." On the other hand, the residents whose evacuation had not been lifted appeared to be exhausted as they continued to face the uncertainty of where and how they would live their lives.

Now that 10 years have passed, many evacuees have returned to their usual daily lives, and the number of disaster-derived consultations from evacuees has steadily decreased. Moreover, specific problems have been gradually left behind, such as the impoverishment caused by the inability to become economically independent, the poor mental health of evacuees who were evacuated outside of Fukushima Prefecture, and the isolation of elderly households. In addition, while we no longer receive consultations about radiation worries, we are concerned that some evacuees may have strong latent radiation concerns but are not ready to disclose them.

Effect of telephone support for evacuees at risk of lifestyle-related diseases

We longitudinally examined the association between the response rate of the MHLS and seeing a doctor or not in FY 2012 based on the outcome of the telephone support provided to those who had not seen a doctor for lifestyle-related diseases, such as hypertension and diabetes, in FY 2011 MHLS. We found that among those who attended a medical institution the following year, there was a higher proportion of individuals who received telephone support compared with those who did not receive telephone support. In addition, those who received telephone support had a significantly higher response rate to the MHLS in the following year than those who did not receive telephone support; therefore, telephone support efforts were considered to be effective in improving the response rate to the survey (Horikoshi, Iwasa, Yasumura, & Maeda, 2017). In addition, a higher proportion of evacuees outside Fukushima Prefecture were eligible for telephone support compared with evacuees in Fukushima Prefecture, suggesting that evacuation life may have affected their lifestyle-related diseases to a higher extent. Furthermore, a high proportion of evacuees outside Fukushima Prefecture suffered from sleep disorders, and it was significant to assess their sleep status soon after the disaster, advise them on how to improve their sleep, and provide them with appropriate support (Horikoshi, Ohira et al., 2017).

Present state of difficulties as described by parents with children

We qualitatively examined the contents of consultations with children in each age group (preschool, elementary school, and junior high school) in the following seven categories: (1) physical discomfort, (2) impact on school life, (3) irritability and violence, (4) anxiety and depression, (5) reactions to earthquakes and radiation, (6) changes in eating behavior, and (7) disturbed sleep.

In particular, preschoolers showed regressions such as acting like a baby, and elementary school students were unable to perform some daily activities that they were previously able to do by themselves. In addition, parents of elementary and junior high school students sometimes talked about the difficulties they faced when changing schools due to evacuation, such as not fitting in at the new school and bullying. These various reactions in children due to PTSD symptoms are confusing and anxiety-inducing for parents. Considering that the parents are also evacuees, their children's reactions seemed to add to the parents' difficulties. We found that the situation of their children further stressed the parents and had reciprocal effects on parents and children (Oikawa et al., 2017).

Usefulness of telephone support through interview survey

To examine the usefulness of telephone support, we conducted a study involving door-to-door interviews with the recipients of telephone support. This study aimed to examine the usefulness of telephone support by focusing on evacuees' subjective evaluations assessed in individual follow-up interviews (Horikoshi et al., 2020). The sample comprised 484 individuals from Fukushima Prefecture evacuated to eight safer regions in and out of Fukushima. We conducted semi-structured interviews with participants receiving telephone support (intervention group) and those not receiving telephone support due to attendance at work, school, or other reasons such as refusal to receive support despite being identified as high-risk (nonintervention group). In the intervention group, the proportion of those reporting satisfaction with the intervention was as high as 74.6% (Fig. 8). Satisfaction was significantly associated with advanced

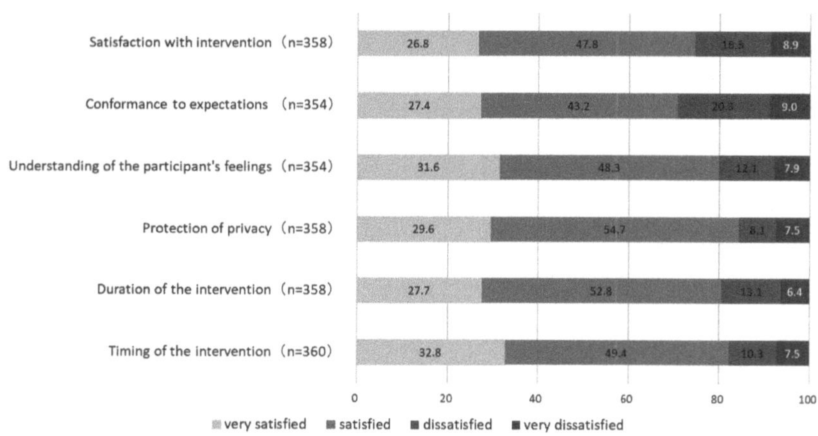

FIG. 8 Subjective thoughts and impressions of brief telephone intervention by the intervention group. It shows a four-point Likert scale: "very satisfied," "satisfied," "dissatisfied," and "very dissatisfied." *(Figure developed by the author with permission from Kamiya, K., Ohto, H., Yasumura, S. (2021). Report of the Fukushima health management survey FY2011-2020. Radiation Medical Science Center for the Fukushima Health Management Survey, Fukushima Medical University, 133–134.)*

knowledge of telephone support availability and receiving advice on health-related practices.

As described above, most respondents who received telephone support positively evaluated their response to such support. In particular, the fact that more than 70% of the respondents were satisfied with the telephone support they received suggests the usefulness of this intervention, which has no precedent in Japan. Moreover, evacuees had high expectations of stress management methods, social resources, information on the results of the MHLS, and advice on lifestyle; we believe that these results should be used as a reference when considering how to provide telephone support in the future. In addition, this telephone support service was suggested to be useful, especially for those evacuated to distant places such as outside Fukushima Prefecture, and we consider it to be an adequate method of support when the ability to support directly is limited, such as in the case of a large-scale natural disaster or COVID-19. Furthermore, its usefulness was confirmed not only for mental health issues but also for physical issues related to lifestyle, and its completion may contribute to improving the response rate to the MHLS (Horikoshi, Iwasa, et al., 2017).

Case study of telephone intervention

The cases described below do not contain any identifiable information.

Case 1: Advice for man living alone

A man in his 60s lived alone in the place of evacuation. When we asked him by telephone about his daily life, he told us that he was drinking all the time because he could not find a job in the evacuation area and had nothing to do. We found that he had problems with drinking every day and drinking large amounts at a time as well as smoking increasingly more per day. Therefore, we checked his sleep status, medical visits, and intention to quit smoking (whether he wanted to quit smoking) and discussed the following topics with him: (1) suggestions for liver rest days, (2) advice on exercise habits, and (3) referral to hospitals that offer outpatient smoking cessation clinics.

Case 2: Continuous telephone support to monitor

A woman in her 1970s lived with her husband in the Tokyo metropolitan area, where she was evacuated. She lived away from her children, and although they sometimes came to visit her for the New Year, they did not usually meet. When we asked her about her current health and daily life over the phone, she simply replied repeatedly, "I'm fine." We introduced her to the local salon activities, but she was reluctant to participate, saying, "I don't want to be known as an evacuee." Therefore, we provided continuous telephone support to monitor her.

Case 3: Coping with a child's anxiety as described by the mother

A woman in her 30s lived with her husband and elementary school-aged children. She reported that she moved from place to place because of forced evacuation a total of seven times. The evacuation was lifted about a year ago, and she had now returned to her former home. In the process of confirming her situation over the phone, we found out that she was worried about her child's reluctance to go to school. In response to our questions, she told us that her child was not chronically absent from school and that she had not yet discussed anything with homeroom teachers. For this reason, we encouraged her to communicate within the family, ask the homeroom teacher to keep an eye on her child, and consult with the teacher when necessary. We also explained how to consult with the school counselor in case of problems and sent her a list of places to consult, such as the Education Center.

Case 4: Consultation with a man suffering from health problems due to stress at work

A man in his 20s lived alone. He was transferred between departments in April and became involved in disaster-related projects. He told us that he was not used to the new project work in addition to his regular duties, and that he could not get support from other staff members because they were also exhausted. Two months later, he started experiencing depressive symptoms such as not being able to sleep well, having no appetite, and not being able to enjoy doing anything. Although he was able to work every day, he often arrived at the office late because he could not sleep well. He also began to experience sleepiness at work, which interfered with his work. We assessed him for depression and suicide using a structured interview method, as we heard him say things like "Sometimes I want to disappear" and "I am worthless." The results showed that he had symptoms suggestive of depression and chronic suicidal ideation, although he had no plan for suicide. We recommended that he should see a doctor immediately, but he was reluctant to do so, saying that he was still fine. However, because he lived alone and had no one to watch him, we decided that we needed to share information with the public health nurse and visit his home for support. Because he agreed to the support visit, the local government public health nurse and the mental health care center provided support, which led to medical consultation.

Discussions

The strength of outreach telephone support is that we can provide support to evacuees in remote areas at a relatively early stage and low cost. In addition, as supporters make phone calls to evacuees who may be unable to seek assistance on their own, we can respond to latent support needs as early as possible. There are many evacuees whose physical and mental health worsened because of the environmental instability caused by the nuclear power plant accident, yet they fell through the various nets of support (Mental Health and Lifestyle Survey for the Fukushima Health Management Survey, 2020). These individuals have been the first target of our telephone support service, and this support itself can play the role of a safety net. In our study on the usefulness of service, we found that evacuees' satisfaction with telephone support was high (over 70%; Horikoshi et al., 2020). This suggests that if we listen carefully to evacuees' thoughts and feelings even over the telephone, they can experience the same level of satisfaction as with visiting support.

However, it is not possible to meet all the support needs using telephone support alone, and its limitations must be assessed. Especially in crisis interventions, it is essential to provide visiting support. As a strategy, we should continue to build and utilize a close network with wide-area local resources (i.e., other support organizations). However, evacuees are affected by disasters over a much longer period owing to prolonged evacuation (Maeda & Oe, 2017). It is not uncommon for evacuees who had been well to later show new symptoms and become eligible for support because of the different phases of postdisaster reconstruction and changes in their life cycle. We need to continue to provide long-term support in the postdisaster reconstruction phase as relief needs change.

Conclusion

The attempt of our center to provide outreach-type telephone support to many evacuees scattered throughout Japan was unprecedented. Hence, we are still testing different aspects while providing telephone support. In Japan, it is well established that outreach supporters visit evacuees' homes during disasters. However, there is a limit to the number of support staff who can visit the affected areas to provide support when access to the areas is poor or areas are extensive. At the time of writing, the COVID-19 epidemic is restricting face-to-face support throughout Japan. As a result, some support organizations that provide in-home support are switching to telephone services as their main support method. However, we feel that outreach telephone support is very significant as an effective form of support, not limited to emergencies such as the current COVID-19 epidemic.

References

Blanchard, E. B., Jones-Alexander, J., Buckley, T. C., & Forneris, C. A. (1996). Psychometric properties of the PTSD Checklist (PCL). *Behaviour Research and Therapy*, *34*(8), 669–673. https://doi.org/10.1016/0005-7967(96)00033-2.

Furukawa, T. A., Kawakami, N., Saitoh, M., Ono, Y., Nakane, Y., Nakamura, Y., et al. (2008). The performance of the Japanese version of the K6 and K10 in the World Mental Health Survey Japan. *International Journal of Methods in Psychiatric Research*, *17*(3), 152–158. https://doi.org/10.1002/mpr.257.

Goodman, R. (1997). The strengths and difficulties questionnaire: A research note. *Journal of Child Psychology and Psychiatry, and Allied Disciplines*, *38*(5), 581–586. https://doi.org/10.1111/j.1469-7610.1997.tb01545.x.

Horikoshi, N., Iwasa, H., Yasumura, S., & Maeda, M. (2017). The characteristics of non-respondents and respondents of a mental health survey among evacuees in a disaster: The Fukushima Health Management Survey. *Fukushima Journal of Medical Science*, *63*(3), 152–159. https://doi.org/10.5387/fms.2017-03.

Horikoshi, N., Maeda, M., Iwasa, H., Momoi, M., Oikawa, Y., Ueda, Y., et al. (2020). The usefulness of brief telephonic intervention after a nuclear crisis: Long-term community-based support for Fukushima Evacuees. *Disaster Medicine and Public Health Preparedness*, *1-9*. https://doi.org/10.1017/dmp.2020.161.

Horikoshi, N., Ohira, T., Yasumura, S., Yabe, H., Maeda, M., & Mental Health Group of the Fukushima Health Management Survey. (2017). Higashinihondaishinsaigo niokeru seikatsusyuukanbyou no risuku gaaru hinansya heno denwashien niyoru chousahyou heno kaitou oyobi iryoukikan jushin no kouka: Fukushima ken kenminkenkouchousa [Effects of telephonic support for evacuees at risk of lifestyle-related diseases on questionnaire responses and visits to medical institutions after the Great East Japan Earthquake: The Fukushima Health Management Survey]. *Japanese Journal of Public Health*, *64*(2), 70–77 (in Japanese) 10.11236/jph.64.2_70.

Kashiwazaki, Y., Maeda, M., Yagi, A., Fujii, S., Takahashi, N., Yabe, H., et al. (2016). Effectiveness of telephone-based intervention for people living in Fukushima disaster area: Fukushima Health Management Survey. *Clinical Psychiatry*, *58*(5), 433–442 (in Japanese).

Kessler, R. C., Andrews, G., Colpe, L. J., Hiripi, E., Mroczek, D. K., Normand, S. L., et al. (2002). Short screening scales to monitor population prevalences and trends in non-specific psychological distress. *Psychological Medicine*, *32*(6), 959–976. https://doi.org/10.1017/s0033291702006074.

Maeda, M., Momoi, M., & Takebayashi, Y. (2020). *Enkakushinrishien sukiru gaido: Dokohedemo tsunagaru enjo [Remote psychological support skills guide: Helping people connect anywhere]*. Seishin Shobo: Tokyo (in Japanese).

Maeda, M., & Oe, M. (2017). Mental health consequences and social issues after the Fukushima Disaster. *Asia-Pacific Journal of Public Health*, *29*(2_suppl), 36S–46S. https://doi.org/10.1177/1010539516689695.

Maeda, M., Ueda, Y., Nagai, M., Fujii, S., & Oe, M. (2016). Diagnostic interview study of the prevalence of depression among public employees engaged in long-term relief work in Fukushima. *Psychiatry and Clinical Neurosciences*, *70*(9), 413–420. https://doi.org/10.1111/pcn.12414.

Member of the Mental Health Survey Working Group in FY 2012 (Chair; Yabe, H.). (2013). *Report on the interview survey of the Mental Health and Lifestyle Survey for adult in FY 2012*. (in Japanese).

Mental Health and Lifestyle Survey for the Fukushima Health Management Survey. (2020). *Heisei 30 nendo "kokoro no kenkoudo/seikatsusyukan ni kansuru chousa" kekka houkoku [Mental Health and Lifestyle Survey for the Fukushima Health Management Survey for FY2018]*. Fukushima Prefecture website. Retrieved February 4, 2021, from: https://www.pref.fukushima.lg.jp/uploaded/attachment/386028.pdf (in Japanese).

Oikawa, Y., Maeda, M., Takahashi, N., Kashiwazaki, Y., Ueda, Y., Hisata, M., et al. (2017). Higashinihondaishinsai niokeru jakunenhisaisya wo motsu oya heno denwashien nitsuite: Fukushima ken "kenminkenkouchousa" kara [Telephonic support for parents of young evacuees after the Great East Japan Earthquake: The Fukushima Health Management Survey]. *Japanese Journal of Traumatic Stress*, *15*(1), 69–75 (in Japanese).

Yabe, H., Suzuki, Y., Mashiko, H., Nakayama, Y., Hisata, M., Niwa, S., et al. (2014). Psychological distress after the Great East Japan Earthquake and Fukushima Daiichi Nuclear Power Plant accident: Results of a mental health and lifestyle survey through the Fukushima Health Management Survey in FY2011 and FY2012. *Fukushima Journal of Medical Science*, *60*(1), 57–67. https://doi.org/10.5387/fms.2014-1.

Yasumura, S., Hosoya, M., Yamashita, S., Kamiya, K., Abe, M., Akashi, M., et al. (2012). Study protocol for the Fukushima Health Management Survey. *Journal of Epidemiology*, *22*(5), 375–383. https://doi.org/10.2188/jea.je20120105.

Chapter 20

The recovery process of the community in Kawauchi Village after the nuclear disaster

Makiko Orita, Yasuyuki Taira, and Noboru Takamura
Atomic Bomb Disease Institute, Nagasaki University, Nagasaki, Japan

Evacuation, return and recovery process of Kawauchi Village

Kawauchi Village is located within 30km of the Tokyo Electric Power Company (TEPCO) Fukushima Daiichi Nuclear Power Plant (FDNPP) (Fig. 1). Before the accident, the total population of the village was approximately 3,000, and its main industry was agriculture.

On March 11, 2011, the Great East Japan Earthquake occurred, resulting in an accident at FDNPP. After the earthquake and subsequent tsunami, at 8:50 p. m. that day, residents who lived within a 2-km radius of the plant were told to evacuate. This order was extended to a 3-km radius at 9:23 p.m. It was extended to a 10-km radius on the morning of March 12, and then to a 20-km radius that afternoon (Nagataki, Takamura, Kamiya, & Akashi, 2013). Due to these evacuation instructions, Tomioka Town, which is located within 20km of FDNPP, decided to evacuate its residents to Kawauchi Village. After March 12, Kawauchi Village accepted almost all of the 12,000 residents of Tomioka Town. Residents of Kawauchi prepared food for the evacuees from Tomioka, but it became difficult because the transportation of supplies was delayed due to the severe damage to roads from the earthquake as well as fear of exposure to radiation among truck drivers. On March 15, residents living within 20–30km were instructed to take shelter inside their houses, and the mayors of Kawauchi and Tomioka decided to evacuate residents to Koriyama City, which is located 60km from FDNPP.

After the accident, it was revealed that contamination levels in Kawauchi were relatively mild compared with other municipalities within 30km of the FDNPP due to the wind direction during the accident. After the prime minister declared the termination of the accident in December 2011, Yuko Endo, the mayor of Kawauchi, decided to return to the town. Since then, the village

Health Effects of the Fukushima Nuclear Disaster. https://doi.org/10.1016/B978-0-12-824098-4.00020-9

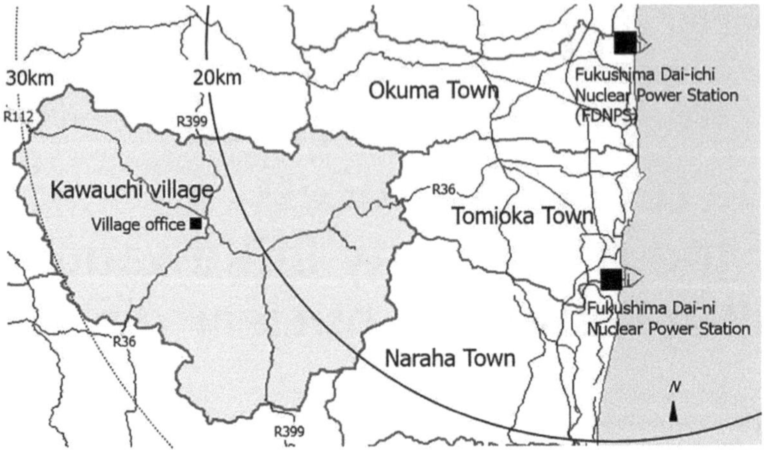

FIG. 1 Map of Kawauchi Village.

has implemented decontamination efforts in the territory to remove radiocesium. In March 2012, the Kawauchi Village office relocated its functions to the village, and the residents started returning. Although the village office reestablished the infrastructure, the number of residents who returned to their hometown within a year was only about 50% of the original population. Fewer members of the younger generation tended to return to the village compared with the older generation, probably due to educational issues of schoolchildren, occupational challenges of parents, and especially anxieties about radiation exposure in the village.

Establishment of a Nagasaki University satellite office and risk communication

Nagasaki University has been assisting in the recovery efforts in Kawauchi Village by evaluating internal and external radiation exposure doses and providing risk communication with the residents according to their individual doses. In December 2011, the University estimated the external radiation exposure doses of the residents of Kawauchi by measuring radiocesium concentrations in soil samples collected from residential areas of the village. When residents started returning to the village, the university dispatched a public health nurse with a masters in radiation medical science to implement risk communication with residents. It was the first case in Fukushima of a radiation medical sciences expert living in the evacuated area for the purposes of risk communication.

In April 2013, Nagasaki University and Kawauchi Village signed an agreement on comprehensive cooperation for the reconstruction efforts of Kawauchi. In the village, the Kawauchi Village Reconstruction Promotion Base was established as a satellite facility of the university. Medical doctors, public health nurses, and pharmacists were stationed permanently, and they worked closely with the village office to measure radioactivity in the soil and food to ensure the safety of residents as well as offer radiation health consultation services based on the measurement data (Takamura, Orita, Yamashita, & Chhem, 2016). Our mission in the satellite office is to evaluate the effectiveness of decontamination through the measurement of radionuclides in soil and internal exposure through the measurement of food and water. In addition, we are providing information and offering individual consultations with residents about radiation and its health effects. In most cases, the university staff conduct door-to-door visits to provide consultation services. In the early days when residents first returned (around 2012 or 2013), they often asked questions about whether the water and rice were safe for consumption, whether children could touch insects, and whether any decontamination work was also carried out after the atomic bomb was dropped on Nagasaki in 1945. The university staff also attended social gatherings organized by the village office to support residents and evacuees. At these gatherings, when the residents asked about the progress of decontamination work and land use, the village office staff answered their questions, and when the residents asked about radiation and health, Nagasaki university staff answered. In this way, consideration has been given to being able to directly respond to the questions and anxieties felt by the residents.

In addition, the university staff has offered radiation health consultation services based on objectively measured data (Fig. 2). After the accident, words such as "microsievert," "becquerel," and "internal exposure," which were not familiar to the general public, became common, and a flood of information and panic occurred. However, with the passage of time, monitoring equipment

FIG. 2 Risk communication with a resident of Kawauchi.

has been installed everywhere in the village, and the ambient dose rates in each area are stated in detail in newspapers every day. Personal dosimeters have also become widespread, and almost all village residents have one or more. As a result, each resident has various types of "dose" information, such as individual dose, internal dose, and ambient dose, around their home. On the other hand, many residents did not understand or were not confident about how to interpret and understand the dose readings. The university staff present the measurement data to residents and explain what each piece of data indicates. In this manner, the university staff address the residents' questions and concerns over radiation exposure in everyday life. The residents often comment that they feel reassured if they can understand the levels of radioactivity in local food or in the environment.

In response to the accident at the FDNPP, the Japanese government created an Evacuation Order Area for residents living within a 20-km radius of the FDNPP. Furthermore, certain areas beyond the 20-km radius, where there remained concerns that cumulative doses of radiation might reach 20 mSv/year, were designated as Deliberate Evacuation Areas (United Nations Scientific Committee on the Effects of Atomic Radiation (UNSCEAR), 2013; Nagataki et al., 2013). Both the Evacuation Order Area and the Deliberate Evacuation Area have been rearranged by the Japanese government, depending on the annual cumulative dose levels. They were reclassified into the following three areas in 2013: Area 1, where evacuation orders are ready to be lifted; Area 2, where residents are not permitted to live; and Area 3, where it is expected that the residents will have difficulties in returning for a long time. Kawauchi Village contained Areas 1 and 2, meaning that residents were able to enter freely during the day, but they were not allowed to stay. Therefore, it was very important for the local office to support the return of the residents of these areas. Before evacuation orders were lifted, from August to December 2013, the university staff evaluated individual doses as measured by personal dosimeters in residents of evacuation order areas who were allowed to return home temporarily, that is, Areas 1 and 2 within the village, which is partially located within a 20-km radius of the plant. We also compared individual doses with the external radiation doses estimated from the ambient dose rates around residential houses. As a result, individual doses were significantly correlated with the ambient doses in front of the entrances to the houses, in the backyards, and in the nearby fields. The maximum cumulative ambient doses in the backyards and fields around the houses were 6.38 and 9.27 mSv/year, respectively. The maximum cumulative individual dose was 3.28 mSv/year, and the median and minimum doses were 1.35 and 0.71 mSv/year, respectively (Orita et al., 2015). Our evaluation confirmed that individual doses are at low levels even in the evacuation order area in the village. Measurement of individual doses is an effective way to support residents of the evacuation order area who want to return home. Long-term follow-up of individual doses as well as internal-exposure doses, environmental monitoring, and reconstruction of infrastructure are needed so that residents may return to their hometowns after a nuclear disaster. In addition, it is important to implement risk communication with residents based on the data of actual measured individual doses to address residents' anxieties about radiation exposure. Kawauchi Village is a model case for the return of residents to their homes and for reconstruction after a nuclear accident.

Although many surveys have shown that the exposure doses of residents are very limited due to the prompt evacuation and food regulation policy (United Nations Scientific Committee on the Effects of Atomic Radiation (UNSCEAR), 2013), there is a gap between residents' risk perception for radiation and their actual exposure doses (Sato, Orita, Taira, & Takamura, 2018). In 2014, we conducted a survey among residents in Kawauchi Village to clarify the factors associated with residents' risk perception for radiation exposure and consequent health effects, and we showed a marked bipolarization of the risk perception after the accident that could have a major impact on social well-being. In 2017, we followed up the risk perception study of residents of the village and found that 37.5% (38.4% in 2014) of residents thought that adverse health effects would occur at radiation exposures from 1 mSv per year [the protection level for the public recommended by the International Commission of Radiological Protection (ICRP)]. Also, 59.0% (57.6% in 2014) of residents stated that adverse effects would occur via their annual intake of mushrooms, including 100 Bq/kg of radiocesium (the current standard value of food regulation in Japan) (Sato et al., 2018). These results suggest that residents did not fully understand the difference between radiation protection policy, which is as low as reasonably achievable, and the actual health effects of radiation based on the results of epidemiological studies. This gap has not significantly changed between 2014 and 2017. For the further recovery of Fukushima in the existing exposure situation, specialists in the radiation health sciences must continue their efforts to communicate with residents in order to narrow the knowledge gap. As of 2020, more than 80% of the residents have returned to Kawauchi Village, and the village has achieved reconstruction by attracting new industrial parks and setting up special nursing homes for the elderly. We will carry out the continuous follow-up of risk perceptions for radiation exposure in residents in order to respond to individual needs.

Evaluation of environmental radioactivity in Kawauchi Village

Nagasaki University has been assisting in the recovery efforts of Kawauchi Village by evaluating internal and external radiation exposure doses and risk communication with the residents according to their individual doses (Fig. 3). As described, in December 2011, we estimated the external radiation exposure doses of Kawauchi residents by measuring radiocesium concentrations in soil samples collected from residential areas of the village. The estimated external effective doses from the soil samples were 0.0011–0.38 μSv/h (0.010–3.3 mSv/year) within 20–30 km of the FDNPP. We could show that

FIG. 3 Measurement of ambient dose in Kawauchi.

the risk of external radiation exposure of the residents was sufficiently low in Kawauchi Village (Taira et al., 2012). One year later, we followed up on the environmental contamination and radiation exposure dose rates in Kawauchi Village and showed that the levels appeared to be decreasing compared to just before the return of the residents to their homes.

With regard to the estimation of internal exposure dose, committed effective doses from local agricultural samples in Kawauchi Village collected from May 1, 2012, to March 31, 2013, were sufficiently low—in the range 18–44 μSv/year for male residents and 20–48 μSv/year for female residents (the range was 18–48 μSv/year for children and 25–43 μSv/year for adults), compared to the public dose limit (1 mSv/year), although the potential for radiation exposure still exists (Taira et al., 2014).

Food monitoring system and mushroom map project in Kawauchi

The Japanese central and prefectural governments began to monitor selected foodstuffs (milk, vegetables, grains, meat, fish, etc.) in March 2011 to minimize the internal radiation exposure of residents of Fukushima (Hamada, Ogino, & Fujimichi, 2012). Those containing radioactive material that exceeded the provisional regulation values (set in the initial phase of the Fukushima accident based on the assumption that the radiation level will decrease in accordance with the half-lives of radioactive materials and that the continuous intake of food contaminated with radiation at this level will not affect health) were prohibited from distribution on March 21, 2011 (Orita et al., 2015), and from consumption on March 23, 2011. Due to this stringent food-control policy, internal radiation from radioactive cesium and radioactive iodine was limited. Fukushima Prefecture reported the results of internal radiation doses measured with a whole-body counter in residents of Fukushima Prefecture and its evacuees from June 2011 to August 2020. Of these, 344,957 (99.9%) showed a committed effective dose of <1 mSv (Fukushima Prefecture, 2020). Considering the experience after the 1986 accident at the Chernobyl Nuclear Power Plant, it is well known that radiocesium tends to concentrate in wild mushrooms (Hoshi et al., 2000; Orita et al., 2018). Hoshi et al. (2000) reported that children consuming mushrooms showed a high radiocesium body burden near Chernobyl, which suggests that mushrooms are one of the main contributors to internal radiation exposure from radiocesium after nuclear disasters.

Before the accident, Kawauchi Village was famous for wild mushrooms, including *Sarcodon aspratus* and *Tricholoma matsutake*. Because the collection and consumption of wild mushrooms is part of the culture of this village, residents are keenly interested in radiocesium in wild mushrooms (Cui, Orita, Taira, & Takamura, 2020). We had a discussion with the residents of Kawauchi and decided to implement a collaboration study called the Mushroom Map Project. Since 2013, we have collected wild mushrooms from the village and evaluated their radiocesium concentrations to estimate the internal radiation exposure of the residents to address anxiety and to reduce future risks of internal exposure. We also started creating a "mushroom map" in 2013 to share results of the radioactivity of mushrooms and the committed effective dose by using understandable terms and figures with the residents (Cui et al., 2020; Nakashima et al., 2015; Orita et al., 2017) (Fig. 4). We received a part of the mushrooms collected by the residents, and at that time, by collecting

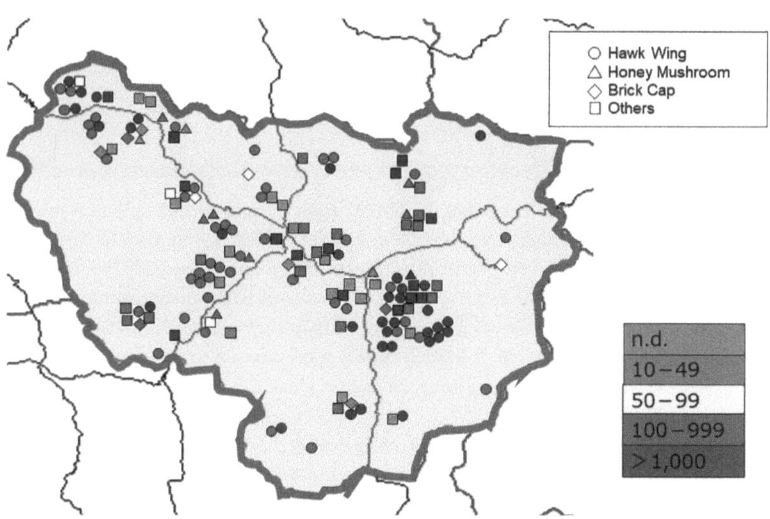

FIG. 4 "Mushroom map" of Kawauchi.

information about the places where they were collected, this map was created so that the type of mushrooms, the concentration of radioactive cesium, and the place of collection could be understood at a glance. This initiative was not well received by the residents at first, mostly because they did not want to disclose the location of their mushroom collection sites that they had kept secret for a long time. However, we proceeded with the mapping by asking the residents to tell us the approximate collection location. During the recovery phase of the nuclear disaster, the engagement of stakeholders, including residents, local authorities, and scientists, is important to decide the future direction of the community. The ICRP emphasizes that stakeholder engagement is key to the development and implementation of radiological protection strategies for most existing exposure situations (International Commission on Radiological Protection, 2009). It is important to continue working in collaboration with local residents, local governments and experts. By measuring the concentration of radiocesium in food and residents being able to visually confirm the results themselves, it is possible to reduce residents' anxiety about radiation exposure and consequent health effects. The relevant activities conducted in Kawauchi Village such as the Mushroom Map Project could serve as a model for further reconstruction efforts.

Conclusion

Based on the experiences in Kawauchi Village, it is clear that the cooperation of residents, local authorities, and experts is essential for the recovery of affected areas after a nuclear disaster. On the other hand, we should note that many evacuees have not yet returned home in some towns/villages where similar efforts have been made. Accumulated experiences and practices should be carefully recorded and evaluated to prepare for unexpected nuclear disasters in the future.

References

Cui, L., Orita, M., Taira, Y., & Takamura, N. (2020). Radiocesium concentrations in mushrooms collected in Kawauchi Village five to eight years after the Fukushima Daiichi Nuclear Power Plant accident. *PLoS One*, *15*(9). https://doi.org/10.1371/journal.pone.0239296, e0239296.

Fukushima Prefecture. (2020). *The result of internal exposure by whole body counter in Fukushima prefecture*. Available: https://www.pref.fukushima.lg.jp/site/portal/ps-wbc-kensa-kekka.html.

Hamada, N., Ogino, H., & Fujimichi, Y. (2012). Safety regulations of food and water implemented in the first year following the Fukushima nuclear accident. *J Radiat Res*, *53*, 641–671. https://doi.org/10.1093/jrr/rrs032.

Hoshi, M., Konstantinov, Y. O., Evdeeva, T. Y., Kovalev, A. I., Aksenov, A. S., Koulikova, N. V., et al. (2000). Radiocesium in children residing in the western districts of the Bryansk Oblast from 1991–1996. *Health Physics*, *79*, 182–186. https://doi.org/10.1097/00004032-200008000-00011.

International Commission on Radiological Protection. (2009). ICRP Publication 111—Application of the Commission's recommendations to the protection of people living in long-term contaminated areas after a nuclear accident or a radiation emergency. *Ann ICRP.*, *39*(1–4), 7–62.

Nagataki, S., Takamura, N., Kamiya, K., & Akashi, M. (2013). Measurements of individual radiation doses in residents living around the Fukushima Nuclear Power Plant. *Radiat Res*, *180*(5), 439–447. https://doi.org/10.1667/RR13351.1.

Nakashima, K., Orita, M., Fukuda, N., Taira, Y., Hayashida, N., Matsuda, N., et al. (2015). Radiocesium concentrations in wild mushrooms collected in Kawauchi Village after the accident at the Fukushima Daiichi Nuclear Power Plant. *PeerJ*, *3*. https://doi.org/10.7717/peerj.1427, e1427.

Orita, M., Hayashida, N., Nakayama, Y., Shinkawa, T., Urata, H., Fukushima, Y., et al. (2015). Bipolarization of risk perception about the health effects of radiation in residents after the accident at Fukushima Nuclear Power Plant. *PLoS One*, *10*(6). https://doi.org/10.1371/journal.pone.0129227, e0129227.

Orita, M., Hayashida, N., Taira, Y., Fukushima, Y., Ide, J., Endo, Y., et al. (2015). Measurement of individual doses of radiation by personal dosimeter is important for the return of residents from evacuation order areas after nuclear disaster. *PLoS One*, *10*(3). https://doi.org/10.1371/journal.pone.0121990, e0121990.

Orita, M., Kimura, Y., Taira, Y., Fukuda, T., Takahashi, J., Gutevych, O., et al. (2018). Activities concentration of radiocesium in wild mushroom collected in Ukraine 30 years after the Chernobyl power plant accident. *Peer J*, e4222. https://doi.org/10.7717/peerj.4222.

Orita, M., Nakashima, K., Taira, Y., Fukuda, T., Fukushima, Y., Kudo, T., et al. (2017). Radiocesium concentrations in wild mushrooms after the accident at the Fukushima Daiichi Nuclear Power Station: Follow-up study in Kawauchi village. *Sci Rep*, *7*(1), 6744. https://doi.org/10.1038/s41598-017-05963-0.

Sato, N., Orita, M., Taira, Y., & Takamura, N. (2018). Seven years post-Fukushima: overcoming the resident-specialist gap. *J Radiat Res*, *59*(4), 526–527. https://doi.org/10.1093/jrr/rry037.

Taira, Y., Hayashida, N., Orita, M., Yamaguchi, H., Ide, J., Endo, Y., et al. (2014). Evaluation of environmental contamination and estimated exposure doses after residents return home in Kawauchi Village, Fukushima Prefecture. *Environ Sci Technol*, *48*(8), 4556–4563. https://doi.org/10.1021/es404534y.

Taira, Y., Hayashida, N., Yamaguchi, H., Yamashita, S., Endo, Y., & Takamura, N. (2012). Evaluation of environmental contamination and estimated radiation doses for the return to residents' homes in Kawauchi Village. *Fukushima prefecture*. *PLoS One*, *7*(9). https://doi.org/10.1371/journal.pone.0045816, e45816.

Takamura, N., Orita, M., Yamashita, S., & Chhem, R. (2016). After Fukushima: Collaboration model. *Science*, *352*(6286), 666. https://doi.org/10.1126/science.352.6286.666-a.

United Nations Scientific Committee on the Effects of Atomic Radiation (UNSCEAR). (2013). *UNSCEAR 2013 Report Volume I. Report to the general assembly scientific annex A: Levels and effects of radiation exposure due to the nuclear accident after the 2011 great east-Japan earthquake and tsunami*. Available: http://www.unscear.org/docs/reports/2013/13-85418_Report_2013_Annex_A.pdf.

Chapter 21

Posttraumatic responses among outpatients in disaster area: Case reports

Arinobu Hori
Department of Psychiatry, Hori Mental Clinic, Minamisoma, Japan

Introduction

Fukushima Prefecture can be divided into three regions: the Aizu, Nakadori, and Hamadori regions from the west. The Hamadori region has the mildest climate, with little snowfall due to its location on the Pacific Ocean. Tokyo Electric Power Company's (TEPCO) Fukushima Daiichi Nuclear Power Plant is located in the middle of this region. In 2011, the Great East Japan Earthquake caused a nuclear accident and a 20-km radius around the plant was designated as a no-entry zone. The area within a range of 20–30 km was initially ordered to be evacuated indoors, and then designated as an emergency evacuation preparation zone where "children, pregnant women, people in need of care, and hospitalized patients, in particular, are not to enter the area" (Fukushima Prefectural Government, 2020). As of November 2020, when this chapter is being written, these evacuation orders are still in the process of being lifted one by one. The number of people who are still evacuating outside Fukushima Prefecture is approximately 29,500 (Fukushima Prefectural Government, 2020). The evacuation order underscores the fact that the Great East Japan Earthquake is a complex disaster that will have a profound long-term impact on the society and culture of a wide region as well as the psychology of the people living there. There is a big difference between the areas where the evacuation orders were not issued outside the 30-km radius and the areas where these orders have made a great impact. For example, Minamisoma is one of the municipalities most affected by the nuclear accident. The city was divided into three parts: about a third of the northern area outside the 30-km zone, the middle third within the 20–30-km zone, and the southern third within the 20-km zone, which made a great difference between the evacuation instructions they received even though they were within the same city (Fukushima Prefectural Government,

Health Effects of the Fukushima Nuclear Disaster. https://doi.org/10.1016/B978-0-12-824098-4.00021-0

2020; Hori et al., 2014; Minamisoma City, 2020a). In addition, the experience of the disaster was quite different in the seaside area from that in the mountainside area. The tsunami seriously damaged the coastal region; on the other hand, radiation contamination was severer in the mountainside region due to the wind direction at the time of the accident. (Hori et al., 2014; Minamisoma City, 2020a). This made it difficult for disaster victims to share their experiences and work together.

At 2:46 p.m. on March 11, 2011, a magnitude six earthquake was recorded in Minamisoma City. At the closest station to Minamisoma City, a 9.3-m wave was recorded, and 40.8 km^2 of the city's area, or about 10% of its total area, was affected by the tsunami. The direct death toll from the disaster was 636. Forty-six evacuation centers were set up in the city, and on the morning of March 15, more than 8,000 people gathered there. By March 20, a mass evacuation of the city was underway, and by March 21, only 171 people remained in the evacuation centers. Although people began moving into temporary housing at the end of June 2011, the shelters were not finally closed until December 28, 2011 (Minamisoma City, 2020a).

The registered population of Minamisoma was approximately 72,000 at the end of February 2011, before the earthquake, but as of the end of August 2020, it was less than 60,000. The rates of people aged 65 and over were 25.8% and 35.6%, respectively. This was due to the outflow of the population, mainly of young people, due to concerns about the health effects of low-dose exposure to radiation and other factors (Minamisoma City, 2020b).

By the end of September 2019, 3,739 disaster-related deaths had been identified in Japan, with Fukushima Prefecture having the highest number at 2,286, of which 516 were residents of Minamisoma City (Reconstruction Agency, 2020). This high number of earthquake-related deaths is thought to have been influenced by the deterioration of living conditions due to evacuation and other factors after the Great East Japan Earthquake and nuclear accident.

Materials and methods

This article discusses the impact of such a disaster on people living there from the perspective of the author's clinical psychiatry practice. The author, a psychiatric clinician who lived in Tokyo before the Great East Japan Earthquake, has been living in Minamisoma since April 2012 and has worked at a local psychiatric hospital for 3 years. The author opened his mental clinic in Minamisoma in 2016. The first case reported here is a patient with whom the author was involved as the primary physician during the period when he was working in a psychiatric hospital. The second case was a patient whom the author encountered after opening the clinic and providing medical treatment by first going to the inpatient ward of the city's general hospital as a house call and then continuing outpatient treatment at his clinic. As part of the Japanese government's support for the victims, they and their families continue to be exempted from paying their medical bills. Consent was obtained from patients to publish their medical history for academic purposes.

Case 1

Case: A woman in her 1960s.

Diagnosis: Severe depressive episode with psychotic feature, complicated grief.

Life history: She was born and raised in the Hamadori region of Fukushima Prefecture. After graduating from high school, she got a job at a clothing company, and continued to work as a sewer. Immediately after she started her job, 10 people competed in sewing a garment as fast as possible, and since she was the fastest, she became the leader of the team. The second fastest person in the same group became her best friend. She got married in her 20s and had a daughter.

Current medical history: The Great East Japan Earthquake occurred in 2011. Immediately after the earthquake, her best friend asked her to drive to her friend's house because her family might be crushed at home. After she responded to the friend, the friend was swept up in the tsunami and died. After the disaster, a large number of employees who had come from abroad returned to their home countries, leaving the office short of staff. Later, to maintain work efficiency, supervisors increasingly reprimanded company employees.

Since she loved her work, she tried to immerse herself in it and forget about the stress, but gradually insomnia and other problems began to appear. When she complained about the burden of stress at work, she was reprimanded instead, and the situation persisted. Her husband also worked as a carpenter in construction and was very busy after the earthquake.

At the time, one of her neighbors committed suicide. The day before she died, the neighbor came to talk to the patient, but the patient did not notice her condition and responded by admonishing her for her sad state. The police contacted the patient after the suicide, as it was in the neighborhood, and asked her to identify the body. Later, the friend who died in the tsunami was laid to rest; afterward, she began to see her friend in her dreams, and her food intake was greatly reduced. She was also sometimes reprimanded publicly about her inability to work efficiently. Gradually, she began to walk around restlessly and began to say things like, "I can't live with myself." At that time, her family took her to the psychiatric hospital, and she was admitted the next day. Her symptoms met the diagnostic criteria for a major depressive episode during the interview on the day of admission, and partly because the author wanted to calm her thoughts of dreadful death, she was told that she would surely get better. She accepted the doctor's explanation with an open mind that "it will get better," and after being admitted to the hospital, her insomnia and loss of appetite soon improved. She began to spend her time in the ward, smiling. However, she gradually began to show signs of a hardening of her facial expressions and often complained about her back pain and constipation. Even at this point, she often told the author that she was "better than before."

When she was asked during the interview, she said that she sometimes felt the presence of her dead best friend in the psychiatric ward. She continued, "I feel like someone is around the corner." "When I wake up in the morning, I get a scary feeling." "I feel like something black is passing in front of me." "I feel like five or six white people are passing. I think they are coming out of the shadows." When the author asked her if she felt that the deceased had left behind some unrequited love for the world and resented the patient, she began to cry. She explained that when she talked to her daughter, she was told that dead people do not resent the patient, but she could not help but think that way, and that it scared her at night. After the author listened to the complaints, the patient seemed to feel better. Small amounts of antipsychotics were prescribed as treatment. After this, she reported, "Since I talked with my doctor, I don't feel any more signs," and started to smile again. However, she started to complain that she was beginning to forget things.

She was discharged from the hospital and continued to visit an outpatient clinic thereafter. After the discharge, she was asked by her company to return to work as soon as possible, but she told the author that she was frequently scolded, which caused her to have palpitations upon recalling. However, she accepted the invitation from her workplace and returned to work on the condition that she would work only half a day. She received a softer response from her superiors. She was not exhausted by the restrictive work schedule and was mentally stable as she returned to the workplace to which she had an emotional attachment. The patient continued to attend outpatient clinics, but suddenly, her anxiety and agitation increased without any specific triggers. She started to talk about her brother and sister-in-law and their two children who had died in the tsunami, which had not been discussed until then.

At the beginning of her second hospitalization, she finally talked about her feelings of being "all alone" as her close friends had disappeared. During this hospitalization, she continued to complain of anxiety and palpitations. She reported dreams of hands reaching out, grabbing her feet, and trying to pull her into the sea. About a month after being admitted to the hospital, she recalled and described a very frightening experience she had had while driving a car at the time of the disaster, when the tsunami was close at hand, and she assumed she was going to die. This interview led to an improvement in her complaints of anxiety and heart palpitations. She also said that she was "tired from work." After the second discharge from the hospital, the patient decided to retire from work. After that, she continued her sick care.

Summary of Case 1: A review of the medical history of this case shows that the patient's condition changed significantly over the course of the postearthquake period. Although the patient herself had experienced great trauma in the earthquake and tsunami, she was initially able to maintain her mental state by immersing herself in work. However, a combination of factors, such as the ongoing turmoil in the entire area (including anxiety about the effects of radiation exposure), the deterioration of the working environment due to an increase in the number of retirees, and suicidal ideations of familiar people, led to the onset of depression and hospitalization for psychiatric care. During this process, it became clear that the patient had depression that arose secondary to trauma responses during the disaster. The content of the trauma reaction was so severe that it went beyond guilt for the deceased friend, such as the feeling of being "wronged and revenged" by the friend, as well as hallucinations that appeared to be dissociative.

As her symptoms improved and her involvement in the workplace became less immersive and more organized, she was able to experience emotions such as loneliness, which she had not been able to experience explicitly before, and was able to think of other relatives who had died beside her friends. However, this temporarily impaired her ability to adjust to the reality of the situation, requiring a second hospitalization, and only then was she able to recount her own traumatic experience of the tsunami, which she feared. This process shows the characteristics of individuals and communities that are insensitive to their own hardships and fears but prioritize concern for others and their organizations, including their workplaces. Although this was a case of considerable severity of symptoms, through this therapeutic process, the patient was able to quit the company and choose a life that prioritized her recuperation.

Case 2

Case: A man in his 1940s.

Diagnosis: Severe depressive episode, PTSD.

Life history: When the patient was in the second grade, his boat fell over and he nearly drowned. From a young age, his parents often cautioned him severely for being the eldest son (of three siblings), and he began to observe the expressions on other people's faces. He felt that his parents' attitude toward him was unfair and he became rebellious and violent from the time he was in junior high school. He graduated from junior high school but did not attend high school. During that time, he was psychologically close to his mother. He worked as a welder in an ironworks. When he was 18 years old, he was involved in a car accident. He was a member of the local firefighting company (Shobo-dan), and due to this, he often saw dead bodies in fires and water-related accidents. Japanese local firefighting companies (Shobo-dan) employ half-volunteer, consenting public servants, and while members may be required to engage in harsh activities to ensure the safety of the community, there is no guarantee of civil servant-like status. In his early thirties, his second brother was killed in a car accident, but he was unable to hold an immediate funeral and spent several days in the vicinity of the accident-damaged body.

Current medical history: At the time of the Great East Japan Earthquake, the patient was working in a manufacturing-related company. After the earthquake, he was ordered to stay in the company for a while and then dismissed. When he returned home, he found the place was in chaos, but he followed the rules of the Shobo-dan and gathered at its camp. He went around his town that day to check the damage and returned home for a while, but slept at the Shobo-dan's camp at night. On the morning of March 12, he started the search operation and, at first, he took a surviving mother and her children to the hospital. Next, he found the body of a girl close in age to his daughter and carried it to a high school where the temporary morgue was located. At first, he found only her hands on the ground and dug them up. He felt that they had killed the girl because they had left her there for a day. He thought hard about blaming himself and the people involved for not searching for them on the night of March 11.

The next move was to go to a senior citizens' home that had been damaged by the tsunami and to retrieve three bodies. On the way there, he heard the news of the explosion in the first reactor of the plant. In the evening of the same day, he returned to his home and spent the morning of March 13 shopping for groceries with his wife. In the afternoon, he joined the Shobo-dan again to act as the deputy head of the search operation. At that time, he retrieved about 15 bodies, but became so emotionally numb that he carelessly handled and accidentally ripped off the arms of a corpse, and blamed himself for treating the bodies as if they were objects. Due to this, he became afraid to touch the body and moved from the group looking for bodies to the group carrying them.

On March 14, his wife cried and begged him to stay at home, but he scolded her and tried to join the Shobo-dan again. However, when he heard the news that the building of the third reactor of TEPCO's Fukushima Daiichi Nuclear Power Plant had exploded and that the second reactor was also in danger, he decided to evacuate to other prefectures, wishing to avoid the effects of radiation on his daughter. He drove his low-gasoline car until about 4 a.m. the next day to reach another prefecture, where he and his family were screened for radiation at the health center and then directed to a gymnasium shelter. His father-in-law, who was with him, was sickly, so he was able to rent an apartment in the city and move there.

Two weeks later, the patient returned to Minamisoma City, where he was later criticized by Shobo-dan officials for evacuating. For about a month, he had a series of hallucinatory experiences, such as hearing Buddhist monks chanting sutras and feeling the sound of people's footsteps and stares. He felt "heartbroken" and was unable to trust anyone. Later on, he resumed assisting the search efforts and disaster clean-up. About three months after the disaster, he moved to Iwaki City, Fukushima Prefecture, to join his wife, daughter, and friends. That was when the nightmares were most active, and he had auditory hallucinations of hearing Buddhist monks chanting sutras all day long. These symptoms were most active at this time and lasted for about 4 years, but he never went to a hospital.

In the summer of 2012, he felt physically and mentally drained and resigned from his job. He later tried to find a job but was not able to continue working. Five years after the disaster, he divorced his wife and left his daughter. His drinking increased and his gastrointestinal condition deteriorated as he preferred to eat spicy food "to his detriment," and was hospitalized twice for ileus. Nevertheless, he was later able to find a new job at a construction company. His wife and children lived in the same city, and he could visit his daughter from time to time.

In the spring of the seventh year after the disaster, he drank too much alcohol at a cherry blossom viewing party and fell and hit his back hard, which caused him to miss 2 days of work. He became remorseful for the trouble he caused at work and attempted to hang himself at home early in the morning. Since he had called his mother just before the suicide, his mother and brother went to his home and found him unconscious and called an ambulance. He was immediately transported to a general hospital in Minamisoma City. Fortunately, he recovered without any physical problems or any after-effects. The surgeon in charge contacted the author, a psychiatrist, and on the day after he was admitted to the hospital, the patient was examined by him. At that time, symptoms such as psychomotor inhibition, depressed mood, insomnia with early morning waking predominance, and suicidal ideation were recognized and diagnosed as severe depressive episodes. At the same time, flashbacks and nightmares about memories of being a member of the Shobo-dan at the time of the Great East Japan Earthquake and searching for the bodies of the deceased were observed to be active. Hence, he was also diagnosed with posttraumatic stress disorder (PTSD).

The author began administering a serotonin selective reuptake inhibitor. Fear, anxiety, and agitation were temporarily heightened when the patient was told about the earthquake, so the author took the time to interview him about his medical history. Little by little, the patient practiced sitting on the first floor of the hospital that he feared because the hospital itself was closely related to his traumatic memories, looking out the hospital window toward the sea, and watching disaster-related videos on YouTube, to familiarize himself with the situation.

More than 3 months after the suicide attempt and after confirming that the suicidal thoughts had disappeared, prolonged exposure therapy, a trauma-focused cognitive behavioral treatment, was started (Foa, Hembree, Rothbaum, & Rauch, 2019). There was a so-called hot spot of traumatic memories that caused a strong sense of fear and emotional confusion, such as when the patient found the body of a girl close in age to his daughter. There were also times when, in the course of treatment, images of dead people reaching out of the sea with regret and trying to drag the patient in became active. At this point, a sense of guilt, including anger that "I killed that girl (because I did not conduct a search operation on the night of March 11)," intensified, which jeopardized the execution of the psychotherapy and led to a two-week interruption. During this period, the president of his current company, whom he admired, reprimanded him for his suicide attempt and urged him to engage in the treatment, and he was able to remotivate himself to continue the treatment.

After resuming psychotherapy, guilt over not being able to help the girl inevitably led to anger at the Shobo-dan officials, which in turn stimulated anger at his parents (especially his mother), making it easier for him to make associations with his parents and his involvement from childhood to adolescence. At the same time, symptoms of PTSD related to the trauma of the earthquake were also active. Therefore, when the author told him, "For the sake of treatment, I want you to focus your attention on the feelings of 'fear' within yourself and not on 'anger and guilt,'" the instructions were abstract, but the patient understood and was able to implement them. We then proceeded to the protocol of prolonged exposure therapy, which enabled him to process traumatic memories related to the disaster effectively. He practiced going to the beach and other places where he searched for bodies. Gradually, he was able to sort out his emotions: "The hardest thing was that there was no one alive, and if there had been one person, it would have given me hope and a sense of purpose, but there was nothing but despair. If people are buried alive, they die after a day. There was no light. It was in the dark. We saw so many dead bodies. In Minamisoma City alone, hundreds of people have died," he said.

The patient spent 4.5 months in the hospital, after which he was switched to home treatment and continued his psychotherapy at the author's clinic. One month after discharge from the hospital, he returned to work. He continued with the psychotherapy, but 3.5 months after he was discharged from the hospital, he recounted for the first time his up-close experience of the tsunami and the fear for his life. After the earthquake occurred in the afternoon of March 11, 2011, he was instructed to go to the coastal villages as part of Shobo-dan's patrol in the local area. While driving along the river road, he saw the tsunami coming up the river. In a quick maneuver, he was able to backtrack and turn down the road and hide in the shadow of the mountains to escape the tsunami. He said, "I had a lot of hard times after that, and I wished I had died then." He later admitted to improving, but after March of the following year, he became unwell again and took another leave of absence from the end of March until August.

During this time, he practiced psychotherapy, which focused on general cognitive behavioral therapy techniques such as stress management, self-monitoring, and mindfulness. Eventually, when his childhood home environment and conflicts with his mother and other family members became the main topic of conversation, the author suggested that he could engage in another psychotherapy program for personality issues derived from childhood experiences, but the patient was not interested, and the treatment was temporarily terminated.

Summary of Case 2: After the Great East Japan Earthquake and Tsunami, the patient was involved in searching for bodies as a member of a Shobo-dan, which resulted in PTSD symptoms. His psychiatric condition was not recognized, and he developed depression after his divorce and other events, which became evident from sudden suicide attempts. What is notable is the severity of the traumatic experiences of the victims of the earthquake and tsunami. However, the course of this case was also greatly affected by two factors: the evacuation due to the nuclear accident and the inherent vulnerability of the personality that existed even before the earthquake. Concerning the former, although he was eager to contribute to the community after the disaster, he was forced to evacuate due to the nuclear accident, which in turn changed his position to be blamed by some of the other Shobo-dan members that he had initially attacked for not starting the search sooner. Due to all this, it is expected that the patient would have had great difficulty in sorting out the psychological events that affected his mental health.

The patient tended to feel very strong remorse and altruism for not being able to help the girl who was found dead, and dealing with this was one of the most difficult parts of his psychotherapy. He was not only full of self-blame but also often highly critical of his superiors and others of the organizations responsible for the rescue operation. To blame his seniors reminded him of the frustration he had felt with his parents since childhood. He identified with, but at the same time felt strong opposition to, the value system imposed on him by his parents, which was to suppress his emotions and contribute to the community.

Similar to Case 1, when the treatment had progressed to the point where the patient was able to organize his relationship with his surroundings to a certain extent, he was able to recall his memories of encountering the tsunami, when he felt his life was in danger. The tendency of personalities to prioritize public relations rather than focusing on the horrors they experienced is a trait found in Japanese people in general, but it is particularly strong in this region.

Discussion

As can be seen from the two cases presented, in areas affected by the Great East Japan Earthquake/Nuclear Power Plant accident such as Minamisoma, there is a marked tendency for both the severity of the traumatic experience as well as the rapid deterioration of the living environment that supports the affected individuals. In Case 1, the patient experienced a tsunami so close to home that she feared for her life, lost her best friend in a way that forced her to feel guilt, and experienced many other losses. However, there was no time for calm, and the working environment became increasingly difficult as the entire region where she lived was in turmoil. Case 2 also experienced the tsunami in a life-threatening way, but in addition to this, he was responsible for the grueling task of searching for bodies as a member of Shobo-dan and experienced conflicts with the community through the evacuation due to the nuclear accident. His mental illnesses led to a poor home and work life, which eventually led to a sudden suicide attempt. In both cases, the nature of the traumatic experience during the earthquake was severe, and the postdisaster environment to support the individual was severely deteriorated.

The main points are enumerated below, focusing on the psychological conflicts that were inevitably imposed on the residents of Minamisoma City, who were forced to live in a deteriorating environment after the earthquake.

(1) *Anxiety and fear of radiation exposure*; Many people did not know the words "becquerel," "sievert," and "cesium" in the first place. Such people were forced to learn about radiation in relation to fundamental issues in their lives, such as "where to live," "what to eat," and "how to deal with anxious family members." Many people were also screened when they entered the shelter, and many were shocked to learn that their bodies were considered "radiation-contaminated material."

(2) *Stigma (prejudice and discrimination) and self-stigma*; People and food products from Fukushima Prefecture may be unfairly discriminated against by those around them. For example, the lowering of prices for agricultural products that have been properly tested and found to be safe is not just a contributing psychological factor, but also a real concern of residents about the future of the region. Even more serious than the stigma is the fact that people living in the affected areas take on the negative opinions of their surroundings as their own. For example, young women may fear that they will not be able to have a normal child in the future. To address this problem, education on radiation should be provided.

(3) *Conflict within the family and community due to differences of opinion on the effects of radiation*; Families and communities often disagreed on matters such as whether they should eat local produce and drink tap water, and to what extent children should be allowed to play outside, all of which are normal aspects of life. In the immediate aftermath of the nuclear accident, it was not uncommon for young mothers with children to leave their husbands and live with their children in evacuation centers.

(4) *Political and ideological conflicts about accepting a nuclear power plant or not directly affect the lives of residents*; One of the effects of the nuclear accident is the increased interest and concern for the disaster area from the outside. Questions such as "Is it safe to eat vegetables from the backfield?" are not closed questions within families and communities, but are linked to the political and ideological conflicts surrounding nuclear power generation that have involved Japan and the rest of the world.

(5) *The tendency for the situation of "ambiguous loss" (*Boss, *1999) to emerge*; In the field of psychology, the experience of loss, the loss of a significant person, job, or property, is a major theme. Facing "loss" is a psychological experience with great pain, but it is also true that confronting it can provide a certain amount of emotional clarity. However, in disasters involving invisible radiation, a situation of ambiguous loss tends to emerge, where it is unclear whether one has lost an important object. For example, a house that was not directly damaged by an earthquake may have a high measured radiation dose. Even if the house becomes uninhabitable, people who have lived there for decades will see their beloved house right in front of them, and there will be little difference in appearance. Some people find it hard to decide to give up their homes while watching them deteriorate. Many people live in high-radiation areas that were deemed "habitable" because of the low radiation levels in their homes while their neighbors' homes may be deemed uninhabitable because of the high radiation levels. However, one such person said, "I cannot live like that. Many people who lived in the former hazardous zone are wondering if they should try to return home. In this case, they are concerned that if they return to their homes, they will not be able to make a living if their neighbors do not return with them. Most people who want to return are elderly, and there is also a concern about whether it is possible to rebuild the community with just the elderly."

(6) *Concerns about the safety of the decommissioning operation of the Fukushima Daiichi Nuclear Power Plant*; Work on decommissioning the accident-plagued plant is expected to continue for decades or more. There is a persistent concern as to whether the work can be safely continued in the future and whether the number of people living in the area can be expected to increase as long as this risk is borne.

(7) *Issues surrounding compensation money*; A lot of money and people were involved in the nuclear accident, including subsidies for the acceptance of nuclear power plants before the accident, postaccident compensation, and a huge budget for earthquake reconstruction. Naturally, people living in the area were also affected by this. Particularly in the case of compensation, subtle differences in disaster situations, such as within or outside the 20-km zone, could make a huge difference in the guarantees, creating an unbridgeable sense of distance between the people in the area that leads leading to psychological division and conflicts.

(8) *Anxiety about the future of the region*; The primary industries in the region (such as agriculture, fisheries, and forestry) were severely damaged by the nuclear accident. Secondary and tertiary industries were also damaged due to the decline in and aging of the population due to the evacuation. In addition, the region will no longer be able to rely on nuclear power, which was one of the major industries in the region. Although reconstruction-related demand has increased since the earthquake through the present, few people are optimistic about the future of the local economy afterward.

When facing these community issues, those who have experienced severe trauma from the earthquake and tsunami are either immersed in action or, conversely, withdraw from social problems in some way, with little opportunity to psychologically process the experience in a receptive atmosphere. It is speculated that there may have been a situation that forced them to react in such ways.

As living conditions deteriorated as listed in (1) through (8), residents lived in an environment where they were expected to contribute actively to support community activities, and where guilt was stimulated by their inability to meet these expectations. In this context, it is assumed that the trauma and grief they experienced had to be downplayed. However, many of the residents may have experienced severe traumatic events during the earthquake. Cases 1 and 2 were exemplified by worsening symptoms that increased the risk of suicide attempts and the need for inpatient treatment, but it is undeniable that potential patients with milder symptoms may have led lives that prioritized contributions to the community over self-care, leaving PTSD symptoms untreated. In a coastal city in the Miyagi Prefecture, which was more severely affected by the tsunami than Fukushima Prefecture, Sakuma et al. (2020) conducted a survey on the long-term trajectories for PTSD symptoms in the local municipality. According to the survey, 6.7% of respondents showed a course of fluctuating or chronic trajectory with the long-term social functioning decline (Sakuma et al., 2020). The author has opened a psychiatric clinic in the disaster-hit area and is conducting prolonged exposure therapy, a cognitive-behavioral therapy, for PTSD (Foa et al., 2019; Hori, Takebayashi, Tsubokura, & Kim, 2019, 2020).

Conclusion

Disasters and nuclear accidents have a significant impact on the psychology of victims. Although the disaster itself is a traumatic situation, it is not uncommon for people to experience traumatic events in the worsened living conditions that follow. Complex disasters create several situations in which survivors must realistically prioritize their responses, and each survivor must expend effort to adapt to difficult living conditions, and, coupled with inadequate dissemination of psychiatric knowledge and practices, may live for more than years without reflecting on their minds and personal traumas. Although we have presented two case studies with a combination of harsh conditions, there may be many more patients with potential PTSD and mood disorders in the community that need to be addressed and aided. In doing so, on the premise that good-quality psychiatric care in a general sense is available, a system that allows trauma-specific care (such as cognitive behavioral therapy for PTSD) is needed. An appeal to the community's values of putting oneself off and neglecting self-care is also necessary.

References

Boss, P. (1999). *Ambiguous loss: Learning to live with unresolved grief.* Harvard University Press.

Foa, E. B., Hembree, E. A., Rothbaum, B. O., & Rauch, S. A. M. (2019). *Prolonged exposure therapy for PTSD: Emotional processing of traumatic experiences—Therapist guide* (2nd ed.). Oxford University Press.

Fukushima Prefectural Government. (2020). *Fukushima Revitalization Station.* Retrieved November 30, 2020, from: http://www.pref.fukushima.lg.jp/site/portal-english/.

Hori, A., Takebayashi, Y., Tsubokura, M., & Kim, Y. (2019). Efficacy of prolonged exposure therapy for a patient with late-onset PTSD affected by evacuation due to the Fukushima nuclear power plant accident. *BMJ Case Reports, 12*(12). https://doi.org/10.1136/bcr-2019-231960.

Hori, A., Takebayashi, Y., Tsubokura, M., & Kim, Y. (2020). PTSD and bipolar II disorder in Fukushima disaster relief workers after the 2011 nuclear accident. *BMJ Case Reports, 13*(9). https://doi.org/10.1136/bcr-2020-236725.

Hori, A., Tsumuraya, K., Kanamori, R., Maeda, M., Yabe, H., & Niwa, S. (2014). Report from Minamisoma City: Diversity and complexity of psychological distress in local residents after a nuclear power plant accident. *Seishin Shinkeigaku Zasshi = Psychiatria et Neurologia Japonica, 116*(3), 212–218.

Minamisoma City. (2020a). *Minamisoma City disaster record.* Retrieved November 30, 2020, from: https://www.city.minamisoma.lg.jp/portal/admin/koho_kocho/1/6/4411.html.

Minamisoma City. (2020b). *Population by age registered in the Basic Resident Registers.* Retrieved November 30, 2020, from: https://www.city.minamisoma.lg.jp/portal/admin/tokeijoho/4203.html.

Reconstruction Agency. (2020). *On the number of deaths related to the Great East Japan Earthquake.* Retrieved November 30, 2020, from: https://www.reconstruction.go.jp/topics/main-cat2/sub-cat2-6/20140526131634.html.

Sakuma, A., Ueda, I., Shoji, W., Tomita, H., Matsuoka, H., & Matsumoto, K. (2020). Trajectories for post-traumatic stress disorder symptoms among local disaster recovery workers following the Great East Japan earthquake: Group-based trajectory modeling. *Journal of Affective Disorders, 274*, 742–751. https://doi.org/10.1016/j.jad.2020.05.152.

Index

Note: Page numbers followed by *f* indicate figures, *t* indicate tables, and *b* indicate boxes.

R

FSC
www.fsc.org
MIX
Papier | Fördert gute Waldnutzung
FSC® C083411

Kolibri 360 GmbH
Ferdinand-Jühlke-Straße 7
99095 Erfurt, Deutschland
produktsicherheit@kolibri360.de